Crack initiation in composites at micro and meso scales
Development and applications of finite fracture mechanics

Universidad de Sevilla
Grupo de Elasticidad y Resistencia de Materiales
DOCTORATE IN ENGINEERING

Crack initiation in composites at micro and meso scales Development and applications of finite fracture mechanics

Doctoral Dissertation of:
Israel García García

Advisor:
Prof. Vladislav Mantič

2014

Acknowledgements

Doctoral study is essentially a time to learn the principles and methods for conducting research. The process of learning generally implies collaboration and interaction with other people. Thus, to be fair, this thesis can be considered the result of the work of all these people who taught, assisted or helped me, whether consciously or not, during this time. The following lines are dedicated to recognize their contribution.

First, I would like to thank my doctoral advisor Dr. Vladislav Mantič for his continuous support, guidance and endless motivation. I would like also to acknowledge his efforts and time invested in inspiring discussions and ideas which are the source of the most relevant results reflected in this thesis. In addition, I am greatly indebted by his meticulous care in reviewing the manuscripts of the articles associated to this thesis.

I also wish to extend my thanks to all my colleagues of the Group of Elasticity and Strength of Materials (GERM) for providing an inspiring working environment and comfortable atmosphere which helped me to develop the work reflected here. A special acknowledgement is deserved by Dr. Federico París, head of the GERM, for promoting this environment. I would also like to thank him along with Dr. Antonio Blázquez for the interesting discussions and suggestions about the problem of transverse cracking in cross-ply laminates. During these years, I have also had the opportunity to take advantage of the experience and computational codes generated by the previous theses developed in the group. Thus, I would like to express my sincere gratitude to Dr. Enrique Graciani and Antonio Blázquez for their BEM codes and Dr. Luis Távara for his parallel implementation of the Graciani's BEM code. Thanks are also due to Dr. Elena Correa for her support with the problem of fiber-matrix debonding and to Dr. Alberto Barroso for assisting me by sharing his knowledge of multimaterial corners.

I would like to extend my sincere appreciation to Antonio Cañas, José Ramón Rodríguez, Isabel del Príncipe and Miguel Muñoz who help and assist me in the laboratory. I am particularly indebted to Jesús Justo and Annaëlle Simon for their collaboration in carrying out the experiments to evaluate the size effect in cross-ply laminates.

Next, I would like to thank my colleagues in the L3 office for their help in those small things that, in spite of being not visible here, have been essential for the adequate development of my doctoral studies: from my first office mates Daniane Vicentini, Afonso Leite, Thomas, L. Barry, Luis Távara, José Reinoso,

i

Christos Panagiotopoulous and Jesús Justo to the new ones: Rafael León, Francisco Moreno, Daniel García and Juan Manuel González. This gratitude is also extendable to our unofficial L3 mate Federico Aguirre.

During these years I had the honor of collaborating with Dr. Dominique Leguillon from the *Université Pierre et Marie Curie (Paris 6)*. This collaboration was particularly strong during my three-months stay in Paris and also during the visits of Dr. Leguillon to Seville. I would like to express my gratitude to him for the interesting discussions and his insightful comments. An entire chapter of this thesis is a result of this collaboration.

In addition, I had the opportunity of collaborating with Dr. Anthony R. Ingraffea from the *Cornell Fracture Group* (CFG) at *Cornell University* during my four-months stay in Ithaca, NY. I appreciated his kindness and interesting suggestions. I would also like to express my gratitude to all my colleagues within the CFG, in particular to Dr. Bruce J. Carter, who assisted me with the use of the software FRANC3D NG.

I had also the privilege of performing a joint research with Dr. Marco Paggi from the *Politecnico di Torino* (now at *IMT Lucca*). I am greatly indebted to him for performing the FEM modeling of the fiber-matrix debonding and providing me with results to compare the models developed here. His motivation, comments and work during our joint research is appreciated very much.

I would like to thank the European tax-paying public as this thesis was supported mainly by the Spanish Ministry of Education by a "Beca de Formación de Profesorado Universitario" (F.P.U. 2009/3968) and the European Social Fund. The partial support by the Spanish Ministry of Science and Innovation (Project MAT2009-14022), the Spanish Ministry of Economy and Competitiveness (Project MAT2012-37387), and the Government of the Autonomous Community of Andalusia (Project TEP-04051) is also acknowledged.

I want to thank to my friends, cousins, aunts and uncles for their warm friendship, care and enthusiasm during these years. A special thanks to Pablo Cárdenas for sharing so many moments which are indispensable in spite of not being reflected directly in this thesis. I would particularly like to thank Dr. Miguel Fosas for our discussions about the sense of the doctoral studies, the analyses of the research and the academia and for being a constant support since I met him.

My sincere thanks to my parents Tomás and María del Carmen and my sister Carla for their patience, care, support and encouragement. In particular, I would like to express my gratitude to my parents for teaching me the passion for knowledge and limitless curiosity. This thesis is undoubtedly a consequence of their influence.

Finally, my very special thanks to Miriam for sharing my moments of frustration and joyful laughter and for taking care of me more than I myself.

Contents

CHAPTER 1

Introduction

1.1 Composite materials

The term "composite materials", commonly shortened to "composites", can be defined, from an etymological point of view, as materials made from several constituent materials with different physical properties, insoluble between them and with the possibility of separation by exclusively mechanical techniques. The adequate combination of the constituents produces a composite material with properties greatly improved with respect to the properties of the constituent materials. This key idea has been employed since prehistoric times for the fabrication of materials which are employed for construction building. The earliest of these man-made materials is considered to be the cob, which is a combination of mud and an organic fibrous material like straw. The mud gives the cob the necessary consistency whereas the straw adds a significant level of toughness and strength, particularly under tension. Other common material conceived following this idea of the combination of several constituents is reinforced concrete. This material combines concrete with a reinforcement, e.g. steel reinforcing bars, counteracting the low tensile strength and the brittleness of the concrete. This combination enables to employ the concrete in structures with regions subjected to high tensile stresses by situating the reinforcing bars in these regions, e.g. in beams.

Although the previous definition of composites is coherent in the etymological sense, this term is traditionally employed by the engineering community to refer to a more specific kind of material developed during the 20th Century which combines unusually high strength and stiff reinforcing materials with weaker materials giving consistency. These materials are also called "advanced composite materials" (ACMs). The most common examples are the materials which combine carbon, aramid or glass fibers with polymer, ceramic or metal matrices. The idea motivating the combination of these constituent materials is the same as described previously for cob and reinforced concrete. For instance, carbon/epoxy

1

combines high-strength and stiff carbon fibers embedded in epoxy resin providing toughness and consistency. The result is a material with a strength-to-weight and stiffness-to-weight ratio significantly higher than those of traditional engineering materials.

In general, the different phases which are combined in a composite can be divided into two categories depending mainly on their function: the reinforcement and the matrix. The reinforcement is usually the material with high mechanical properties. This is normally shaped as microscopic particles or fibers. The matrix is the material giving mainly consistency to the composite, which is necessary since the reinforcement is often discontinuous. The matrix can also add other kind of properties to the composite. The interface between the reinforcement and the matrix is often considered a third type of constituent given its high influence on the composite properties, in particular on those which govern the failure. In addition, due to the small size of the reinforcement, the area of reinforcement-matrix interface per unit volume is huge which increases its relevance.

The relevance of these materials in some industries where the lightweightness is a key aspect of design has increased drastically during the last decades as a consequence of their high properties per unit weight. This is the case in the aerospace industry, which is motivated by the fact that a small decrease of the aircraft weight results in a significant fuel saving during its entire lifetime. In the aeronautical industry, this evolution is particularly observed in airliners (also called commercial aircrafts) where the aluminum alloys have been replaced in by composites as the main material. Initially, composites were restricted to non-structural applications given the lack of knowledge about their structural performance. The first uses in structural parts in the 1950s were limited to secondary parts with a very low responsibility in the global structural integrity, and representing about 2% of the total structural weight. A slow but continuous increase in the use of composites followed during the next decades. These first applications were very limited by the restrictive safety standards in the commercial aviation. In the 1980s and 1990s the two largest airliner manufacturers followed two different strategies with respect to the evolution of the use of composites. During these decades Airbus increased continuously the structural weight of composites in each new model whereas Boeing was much slower, their percentage in the structural weight staying approximately fixed in the models of these decades. However, at the beginning of the 21st century, Boeing launched the development of the B787 with more than 50% of structural weight in composites, well ahead of the aluminium alloys which was the 20% of the structural weight. Moreover, in volume the B787 structure is 80% composite. This has represented a jump for the aeronautic industry and in particular for Boeing due to its strategy during the previous decades. Thus, currently the use of composites spans the majority of the aerostructures of the airframe, including primary ones. Simultaneously to the development of the B787, Airbus launched the development of the A350, an airliner of a similar type to the B787. This was conceived initially as a redesign of the A330 but later it was announced to have an almost totally new design. The main reason for this change was the announcement of the percentage of structural weight in composites of the B787 which would make the A350 (as a redesign of the A330) becomes obsolete in terms of fuel-efficiency even before its maiden flight. Thus, the new A350 is made out of 53% composites which is the most prominnent structural material, the next highest being the aluminium alloys with 19% of the structural weight. Observe that percentage are very similar in both models. These last developments show that composites are already the main material in commercial aviation. Composites are also extended as structural material in military aircrafts. In fact, the weight reduction caused by the use of composites has other more relevant advantages above and beyond the increase of fuel-efficiency for some types of military aircrafts such as fighters. These aircrafts require a high level of maneuverability in dogfight situations which is greatly improved by the weight reduction as, e.g., the increase of the thrust-to-weight ratio. Composites are also extended to space applications where the weight reduction is a key factor as well. Besides their high strength-to-weight ratio which is a common quality for the whole aerospace industry, in the case of space applications, the high stiffness-to-weight ratio of composites is also a great advantage given the high requirements on stiffness. This need is based on the fact that the natural frequency of the structures is limited to a certain value during the liftoff to prevent breakup. Moreover, this is also an advantage for specific applications such as communications satellites which require the deformation to be kept within a small margin to assure the correct operation of functional parts like antennas.

The use of composites is also particularly useful in other transport industries since the qualitative relation between weight reduction and fuel saving can be extrapolated to other modes of transport. The most relevant example is the boatbuilding industry. Initially, their use was limited to small pleasure yachts and fishing boats. The earliest application for large structures was military minesweepers employing glass fibers embedded in a polymeric resin. The main advantage of composites in this particular application was their low magnetic signature. Their use is already extended to superstructures in the largest ships. Besides the increase of fuel-efficiency, composites present other advantages for particular problems of the boatbuilding industry, e.g. a high resistance to corrosion and the ability to improve stability. Their use has also extended to the automobile industry. Although the application of composites to this industry presents the common advantages described for the transport industry, their use is still limited mainly due to two reasons: their high cost and the difficult tayloring of their manufacturing, both aspects being key in the automobile industry. As a consequence, their application to primary structures is limited to certain luxury sport cars. In fact, these cars could be included in the frame of high-performance products for sports where composites are broadly used, e.g. bicycle frames, baseball bats, golf clubs and rackets.

The previous summary of the main applications of composites in the industry has enabled envisagement of the main advantages of composites for structural applications with respect to the traditional materials employed in each industry. In what follows, these advantages are discussed focusing on the main composites employed in the aerospace industry: long-fiber-reinforced composites based on fiber, glass or aramid embedded in polymeric resins. Since the properties referred to in the following are highly directional, the properties corresponding to the fiber direction are taken.

- High strength-to-weight ratio. This quality is due to the combination of the low density of the constituent materials and the high strength of the fibers along their axis direction. The low density is a direct consequence of combining polymeric resins and relatively light fibers. Both constituent materials have typical densities ρ ranging from about 1000 kg/m^3 to 2500 kg/m^3 which is significantly lower than the densitiy of the typical materials employed for structures as aluminium ($\rho \approx 2700$ kg/m^3) or steel ($\rho \approx 7800$ kg/m^3). With regards to the strength, in general the fibers have a high strength along their axis direction since the small dimensions along the transverse direction (typically measured in micrometers) minimizes the scale of possible defects, thus increasing the strength. In addition to this common feature, each type of fiber presents additional reasons for the high strength as e.g. strong chemical bonds. The tensile strength of isolated fibers employed in the aerospace industry ranges from 2 GPa to 7 GPa which is much higher than the typical values for aerospatial aluminium (0.3 GPa) or the typical civil-industry steels ($0.4 - 1$ GPa). In spite of the much weaker properties of the polymeric resin, these differences remain significantly high even when the whole composite material is taken into account. The high strength-to-weight ratio is the most relevant property of the composites studied here and justifies their extension to structures in the transport industry where the weight has always been a variable to be minimized.

- High stiffness-to-weight ratio. This quality is due to the low density of the constituent materials and, in some cases, is also combined with the high stiffness of the fibers. The stiffness is affected mainly by the chemical bonds between the atoms or molecules forming the constituent materials. Typically, the polymeric resins have a low stiffness with Young modulus E ranging from 1-8 GPa. Regarding the fiber stiffness, carbon fibers present a higher E along the axis direction (200-700 GPa) than typical aluminium alloys (70 GPa) and even higher than steel (210 GPa). However, glass or aramid fibers present a much lower stiffness (70-150 GPa). These values are also reduced when the whole material is taken into account. Combining these values with the lower weight, composites made with carbon fibers are much more suitable for application with high requirements of stiffness-to-weight ratio. This is typically required when deformation is highly restricted, such as in optical or telecom applications in satellites.

- High level of anisotropy in their elastic properties. This quality is mainly due to the presence of fibers and the great difference in stiffness between the typical fibers and the polymeric resins. As a consequence, in unidirectional laminates the contrast between longitudinal and transverse Young's moduli can become several orders of magnitude. This directional stiffness allows designing the laminate to meet certain objectives to improve the global performance. The directional stiffness is a characteristic which has not been fully utilized yet. An example of how the use of the directional stiffness can be used in a beneficial manner is aeroelastic tailoring. This is based on controlling the aeroelastic deformation by designing the directional global stiffness of the laminate and the structural part through its layup to affect the structural and aerodynamic performance in an advantageous way.

- High fracture toughness. Both polymeric resins and carbon, glass and aramid fibers have a very low fracture toughness when they are isolated. However, composites formed from them have a high fracture toughness, its value even reaching the fracture toughness of some metals. This increase with respect to the fracture toughness of the constituent materials is due to the complex mechanisms associated to the cracking, which are promoted by their complex microstructure. The crack propagation does not only require the fibers or resin to break but also it needs to activate a complex variety of microscopic damage mechanisms, which imply an additional energy dissipation per unit broken area. The contribution of these additional terms increase the global fracture toughness by several orders of magnitude. Thanks to this increase, these composites are tough enough to be used in large primary structures.

- Good fatigue performance. This is very similar to the previous property. In this case, the fatigue response is also strongly affected by the complex microstructure of these composites. In particular, the fatigue propagation of cracks is continuously arrested by microstructural barriers at different scales. e.g., fiber-matrix interface at a micro scale or the interface between plies with a different orientation. As a consequence, composite structures usually have a good fatigue behavior. This property is particularly relevant for the aerospace industry taking into account that fatigue is a key issue in the performance of aeroestructures, which are normally subjected to cyclic loading.

1.2   Failure in composites

In the previous section, the high strength-to-weight ratio of composites has been highlighted. However, this advantage can be partially reduced or even totally missed if higher safety factors have to be employed as a consequence of a certain lack of knowledge on their particular failure mechanisms. This is a clear disadvantage with respect to other traditional materials such as aluminium alloys, which have been studied and tested for decades, their failure mechanisms being known. This lack of knowledge about the failure also affects the accurate estimation of the true reserve factor, i.e. the ratio between the load leading to failure and the design load. These facts combined with the inability of the traditional failure criteria employed in metals to be applied in composites have motivated a huge effort of studying the failure in composites during the last decades.

In spite of having motivated intense research during the last decades, failure in composites still has not been studied enough. The reason is mainly the high complexity of the mechanisms governing the global failure. These mechanisms are associated to the microstructure at different scales as well as the interaction between them. This is a direct consequence of the defining feature of composites: the presence of several constituent materials which are combined on different scales. As a consequence, the presence of interfaces between dissimilar materials and a high variability in the elastic, strength and fracture properties corresponding to the different constituent materials can be found at different scales. This strongly affects the failure mechanisms at these scales adding a high complexity to these mechanisms seen from a macro scale perspective.

These failure mechanisms are particularly complex in laminates composed by plies with different orientations, which are intensely employed in the majority of applications described in the previous section. A large number of failure criteria based on a variety of assumptions have been proposed for these laminates. The coexistence of several failure criteria has motivated the organization of world wide exercises to evaluate the accuracy of the predictions given by the criteria proposed by comparing with a set of experiments, see Soden et al. (1998); Hinton et al. (2002); Kaddour and Hinton (2013). Results showed a great margin for accuracy improvement.

In spite of the large number of failure criteria proposed, according to the estimation by Sun et al. (1996), the most used criteria in the 1990s were the "maximum strain" and the "maximum stress" criteria, being used by about 52% of composite designers. This estimation also shows a great diversity in the failure criteria employed. The maximum stress or strain criteria are based on assuming a set of critical values for the stress or the strain, respectively, corresponding to certain relevant orientations. The failure is predicted when the stress or the strain along one of these directions reaches its corresponding critical value. Note that the most used criteria are based on the simple assumption that a finite number of different failure mechanisms exist and are activated exclusively by a certain level of either strain or stress along a certain direction. Therefore, they neglect the existence of other failure mechanisms and even the interaction between the mechanisms considered. On the contrary, the majority of the other failure mechanisms predict an interaction between them. Thus, the preferential use of the simplest ones evidences a low level of trust of the designer community in the prediction given by the rest of the failure criteria, particularly in the form that the failure mechanisms are predicted to interact. Previous conclusions are based on estimations published in 1996 which could have changed given the fast and continuous evolution of the discipline.

According to París (2001), a comprehensive understanding of the physical basis which governs the failure in composites is necessary in order to improve and generate new physical-based failure criteria. These criteria have to take into account the inherent complexity of the micro and meso structure of these types of material systems. This has motivated the study of the mechanisms of crack and damage propagation at these scales. Enormous progress has been made. In fact, many results can be found in the literature characterizing the crack propagation at different scales. A revision of the previous works dealing with the problems studied in this thesis can be found in the corresponding chapters.

In contrast with the number of works dealing with the crack propagation at these scales, the crack initiation at micro and meso scales is still an open matter which has been relatively under-developed. The main reason for this is the classical inability of the linear elastic fracture mechanics to predict the initiation of a crack in an intact material or interface. Some works have been developed in order to study the initiation by computational methods as cohesive zone models requiring large computational resources. Although crack initiation is a controversial concept, as discussed in Chapter 2, adequately interpreted it can be considered as the first step in a wide variety of failure mechanisms. Therefore, its prediction is very relevant for the full characterization of a failure mechanism.

This thesis focuses on studying the crack initiation as the first step of several failure mechanisms. This study is based on the application of the coupled stress and energy criteria, proposed by Leguillon (2002) in the framework of the finite fracture mechanics introduced by Hashin (1996), see Taylor (2007) or Chapter 3 of this thesis for a review. This criterion often enables the prediction of crack initiation by semianalytical expressions, which is very useful to the study of the influence of the different parameters governing the problem or to be employed as part of a future global failure criterion.

1.3 Objectives

This thesis is part of a number of works carried out within the "Grupo de Elasticidad y Resistencia de Materiales" (GERM) with the objective of understanding the failure mechanisms in composites from the physical point of view. The long term objective is the development of a failure criteria established on a well understood physical basis. Previous works focused on studying the first steps of the main failure mechanisms by means of micro and meso scale models. However, as discussed previously, in spite of the fact that crack initiation is considered the first step in some of these failure mechanisms, it is still not fully understood. In view of these previous studies the objectives of this thesis can be divided into two groups: a set of general objectives and another set of objectives wich are common to each failure mechanism studied, with their application being individually particularized.

The general objectives are:

- Contribute to the development of the coupled stress and energy criterion of the finite fracture mechanics because it is the main tool employed for the analysis of the crack initiation in this thesis. The particular characteristics of the problems here for which this criterion is applied require a further development of this criterion. In this sense, the contribution of this thesis is twofold. First, the original theoretical formulation is generalized in order to take into account certain phenomena, which are relevant for the problems studied in this thesis but were neglected in the original formulation. Second, a set of methodologies are developed with the purpose of enhancing the application of this criterion.

- Evidence that the coupled stress and energy criterion is suitable for the prediction of crack initiation at several scales in composites in the context of the definition of a new failure criterion for composites in the terms discussed previously. Thus, it is essential to show that the coupled criterion is able to obtain analytical or semianalaytical expressions for the prediction of crack initiation, these expressions being simple enough to be implemented in a global analysis. The simplicity of these expressions is also required for the physical interpretation of the results. Even more important, it is necessary to verify that the predictions given by the coupled criterion agrees with experiments or other accepted techniques.

- Generate new knowledge about the nature of the first step in the failure mechanisms studied. The results must enable the ability of predicting the conditions under which crack initiation occurs, i.e. the critical loads leading to it as well as the characteristics of the crack immediately after the initiation. In addition, it is necessary to connect it to the previous works in order to integrate the new knowledge.

The particular objectives, enumerated in what follows, are actually a detailed description of the last objectives of generating new knowledge about the failure mechanisms. They are discussed here because although they are particularized for each failure mechanism, their formulation is common in several of the problems studied in this thesis

- Obtain theoretical models with analytical or semianalytical expressions to characterize the crack initiation as a function of the elastic, strength, fracture and geometric properties. It is particularly relevant that these properties correspond to well-known physical properties which can be obtained by well established experiments. As described in Chapter 3, the characterization of the crack initiation from the perspective of the coupled criterion consists of obtaining the critical load for the initiation and the crack geometry immediately after its initiation.

- Study the variation of the previous results with the main parameters of the problem. In addition, given the large number of parameters, it is necessary to evaluate the relevance of different parameters for this variation with the aim of identifying the main parameters governing the problem.

- Propose indirect experimental procedures to obtain the fracture and strength properties based on the theoretical models developed. Some fracture and strength properties governing the problems studied are difficult to be obtained experimentally. The difficulty originates from the small scales of the necessary tests.

1.4   Outline of the thesis

This thesis is structured in 8 chapters in which the previous objectives are developed. Excluding this first chapter of introduction and the last one describing the conclusions and future developments, the thesis outline can be divided into two parts: the first part studies the phenomenon of crack initiation and the criterion employed here for its prediction; in the second part the application of this criterion to the different failure mechanisms and scales studied is described. The general objectives described previously are not equivalently developed in the two parts. The first objective is mainly developed in the first part, whereas the third objective is developed in the second part. The second objective is present in both parts.

The first part is composed of Chapters 2 and 3. In Chapter 2, the controversial concept of crack initiation is discussed and an interpretation in the context of this thesis is provided. In addition, this chapter describes briefly the main tools employed to predict crack initiation. In Chapter 3, the coupled criterion of the finite fracture mechanics is discussed in the context developed in the previous chapter. Both the original formulation and the new contributions are described from a practical perspective leading to a global generalized formulation. The different techniques associated to the application of the coupled criterion are also detailed, including some originally proposed in this thesis.

The second part is composed of Chapters 4 to 7 and deals with the application of the coupled criterion in the different problems studied. The chapters are sorted by the scale of the problem studied, from the smallest to the largest. In Chapter 4, the problem of crack initiation at the fiber-matrix interface when composite material is subjected to transverse load is analyzed. First, the preference of the different forms of the crack initiation predicted by other methods in the literature is analyzed from the perspective of the coupled criterion. Next, the results predicted for the case of uniaxial transverse loading are compared with the results given by a well-accepted method such as the cohesive zone model. Finally the influence of a secondary transverse load is analyzed focusing on the variation of this influence with the value of the main parameters. In Chapter 5, a similar problem is studied: the crack initiation at the interface between a spherical inclusion and the surrounding matrix when the material is subjected to tension. The results found are compared with preliminary experiments found in the literature. Chapter 6 deals with the problem of crack initiation in the off-axis ply in cross-ply laminates. This is a classical problem and a large number of particular criteria have been proposed to interpret some phenomena typically found in experiments. The objective here is the interpretation from the perspective of a general criterion such as the coupled criterion. First, a simple but reasonable 2D model of the problem is proposed, which is later generalized to take into account 3D phenomena as well as thermal effects. Finally, some experiments with cross-ply laminates carried out in the context of the thesis are described and the results obtained are compared with the prediction of the theoretical model. In Chapter 7, the crack initiation at a reentrant corner in the presence of an interface is analyzed. The theoretical model enables an analysis of the variation of the failure typology as a function of different problem parameters.

CHAPTER 2

From damage to fracture: crack initiation

Crack initiation under static loads has been a controversial topic for a long time in spite of being, *a priori*, the first step of some failure mechanisms. The reason can be found in the combination of historical causes and the difficulties inherent to the physical characterization of the process. This chapter is devoted to introduce how the concept of crack initiation is understood and predicted in the context of this thesis. First, some historical notes on the fracture mechanics will be reviewed in Section 2.1 to present the main problems that the study of crack initiation has from the point of view of the classical fracture mechanics. Subsequently, the two main problems of the study of crack initiation will be discussed in Sections 2.2 and 2.3, describing how they have been solved.

2.1 Crack initiation from the classical fracture mechanics perspective

The conditions under which a crack in a body propagates have been studied for near a century leading to a new field: the fracture mechanics. Traditionally, the criterion proposed by Griffith (1921), which is widely used yet, is considered the first[1] milestone of this field. This criterion is motivated by the works carried out by Griffith and his colleagues about the lifetime of aircraft elements at the Royal Aircraft Factory in the context of the World War I. They observed that lifetime of rods at the aircraft engines was strongly increased when their external surfaces were polished. An attempt to explain this observation was made by the stress

---

[1]Currently it is known, thanks to Rossmanith (1995), that, some years before, Wieghardt (1907) proposed a theory for crack propagation in terms of the stress intensity factor equivalent to that proposed by Griffith (1921). The reference to Griffith's works is kept here by their historical relevance and the importance of the Griffith's energy perspective in the context of this thesis

analysis of scratches in the rod surfaces, modeling them as elliptical flaws. Using the maximum stress criterion of fracture, widely used by his contemporaries, this analysis led to the conclusion of that the maximum stress expected was almost independent of the depth of scratches beyond a certain depth. However, this result was contrary to the facts observed by the engineers when rods were polished.

The result of this first analysis and his subsequent experiments with glass fibers containing artificial flaws motivated Griffith to reject the maximum stress criterion and to propose his new criterion. The idea behind the Griffith criterion is simple and revolutionary: a certain amount of energy is necessary to break the bonds joining the atoms or molecules, and this energy comes from the potential elastic energy released in the elastic body and from the work done by the external forces due to the propagation of the crack. Thus, a crack is expected to grow when the energy released by an infinitesimal increase of the crack length reaches the energy necessary to break the inter-atomic or inter-molecular bonds at an infinitesimal distance of the crack tip[2]. In addition, Griffith assumed that the energy necessary to open a new surface is proportional to the amount of bonds to break, hence proportional to the area of the new surface. The energy necessary to generate a new surface per unit area is known as "fracture energy" γ_f, whereas the energy necessary to open a new crack per unit area is the "fracture toughness" $G_c = 2\gamma_f$. The elastic potential energy released per unit of new crack area is the "energy release rate" (ERR) G. Thus, the Griffith criterion for propagation is written as,

$$G \geq G_c, \tag{2.1}$$

which summarizes the Griffith criterion. The application of this new criterion to a straight crack of length $2a$, within a semi-infinite plate subjected to a constant and remote tension σ^∞, gives that the crack is propagated when σ^∞ reaches a critical value σ_c^∞,

$$\sigma_c^\infty \propto \sqrt{1/a}. \tag{2.2}$$

This dependence allowed Griffith to explain the influence of the scratches on the lifetime of engines rods such as the experimental results obtained in glass fibers.

After the seminal papers by Griffith (1921, 1924), a large period followed without a significant development of the Griffith criterion because his works were ignored by the engineering community for almost 30 years coincident in part with the decrease of attention to the problem during the first part of the interbellum, see Figure 2.1. In the meantime Westergaard (1939) obtained the power of singularity of the stresses around the crack tip,

$$\sigma \propto \frac{1}{\sqrt{r}} \tag{2.3}$$

where r is the distance to the crack tip.

---

[2]Actually, Griffith (1921) proposed his criterion as a generalization of the "theorem of minimum energy" by taking into account the surface potential energy due to the inter-molecular or inter-atomic bonds. To keep these historical notes as simple and self-contained as possible, the equivalent explanation given was chosen.

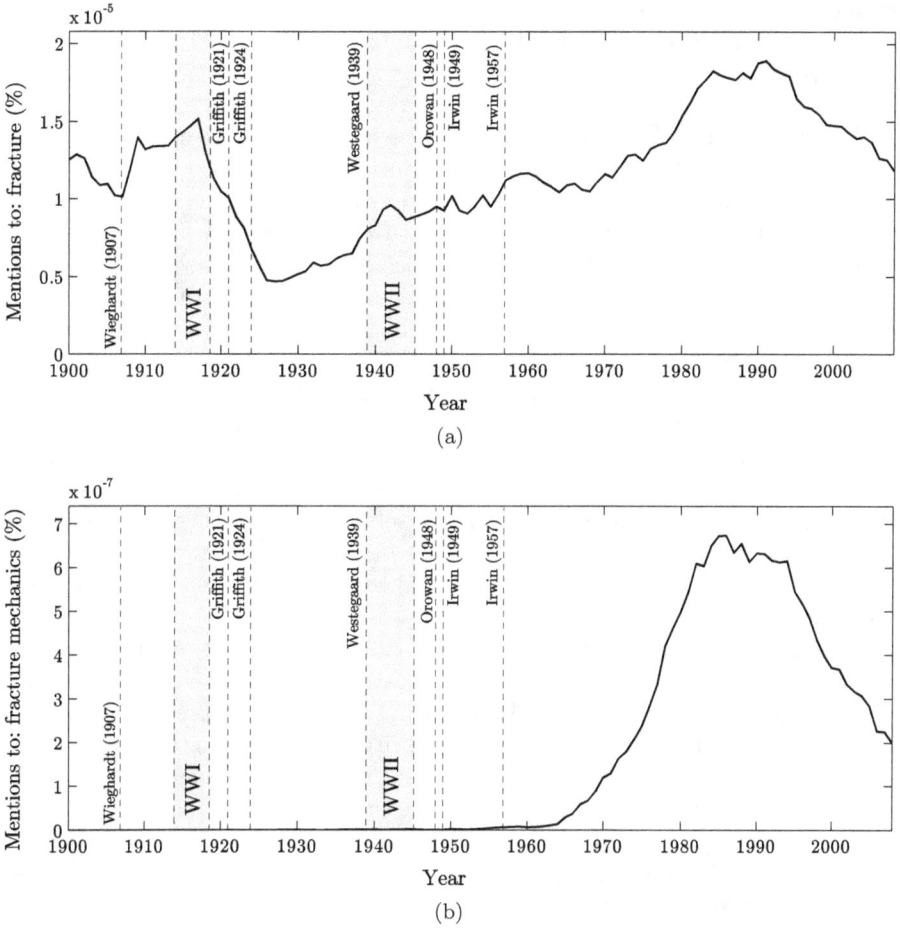

Figure 2.1: Use over time of the terms (a) "fracture" and (b) "fracture mechanics" normalized with the total number of mentions. Data are extracted from the Google Books Ngram Viewer developed by Michel et al. (2011) and setting smoothing to 2.

According to Erdogan (2000), the Griffith criterion was not immediately accepted in the engineering community mainly due to two reasons: a) In metals, which is, along with concrete, the main material family used in large structures, the fracture energy due to inter-atomic bonds is several orders of magnitude lower that the actual fracture energy due to energy dissipation in other irreversible processes as yielding and b) the stresses obtained by a linear analysis around the crack tip are very unrealistic for metals because yielding is expected at these points. Both problems were solved at the end of 1940's and the 1950's. On the one hand, Irwin (1948); Orowan (1949) proposed to correct the value of fracture toughness by adding the terms of the dissipated energy due to other irreversible processes. On the other hand, Irwin (1957) demonstrated that the energy release rate is well approximated by a linear elastic analysis if the plastic zone around the tip is small enough even though the stresses near the crack tip are unrealistic.

The resolution of these problems allowed to apply the Griffith criterion to engineering materials. The combination of this fact with a succession of several catastrophic failures of large metallic structures motivated an extensive development of the linear elastic fracture mechanics from then on. A large and intense amount of contributions followed dealing with the characterization of the stress tensor around the crack tip, the definition of the "stress intensity factor" as the coefficient of the singular term of the asymptotic solution near the crack tip, the relation between this factor and the energy release rate. In addition, the first methods to obtain experimentally the fracture toughness were developed in order to measure the main magnitude of the new field. Next decades[3], the efforts were devoted to develop the non-linear fracture mechanics which provides with tools designed to study problems with nonlinear effects which cannot be neglected. A huge effort was also spent to obtain elastic solutions for geometries with cracks of practical importance. Subsequently, the development of the computational methods enabled to solve more complex problems such as to simulate nonlinear phenomena.

As shown in the previous historical notes, the answer to the question of which are the conditions necessary for crack propagation has been studied more or less intensely during a whole century. However the very related question of how and under which conditions a crack initiates has drawn much less attention. In fact, at least in the case of static crack initiation, few pages dedicated to this topic can be found in the majority of books of fracture mechanics. Crack initiation is often studied as a part of damage mechanics.

This traditional undervaluing of this, *a priori*, step on the fracture process can be explained by the addition of two reasons:

- Many mechanicians claim that crack initiation has not any sense since flaws, scratches, voids and/or other types of defects exist within any material system right after its manufacturing. Thus, according to this perspective, in brittle and quasi-brittle materials, the failure initiation would be given by the statistical distribution and the lengths of these defects, which are very linked to the manufacturing processes and material nature. Hence, the study of crack initiation under static loads would be spurious and of no relevance.

- Griffith's criterion is not able to predict crack initiation at least according to its classical formulation. For the example of a crack in a semi-infinite plate, this result can be directly derived from (2.2). For the crack length

---

[3]These lines are not claimed to be an exhaustive historical review of fracture mechanics but a brief introduction to provide a context to the subject of crack initiation. The reader interested in the subjects can find excellent works covering it in the literature, see e.g. Erdogan (2000); Cotterell (2010)

corresponding to crack initiation ($a \rightarrow 0^+$), the critical remote tension $\sigma_c^\infty \rightarrow \infty$. Hence, no crack will appear in the lack of an initial one according to the classical Griffith criterion. This result is general for any problem.

Note that both reasons are of a very different nature. Whereas the first one discusses the physical and practical sense of studying the crack initiation, the second one shows a limitation of the Griffith criterion. In what follows, this chapter is consecrated to delimit in which sense crack initiation is going to be understood in the context of this thesis and to briefly present how both problems have been solved previously.

2.2   Do cracks initiate or exist before loading?

Griffith verified its theory by testing glass with artificial flaws larger than those existing in the material according to his assumptions. Next decades, an effort followed dealing with the observation of the "natural" flaws and its characterization. Unfortunately, observing directly these flaws was impossible in the glass used in the experiments by Griffith (1921). The reason was not that they do not exist but the limited means of the optical science and the typical small size of flaws in glass. As pointed out by e.g. Lawn (1993) the typical flaw sizes depends strongly on the material, ranging from nanometric in pristine optical fibers and whiskers, submicroscopic in glass and milimetric in rocks. With the development of the modern electron microscopy, flaws have been observed in the majority of materials of interest, see e.g. micrometric flaws in composites by Miller and Wingert (1979), submicroscopic flaws in Kevlar by Wagner (1986) and nanometric voids in alumina films by Huang et al. (2004).

The origin of these flaws is very variated, some of the most important are:

- Chemical origin: The exposition of the external surfaces to chemical spices generates chemical micro-reactions. This can lead to generate surface flaws. In the case of materials composed by several phases as composites or concrete, the micro-reaction between the aggregates and the stochastic characteristic of this reaction can also generate micro-flaws at the interface between phases or in the vicinity. In the majority of cases, these reactions are highly promoted by temperature, hence a large part of these flaws appear during the manufacturing process.

- Thermal origin: Temperature change and gradients generates high stresses, which promotes the incompatibility of deformations promoting the appearance of flaws. In addition, the solidification of granular materials as metals generates the appearance of voids and bubbles.

- Radiation origin: Low-energy radiation favors chemical micro-reactions at the external surfaces, generating new phases or even voids. This is particularly important in polymers since it is well known that radiation can generate chain scission or crosslinking. In the case of chain scission, a weaker zone is generated, which is equivalent to a flaw in some manner.

In the case of crosslinking, the gradient of properties can lead to a stress concentration leading to flaws in the first step of the loading process.

- Micromechanical origin: Multiple mechanical effects promote the generation of flaws, e.g., microcontact between phases or boundary between phases sliding.

Once the existence of flaws have been shown for the majority of materials of interest, the failure of brittle or quasi-brittle materials should be studied, *a priori*, as a statistic phenomenon. According to Weibull (1939), a statistic distribution can be defined as the probability of finding a flaw of a certain effective length per unit volume. On the other hand, it can be assumed a direct relation between the effective length of a flaw and the critical stress leading to the failure through this flaw. This theory is able to explain several observed phenomena, e.g., the lower strength of larger structures since if the flaw effective-length follows a distribution per unit volume, a larger volume increases the probability of having a large flaw and hence the breaking for a lower load. Although the failure prediction becomes a statistic matter, this theory remains compatible with predicting that structures fails at these points with strong stress concentrations, at least in terms of probability. The reason is that at those points with high stresses, the critical flaw size required for failure is reduced in comparison with the zones with lower stresses, therefore the probability of failure increases at the stress concentration points.

Studying crack initiation from a deterministic point of view is apparently contradictory with the demonstrated fact of that the majority of materials contains flaws, microcracks or voids since its manufacturing. However, both perspectives became compatible if they are confined to different scales. Thus, if the inhomogeneity introduced by the presence of flaws, microcracks and other defects is assumed to be enclosed to a certain (small) scale, crack initiation can be studied at a larger scale, for which the material can be assumed to be continuum and homogeneous. At this larger scale, a continuum and intensive variable denoted damage can be defined measuring the growth of these flaws, microcracks or in general any irreversible process. These processes are strongly associated to the material microstructure. They correspond to coalescence of, e.g., voids in metals, crazes in polymers and microcracks in concrete. In addition, at the scale for which the continuum is assumed, the elastic properties can vary with the damage variable. Since this continuum perspective, crack initiation can be associated with the abrupt localization of damage at a certain surface. The analysis derived from this vision is the subject of study of a whole field denoted damage mechanics, see e.g. Lemaitre (1985).

This approach is coherent with the continuum mechanics and the damage variable is analogous with other more established variables, e.g., Young's modulus for grained materials as usual metals. In this case, the Young's modulus measured at a macro scale only has sense for scales sufficiently larger than the typical grain size. Analogously, the representativity of the damage variable is limited to the scale studied and has not sense below a certain scale where the homogenization of the effect of the presence of flaws cannot be assumed.

In the framework of this thesis, crack initiation will be understood following this perspective, the initial flaws being assumed to be confined to a scale sufficiently smaller than the scale chosen for the study of each problem. This idea is schematized in Figure 2.2. For the sake of illustration, a classical example of a geometry with a stress concentration is represented. Initially, and prior to any loading, a distribution of flaws exists at the solid with varied size and origin as described previously. These defects are visible below a certain scale which is on the right row of Figure 2.2. Thus, above a certain scale, the solid can be assumed as homogeneous having some equivalent elastic properties for a certain scale. When the solid is loaded monotonically with, e.g, a tension σ^∞ and it exceeds a certain value, the flaws and other defects begins to grow stably or generate plastification around them or other irreversible processes. These processes are accounted by the increase of the damage variable defined at a larger scale, which can modify the effective elastic properties at this scale. In some cases, this damage, understood as continuum, can be "visible" at larger scales. This occurs for example in some polymers which present "stress whitening", see e.g. G'Sell et al. (2002). In these polymers, regions with high levels of damage become white and translucent as a consequence of certain types of damage. In general, if the tension σ^∞ remains increasing, an abrupt coalescence of flaws, microcracks and other defects occurs for a critical value of the tension σ_c^∞. This process leads to the apparition of a crack at an upper scale, at which the damage variable was defined. Note that, according to this analysis, crack initiation can be understood as an inter-scale jump of the discontinuities, from the scale with heterogeneities to the scale initially assumed with homogeneous properties. In what follows in this thesis, the concepts of crack initiation and crack onset will be understood in this sense. Since the problems analyzed here are studied at different scales, heterogeneities will be assumed to be confined to a scale sufficiently smaller than the scale used for each problem.

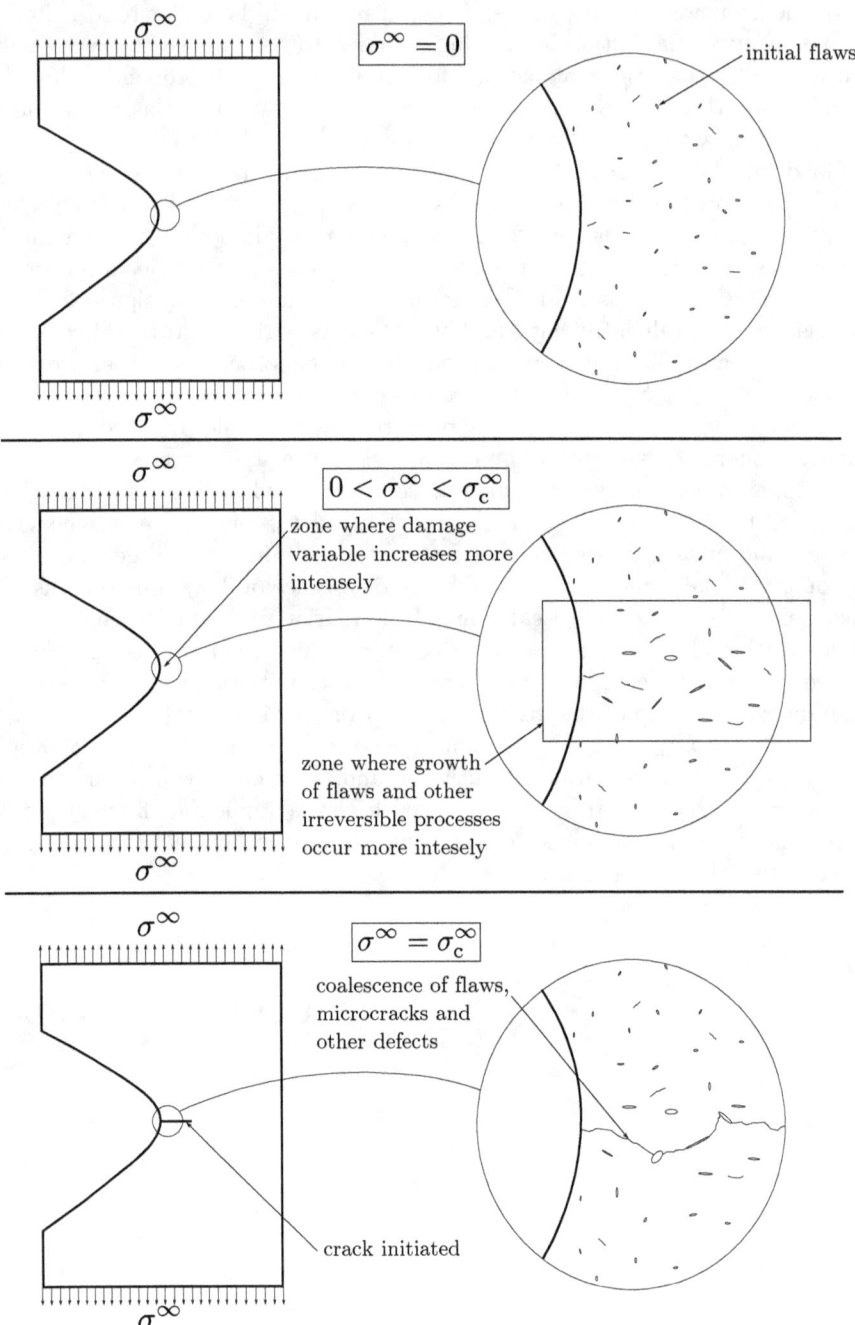

Figure 2.2: Schematic of the crack initiation process understood as the inter-scale jump of the discontinuities

2.3 Predicting crack initiation

Once the interpretation described in previous section for accepting the study
of crack initiation is assumed, it remains the question of how to predict the
apparition of the crack. In particular, it is of great practical interest to obtain
the critical load for which the crack is initiated, along with the situation and
size of the new crack. In the following, the main methods employed to predict
crack initiation are briefly introduced. Some of these methods were not proposed
with the main motivation of predicting crack initiation but motivated by other
problems.

2.3.1 Damage mechanics

Within the continuum mechanics, the more natural approach in accordance to
the previous discussion is to use a constitutive law which contains the information
about the evolution of the damage variable as a function of the other state variables, e.g., stresses and strains, see Lemaitre (1985, 1990); Benallal et al. (1992).
The damage evolution will be given by phenomenological laws, which model the
evolution of the flaws at the microstructure scale in a homogenized form. When
the damage evolution at the microstructure is unknown, these laws are normally
obtained by complex and numerous experiments of the material to fully characterize the irreversible processes leading to the evolution of the damage and its
influence on the material properties. In other cases with more knowledge about
the defects evolution at the microstructure, damage models are proposed based
on this knowledge, see e.g. Maimí et al. (2007). In general, the Thermodynamic
of irreversible processes represents a obliged framework for the development of
these laws. The problem resulting from introducing these constitutive relations
within the usual framework of the elasticity is, in general, nonlinear. The solution of these models requires the use of the typical computational procedures
in applied mechanics. These models can have a high computational cost due
to its nonlinearity and complexity, in particular, if the nonlinearity affects large
regions of the domain. Moreover, they have well known problems as inherent
mesh sensitivity above a certain damage level, due to the local loss of the elliptical characteristic of the partial differential equations governing the problem, see
e.g. de Borst (2004) for a full description of the problems of these models.

2.3.2 Atomistic modeling

Out of the continuum mechanics framework, the atomistic models study the fundamental materials behavior and deformation phenomena at the molecular and
atomic scale. These models are able to predict crack initiation along with all
the steps of the failure process, see e.g. Buehler (2008) for a full review. In
the atomistic models, the true discreteness of the materials is explicitly modeled. For example, in metals, each atom is considered as an individual particle
which is glued to the other atoms in the crystal structure. In general, atoms and
molecules are taken as indivisible particles and a set of forces is applied between

them, representing, e.g., chemical bonds. The results of atomistic simulations are
the positions and velocities of the particles at a certain time. Thus, the resulting
system grows in complexity as the amount of particles modeled increases. The
vast majority of these systems have to be solved computationally. Current computational technology is able to simulate billions of atoms (10^9). Unfortunately,
engineering scales go far beyond this amount of atoms since a cubic millime-
ter contains about 10^{20} atoms. Even if atomistic models are not (yet) able to
study entire solids at the traditional engineering scales, this is a powerful tool to
study phenomena which occurs at submicroscopic scale, which is very common in
failure of many material systems, see e.g. the atomistic study of dislocations by
Buehler et al. (2004). Atomistic modeling can also be used to obtain constitutive
models for the continuum mechanics, see e.g. Krull and Yuan (2011) and as a
part of multiscale models for failure, see e.g. Zeng and Li (2010). Moreover, this
is suitable to study failure in nanometric materials as graphene, see e.g. Cao and
Qu (2013).

2.3.3   Cohesive zone models

Other methods being used to study crack initiation have been proposed as a
generalization of the Griffith criterion. In the case of the two methods studied in
the following, the generalization introduced enables the study of crack initiation
remaining coherent with the Griffith criterion in the presence of a crack when
nonlinear effects can be neglected. The form in which each generalization solves
the inability of the Griffith criterion for crack initiation is based on moderating
different hypotheses used by Griffith.

One of the strong hypotheses behind the Griffith criterion is the assumption
of an infinite jump in the stresses supported at points along the crack path when
the crack propagates. According to the Griffith theory, when a crack propagates,
points situated along its path go from fully intact to fully broken in a discreet
manner. This is one of the reasons which avoids the Griffith criterion to be
able to predict crack initiation. If the stresses supported by a point have an
infinite jump, the dissipated energy necessary for the process has a finite value
which has to be fulfilled immediately being extracted from the energy released.
This condition can be fulfilled in the presence of a crack with finite length. On
the contrary, due to the different scaling of energy dissipation and release, the
Griffith criterion cannot be fulfilled for a crack with vanishing length. Hence,
if this hypothesis is released, i.e. the stresses can evolute in a more continuous
manner, the energy necessary to open a new crack surface with infinitesimal
area can be fulfilled continuously according to a prescribed evolution. Thus, the
cohesive zone models (CZM), introduced by Dugdale (1960); Barenblatt (1962),
implement laws which relate the separation of the cracks faces and the stresses
transmitted through them.

In general, cohesive zone models prescribe normal tractions σ along the cohesive zone as functions of the relative displacements (separation) δ between the
two surfaces, see Figure 2.3(a). Tractions vanish when the relative displacements
reach a critical value δ_c. In general, the cohesive law can be expressed using a

law as

$$\sigma = \sigma(\delta, f), \tag{2.4}$$

where f is a history parameter which contains the information about the loading history. This parameter is particularly relevant for problems with cyclic loading. In the case of monotonic loading, the influence of the loading history on the cohesive law is generally neglected. Cohesive laws which are history-independent are named holonomic. Actually, this distinction is more associated to the implementation since the majority of cohesive laws can be implemented either in a holonomic way or introducing a history-dependence. In what follows, only holonomic cohesive laws will be described since only crack initiation under monotonic loading is studied in the context of this thesis.

Figure 2.3(b) represents an example of cohesive law. The most important parameters defining the cohesive law are δ_c and the tensile strength σ_c which is the maximum normal traction which can be supported. The fracture energy G_c is given by the integral of the cohesive law,

$$G_{1c} = \int_0^{\delta_c} \sigma(\delta) \mathrm{d}\delta. \tag{2.5}$$

This expression for the fracture energy can be rewritten in terms of dimensionless tractions $\hat\sigma(\delta/\delta_c) = \sigma(\delta/\delta_c)/\sigma_c$ and relative displacements $\hat\delta = \delta/\delta_c$ as,

$$G_{1c} = \sigma_c \delta_c \int_0^1 \hat\sigma\left(\hat\delta^t\right) \mathrm{d}\hat\delta. \tag{2.6}$$

Note that according to this expression, G_c is directly proportional to σ_c, δ_c and a dimensionless factor which depends only on the form of the cohesive law. Most fracture problems are well modeled by assuming pure mode I since cracks usually follow paths which keep the crack in pure mode 1, in accordance with the criterion of local symmetry by Barenblatt and Cherepanov (1961); Erdogan and Sih (1963). However, the mode II can be relevant in some problems, e.g., crack initiation at interfaces. The reason is that interfaces can have weaker fracture and strength properties than the bulk, therefore the crack propagation along the interface can be preferential even if it does not occur in pure mode I. To account for this, cohesive laws containing also a dependence between the sliding separation δ^t and the shear stress τ can be defined. In this case, the sliding component of the cohesive law will be determined by analogous properties as shear strength τ_c and critical sliding separation δ_c^t. The fracture mode in mode II will be given by the analogous expression to (2.6),

$$G_{2c} = \tau_c \delta_c^t \int_0^1 \hat\tau\left(\hat\delta\right) \mathrm{d}\hat\delta^t, \tag{2.7}$$

where $\hat\tau = \tau/\tau_c$ and $\hat\delta^t = \delta^t/\delta_c^t$ are the analogous dimensionless functions and variables. Both fracture modes can also be coupled prescribing an interaction between modes, see e.g. Tvergaard and Hutchinson (1993) for a cohesive law proposed with both modes being coupled.

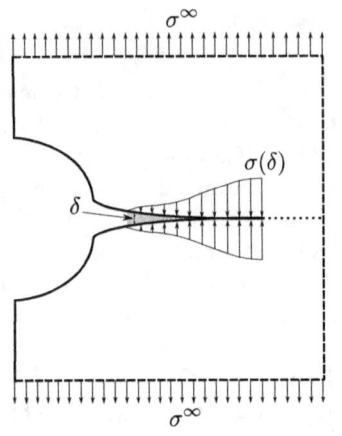

(a) Example of a body with a cohesive law prescribed along a line.

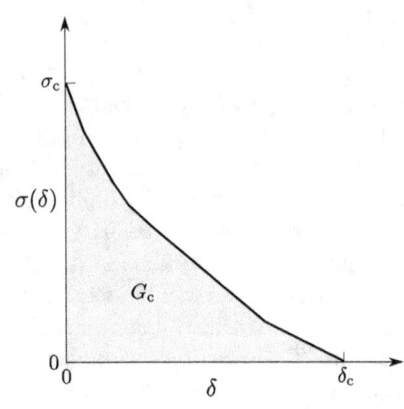

(b) Example of cohesive law

Figure 2.3: Cohesive zone model

Actually, CZM can be considered, with some conceptual differences, to be an application of the damage mechanics described previously with the damage confined into a finite number of surfaces. The dimensionless separation $\hat{\delta}$ can be seen as a damage variable which governs the constitutive law. In CZM, the damaged zone is denoted process zone since for cracked bodies this is the zone where the irreversible processes occur, see Figure 2.4. This figure shows the typical zones of a cracked body modeled with CZM and the physical sense for two materials for which CZM is extensively employed. Observe the particular microstructural processes which justify the assumption of a continuous separation law and the existence of cohesive tractions. In concrete, the interaction between the constituents promotes basically the generation of unconnected microcracks among other phenomena, which reduces partially the ability of supporting tractions, see e.g. Otsuka and Date (2000) for a full description of the process zone in concrete. In composites, the reason for a process zone is also due to the different behavior of the constituents. In addition to the reason described for concrete, in composites, long fibers can act as traction bridges maintaining an important part of the supporting ability.

Previous examples to justify the existence of cohesive tractions show that the cohesive law depends strongly on the material through the microstructure and its failure behavior. Thus, apart from the failure properties as strength and fracture energy, the form of the law depends also on the material. Hence, an extensive amount of forms have been proposed to be used for certain materials or problems. For the sake of simplicity, in what follows, the discussion about the forms is reduced to pure mode I. Some of the most typical examples are:

- Constant: This is surely the simplest form for a cohesive law. Dugdale (1960) proposed to model the plastic zone in the vicinity of the crack tip

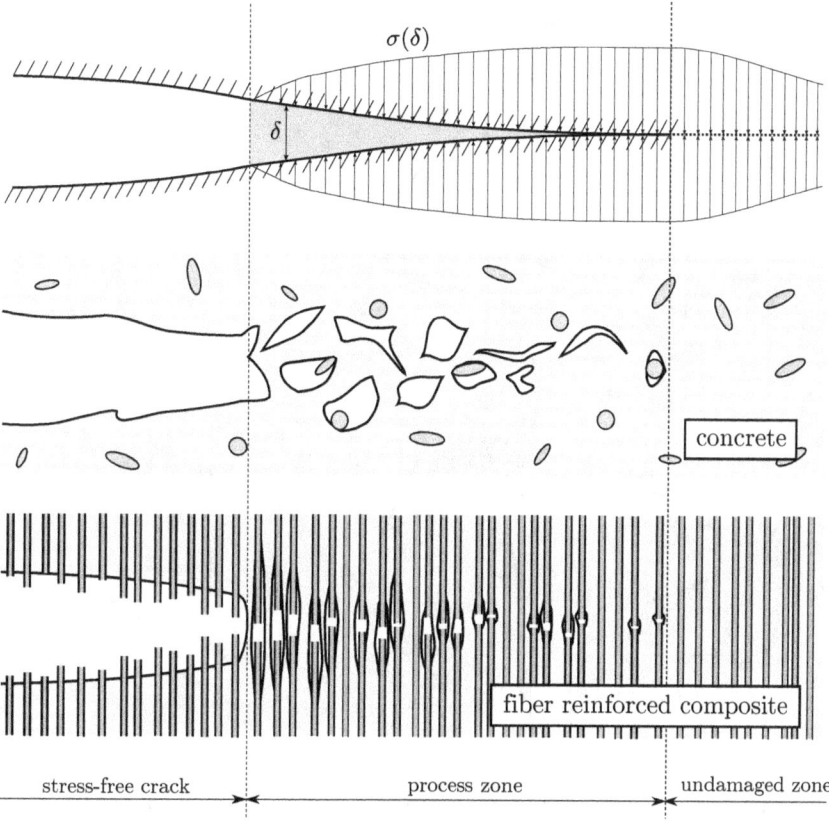

Figure 2.4: Different zones which can be found along a surface with a cohesive law prescribed compared to examples of the failure at the microstructure scale for concrete and long fiber reinforced composites.

by limiting tractions to the yield strength σ_y. If this is prescribed within a range of relative displacements $[0, \delta_c]$ between faces, this is *de facto* a cohesive law with the form represented in Figure 2.5(a). This idea enabled to avoid the unrealistic singularity of stresses at the crack tip in metals. The law is fully defined by σ_y and the fracture toughness G_{1c} because $\delta_c = G_{1c}/\sigma_y$.

- Linear softening: It was proposed by Hillerborg et al. (1976) for its use to model fracture in concrete inspired in experimental results. According to this law, see Figure 2.5(b), no separation occurs for tractions below the tensile strength. Once it is reached, a linear softening follows up to δ_c. Similarly to the previous cohesive law, this is fully defined by σ_c and G_{1c} since $\delta_c = 2G_{1c}/\sigma_c$.

- Triangular: This is essentially similar to the previous cohesive law with the

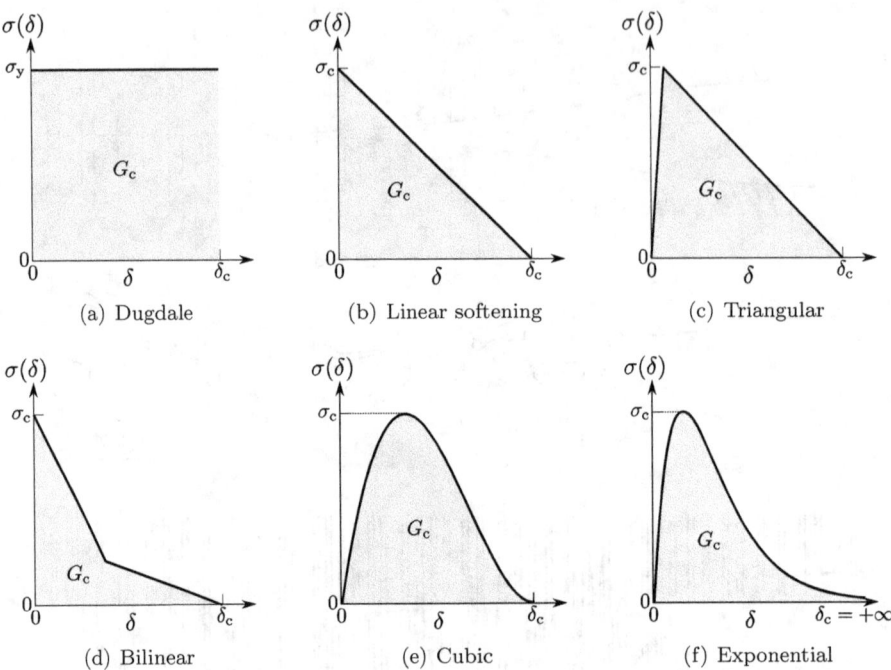

Figure 2.5: Examples of cohesive laws

addition of a linear hardening part with a finite slope instead of the vertical first part of the previous one, see Figure 2.5(c). In general, the finite initial slope introduces a new independent parameter, which has to be defined. In many practical problems, this initial slope is introduced only because of computational reasons and it is set high enough to neglect the influence of the value on the results. However, the initial slope has been reported to be important on the tendency of some results, see Section 4.2 for a discussion in the context of one of the problems studied in this thesis.

- Bilinear softening: This cohesive law, represented in Figure 2.5(d), is a generalization of the linear softening by defining two different slopes for the softening part. These two zones are inspired in experimental results in concrete, see e.g. Petersson (1981). An additional parameter is necessary to fully define this cohesive law in comparison with the linear softening law. Hence, the traditional fracture energy is divided into initial fracture energy and total fracture energy, see Jenq and Shah (1985).

- Cubic polynomial: The form of this cohesive law is given by a cubic polynomial which vanishes for $\delta = 0$ and $\delta = \delta_c$, the last one being a double root, see Figure 2.5(e). This law was proposed by Needleman (1987) to study crack initiation at interfaces. Only two parameters, σ_c and G_{1c}, have to be defined to fully characterize this law. A wide variety of generalizations of this law to take into account mixed mode has been proposed, due to the relevance of fracture mode-mixity in interface cracks.

- Exponential: This law is represented in Figure 2.5(f) and is characterized by an exponential form. This was proposed by Needleman (1990) to model crack initiation at interfaces. The exponential curve is derived from the comept of universal binding energy correlation reported by Rose et al. (1981).

The suitability of each cohesive law has been extensively studied, see e.g. Alfano (2006) for a critical review of cohesive laws used in composites and Park and Paulino (2011) for a general recent review.

2.3.4 Finite fracture mechanics

Finite fracture mechanics is based on releasing other hypothesis of the Griffith criterion: the assumption of that crack growth is infinitesimal. If this assumption is released and the energy balance is formulated in an incremental manner, the resulting balance is able to predict initiation of cracks with finite length. Based on this idea, several methods, either based on this energetic balance or on stress conditions, have been proposed and used in a intermittent manner along the last half-century, see e.g. Neuber (1958); Tszeng (1993); Hashin (1996); Leguillon (2002); Taylor et al. (2005); Cornetti et al. (2006). The common framework of these methods is also named "Theory of critical distances" by other authors.

Under the term finite fracture mechanics, a large set of criteria for crack initiation are framed. They all are based on assuming that a critical length, either material- or structural-dependent, governs the crack initiation. A large part of them have been used in an intermittent manner for a long time. Actually, the term finite fracture mechanics was initially proposed by Hashin (1996) to refer to an energy-based criterion which assumes a crack onset of a finite length. However, the term has been also employed since then by the engineering community to refer to either other criteria based on a critical length or to the complete framework. In the context of this thesis, the original criterion used by Hashin (1996) is denoted as "incremental energy criterion", being described in Section 2.3.4.1, whereas the term "finite fracture mechanics" is reserved for the general framework.

Although every criterion within this framework uses a critical length one way or another to predict crack initiation, this is not necessarily associated to a length with a physical sense. Only in some cases, this critical length is associated to the crack length predicted immediately after the onset.

The main general advantage of the criteria for crack initiation proposed within this framework, in comparison with the methods described previously, is that they are based generally on linear analyses. Thus, the criteria within the finite fracture mechanics are more adequated for applications which require the combination of a high level of simplicity, e.g., industrial applications or multiscale analyses, along with a good agreement with experiments, as reported in the majority of problems. However, as pointed out by some authors, see e.g. Taylor (2007), the other side of the coin is that, at least at the moment of writing these lines, the majority of these criteria are not fully physically fonded. They are well based on typical physical laws and its generalization, but a rigorous interpretation of why they works has not been provided, due to the inherent difficulty on deducting general conclusions of the complex nonlinear and stochastic nature of the physics governing the crack initiation. In spite of this fact, these criteria are claimed as suitable tools to provide useful and simple expressions to predict crack initiation.

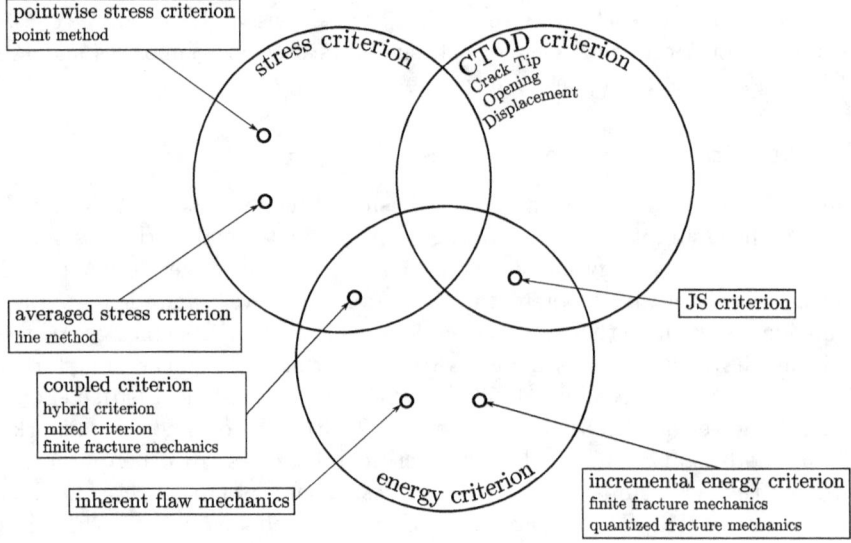

Figure 2.6: Schematic of the main criteria in the framework on the finite fracture mechanics

Figure 2.6 shows a general outline of the different criteria within finite fracture mechanics. They are classified by which type of condition or conditions implement. Due to the recent and intense development of these criteria, each criterion have been referred to with several names in the literature. Alternative names found in the literature are written in italics. In the following, the main criteria are described following the structure proposed by Maimí et al. (2013) by splitting the criteria into two groups: those whose critical length is either fixed or variable.

2.3.4.1 Fixed-critical-length criteria

This category includes a diverse series of criteria with a common point: they are based on a critical length l_c, which depends on the material properties. The form in which l_c is obtained differs for each criterion. In spite of the inherent diversity on the nature of each criteria, in the majority of cases the estimation of l_c is based on setting it so that the criterion matches with the classical criteria for the problems for which they are assumed to work. Basically, these problems correspond to a body with a sufficiently large crack on the one hand and a tensile specimen on the other. The classical criteria to match with, are, respectively, the Griffith criterion and the tensile stress criterion. Thus, in these cases the critical length depends mainly on the typical material properties governing both problems: the fracture toughness G_c and the tensile strength σ_c.

In the following, the main criteria based on a fixed l_c are divided into two groups depending on if they are founded on either a stress or an energy analysis.

Stress-based criteria The stress-based criteria were the first criteria to be used to predict failure and crack initiation. However, they were rejected by Griffith (1921) because they contradict the experimental evidences of a strong size effect on failure of brittle materials with initial defects. The reason is that initial stress criteria were based on the maximum value of stresses found in the whole body. Thus, the scaling of the geometry does not modify the stress state at the point where the maximum is found, and as a direct consequence any size effect cannot be predicted. However, if the stresses are regarded along a fixed length and this length is not scaled with the geometry, because, e.g., it is considered a material property, the stresses along the critical length l_c do change with the geometry scaling, hence a size effect can be predicted.

Once it is accepted that it is necessary to examine stresses along a critical length, the next question arises: which stress value of the whole length has to be compared with the strength material properties. Different answers to this question produce different criteria. The two most relevant criteria are described in the following.

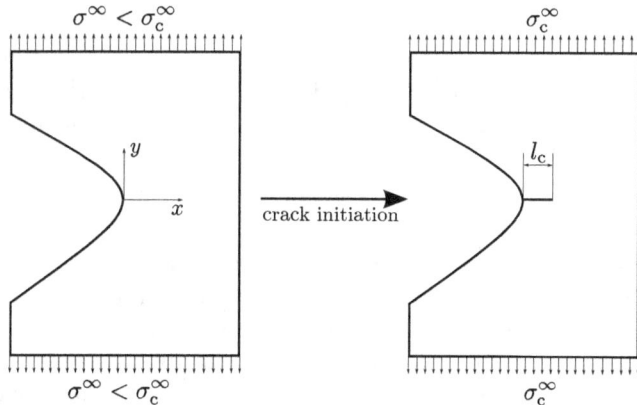

Figure 2.7: Schematic of the process of crack initiation according to the criteria in the framework on the finite fracture mechanics

Pointwise stress criterion According to the review by Taylor (2007), the pointwise stress criterion, also known as point method, was initially proposed by Peterson (1938, 1959) based on his works about fatigue in the 1930's. Unlike the

usual strategies to obtain l_c described previously, these authors initially related
the critical length with microstructural material distances as grain size or inter-
atomic distances.

This criterion is based on the condition for the crack onset of that the stresses
at all the points situated at a segment of length l_c have to exceed or equal the
tensile strength σ_c,

$$\sigma(x) \geq \sigma_c \; \forall x \in (0, l_c), \tag{2.8}$$

where x is the coordinate which varies along the segment, see Figure 2.7. When
this condition is fulfilled a crack onset occurs. This criterion is usually applied along segments where stress field is monotonic, e.g., stress concentrators
as notches or cracks. In this case it is only necessary to verify this condition at
the point situated at a distance l_c of the stress concentration point. Thus, the
condition (2.8) can be rewritten more simply as,

$$\sigma(l_c) = \sigma_c. \tag{2.9}$$

This form of expressing this criterion causes that this criterion is also known
as point method (PM). Note that only the solution at one point of a linear-
elastic problem is required. This value can be easily obtained by means of either
analytical methods or, if required, well established computational methods as
finite elements.

The critical length corresponding to this criterion l_c^{PSC} can be derived by
imposing the agreement with the Griffith criterion in the case of large cracks,

$$l_c^{\mathrm{PSC}} = \frac{EG_c}{2\pi\sigma_c^2}, \tag{2.10}$$

which can be obtained by evaluating the stresses at $r = l_c^{\mathrm{PSC}}$ given by the
Westergaard (1939) solution for a cracked infinite plate, applying the Griffith
criterion based on the Irwin relation between G_c and the critical stress intensity
factor K_c, and equaling the resulting stress to the tensile strength σ_c.

Averaged stress criterion   The averaged stress criterion was introduced
by Neuber (1958) based on his studies in Germany contemporaries to Peterson's.
Similarly to the previous criterion, this was proposed initially for fatigue and
subsequently reinvented for problems with static loading by Novozhilov (1969).
Instead of imposing that every value of stresses along the segment has to exceed
or equal to σ_c, this criterion compares the average of stresses along the segment
with σ_c. Thus, the condition for the crack onset is

$$\frac{1}{l_c} \int_0^{l_c} \sigma(x) \mathrm{d}x = \sigma_c. \tag{2.11}$$

This expression has been extended to three-dimensional problems by Nazarov
(2004).

The critical length corresponding to this criterion l_c^{ASC} can be obtained taking the averaged stress of the Westergaard (1939) solution giving

$$l_c^{ASC} = \frac{2EG_c}{\pi\sigma_c^2}. \tag{2.12}$$

Energy-based criteria This category includes those criteria based on the combination of an energy condition and a fixed length.

Incremental energy criterion This criterion proposes a condition based on the same energy balance used by Griffith (1921) but formulated in a incremental manner. The critical length l_c is the incremental jump of crack length and is normally associated to a certain material characteristic length at the scale employed for the study. The condition can be written as,

$$\int_0^{l_c} G(a)\mathrm{d}a = G_c l_c, \tag{2.13}$$

where a is an intermediate crack length used to obtain the integral term. This criterion has been used in an intermittent manner for a long time without an apparent connection between the authors employing it. To the author's knowledge, it was initially proposed by Aveston et al. (1971) in the context of micromechanics of unidirectional fibre composites. Some years later, the criterion was implemented by Parvizi et al. (1978) to explain the size effect found in the transverse crack onset in cross-ply laminates. They identified l_c with the inner-ply thickness, see Chapter 6 for an analysis of the problem. Subsequently, Tszeng (1993) applied the same criterion to the crack onset at the interface between a stiff spherical inclusion and the surrounding matrix. In this case, l_c was associated to an arbitrary fixed debond angle. Three years later, Hashin (1996) examined the evolution of the crack density as function of the external load in the problem of transverse cracking studied by Parvizi et al. (1978). This analysis provided an analytic expression to predict the crack density based on this criterion. Similarly to Parvizi et al. (1978), Hashin (1996) identified l_c with the inner-ply thickness. In addition, he proposed the name of finite fracture mechanics for this criterion, now used to denote the complete category of methods as discussed previously. More recently, Pugno (2004) proposed a similar method, called quantized fracture mechanics, and applied it to the atomistic scale. This application leads to the interesting result of the existence of prohibited strength bands if l_c is associated to multiples of the inter-atomic distance.

Inherent flaw model This criterion is based on assuming the existence of flaws which grow in a stable manner as "non-Griffith cracks", according to the exact words used by Dvorak and Laws (1986). This stable growth ends when flaws reach a critical length l_c, which is a material property. Above this length, flaws become governed by the Griffith criterion and if its conditions are fulfilled, the crack grows unstably. Note that, from the point of view of the result obtained, this is often equivalent to assume the existence of a crack in the material with length l_c. This criterion was initially proposed by El Haddad et al. (1979) for fatigue problems based on the length for which a crack arising from a notch has a minimum value of G. Subsequently to this work, Dvorak and Laws (1986) presented a theory to predict transverse crack initiation in cross-ply laminates, which is based on an equivalent idea. Next decades, many works based on the vision of Dvorak and Laws (1986) of transverse cracking have been presented.

In summary, this criterion for crack initiation can be written as,

$$G(\sigma_c^\infty, l_c) = G_c. \tag{2.14}$$

2.3.4.2 Variable-critical-length criteria

Many authors have claimed that the critical length is not a material property, see Maimí et al. (2013) for a review, but it also depends on the other problem parameters, being, in general, a structural parameter. In this sense, the criteria included in this category are based on a critical length which is a result of the analysis and not a material property.

JS criterion   This criterion was proposed by Jenq and Shah (1985) to study crack initiation in concrete. It is essentially similar to the inherent flaw model described previously but with a variable critical length. The basic assumption is similar: a defect or damaged zone grow stably (even tough the evaluation of the Griffith criterion predicts an unstable growth) up to a critical length for which the defect propagation becomes governed by the Griffith criterion. The critical length is given by the combination of the Griffith criterion and the crack tip opening displacement (CTOD) criterion proposed by Wells (1961) and Cottrell (1961). The CTOD criterion was proposed to account for the nonlinear processes near stress singular points when their effect cannot be neglected. The combination of both criteria leads to a system of two equations with two unknowns: the critical length l_c and load σ_c^∞.

This criterion can be written as,

$$G(\sigma_c^\infty, l_c) = G_c, \tag{2.15a}$$

$$\text{CTOD}\,(\sigma_c^\infty, l_c) = \text{CTOD}_c. \tag{2.15b}$$

Coupled stress and energy criterion   Multiples evidences exist in the literature of that in many problems, a simultaneous fulfillment of stress and energy criteria is necessary to initiate a crack, see Chapter 3 for a revision. Based on these evidences and within the context of the finite fracture mechanics, Leguillon (2002) proposed a coupled stress and energy criterion to predict crack initiation. The two criteria impose two conditions for two unknowns: the critical length and load, hence the problem is determinated. If the pointwise criterion is taken as the stress criterion following the original proposal by Leguillon (2002), the system given by this criterion can be written as,

$$\sigma(\sigma_c^\infty, x) \geq \sigma_c \ \forall x \in (0, l_c), \tag{2.16a}$$

$$\int_0^{l_c} G(\sigma_c^\infty, a)\mathrm{d}a = G_c l_c. \tag{2.16b}$$

Note that the two equations correspond to the previous equations defining the pointwise (2.8) and energy criteria (2.13). However, in this case l_c is an unknown. Many modifications of this criterion have been proposed to study problems of a very different nature. Since this is the criterion employed in this thesis, Chapter 3 is consecrated to present it in more detail.

The coupled criterion of the finite fracture mechanics

From the point of view of the classical theory of fracture, the crack growth seen at the smallest scales is interpreted as the simplest case of fracture, known as cleavage fracture, i.e. the separation of atomic planes. However, in the majority of materials of interest the actual processes of crack initiation and growth are much more complex that the cleavage fracture since they are associated to complex and varied processes at different scales. These processes are strongly affected by local anisotropy or inhomogeneities as inclusions or porosity, which strongly depends on the material nature. All these factors complicate considerably the general formulation of a fracture theory based on the micromechanism. Moreover, the case of crack initiation is more complex than the case of crack growth because the presence of a crack localizes the processes associated to the fracture to a relatively small region around the crack tip whereas in the case of crack initiation these processes are associated to a larger region which increases the interaction between scales.

The complexity inherent to the process of crack growth has been successfully avoided for a wide variety of materials by the linear elastic fracture mechanics (LEFM). LEFM enables the analysis of this process in the terms of the continuum mechanics leading to relatively simple analyses. One of the reason for this success has been that the theory is based on well-known general laws as the conservation of the energy. The use of general laws makes possible that, within a certain range of crack length, the crack growth can be predicted using the same theoretical tool for materials so different as glass and steel, in spite of the dissipative phenomena associated to the fracture of each one are very different. The key is that the conservation of energy has to be fulfilled independently if the energy is dissipated on the separation of atomic planes in glass or mostly in permanent plastic strain in the vicinity of the crack tip in steel. From a practical and engineering point of view, it is clear that this lack of distinction between the actual process of dissipation is not relevant since the LEFM is a common and useful tool on the current engineering community.

Following the previous discussion for the case of crack growth, from the engineering point of view, the coupled criterion represents a good equilibrium between simplicity and physically foundation of the theory for the study of crack initiation. The coupled stress and energy criterion was proposed in the context of the finite fracture mechanics and is utilized to predict the conditions which lead to crack initiation. This criterion assumes that a crack onset of a finite extension initiates when the next two classical criteria are simultaneously fulfilled: on the one hand, a stress criterion prescribes a condition over the stresses, prior to the crack onset, at the surface where crack opens. On the other hand, an energy criterion is imposed, being based on the energetic balance between the states prior and after the crack onset.

The physical foundation of the coupled criterion is inherent to the two criteria which are combined. The stress and energy criteria have been justified and utilized separately for a wide variety of materials with a relative agreement within the range of validity classically assumed for each one. Therefore, it can justify the combination of both criteria in intermediate situations between the classical range of applications of each criteria. Even though the coupled criterion is not as rigorous and physically fonded as the atomistic modeling, this lack is balanced out by the possibility of keeping in the continuum mechanics approach and solving problems at the macro scale, which would be computationally impossible to solve by an atomistic model currently. Moreover, in a great amount of problems, semianalytical expressions for the critical load leading to the crack initiation can be obtained. Regarding to other approaches based on the continuum hypothesis as cohesive zone models or damage mechanics, the comparison is qualitatively similar, these approaches requiring a much lower computational cost than the atomistic modeling but much higher than the coupled criterion. On the other extreme, the stress criteria based on a fixed critical length, see Section 2.3.4.1, requires even a lower computational cost than the coupled criterion. However, the introduction of a material-dependent critical length is difficult to be physically justified.

This chapter presents a review of the state of art of the coupled criterion and is organized as follows. After the description of the historical background in Section 3.1, the key hypothesis of the coupled criterion is discussed in Section 3.2. The stress and energy criteria are analyzed in Sections 3.3 and 3.4, respectively, focusing on the theoretical definition on the one hand and on the practical application on the other. Finally the combination of both criteria is studied in Section 3.5.

3.1 Background

The idea of the need of fulfilling both stress and a energy conditions for the crack initiation has been suggested and even used in a intermittent manner for a long time. The pioneering works circumscribes the discussion or application of this criterion to the particular problem studied by each work. As a consequence of the lack of a general discussion, this criterion has been proposed several times by different authors without an apparent connection between them up to the general discussion by Leguillon (2002).

To the knowledge of the author, the first work suggesting the need of fulfilling both stress and energy criteria for the crack initiation is due to Bader et al. (1980) who, in the context of discussion of some experimental results obtained for transverse cracking in cross-ply laminates, wrote,

> *The basis of our present understanding is that in order for a crack to form it must be both mechanistically possible and energetically favorable.*

This sentence can be understood as a suggestion of the need of fulfilling two conditions, the mechanistic condition, which could correspond to a stress criterion, and the need of being energetically favorable, which corresponds to an energy condition. However, Bader et al. (1980) and some coworkers in other papers as Garrett and Bailey (1977); Parvizi et al. (1978); Bader et al. (1980) applied exclusively an energy criterion with a fixed critical crack length. They identified this critical length with the inner-ply thickness by assuming that the crack spans the whole thickness of the transverse ply.

Almost simultaneously to this work, Goods and Brown (1979) utilized the idea of combining both stress and energy criteria for the study of crack initiation at the interface between a spherical particle and the surrounding matrix. Subsequently to the formulation of an energy criterion, the authors wrote

> *It is of course not sufficient for just the energetics to be favorable: in order to nucleate cavities the interface strength σ_t must also be reached.*

Note that the need of fulfilling both criteria is presented as a well known condition for cavity nucleation, which can be identified with crack initiation. This also occurs in subsequent papers dealing with the same problem as Fisher and Gurland (1981) who claimed, after formulating an energy criterion,

> *[...] it is additionally required for void nucleation that the normal stresses σ_n at the interface satisfy the critical conditions that σ_n be greater than or equal to the critical bonding stress at the interface σ_c over the area of the nucleated crack.*

To remark from these words that the authors assume that the idea of both stress and energy conditions have to be fulfilled is well established, since identically to the previous work, they neither explain this assumption nor cite any work doing it. It is worth highlighting that these authors denote it as "double criterion". These two works were revisited in a review consecrated to the mechanisms of plastic flow in materials hardened with particles by Emburi (1986). He cited the last two works along with other similar ones, remarking explicitly the condition

of fulfilling both criteria for crack initiation. Tszeng (1993) studied the problem of crack initiation at the particle-matrix interface for an ellipsoidal particle
applying the coupled criterion and citing the previous works to justify the condition of satisfying both stress and energy criteria. He also presented an interesting
discussion about the critical angle of debond at the crack onset.

In view of the previous cited words, the condition of satisfying both criteria
seems to be a well established law. However, no general and formal discussion
has been found up to the paper by Leguillon (2002). He discussed the condition
of fulfilling both conditions by presenting evidences corresponding to classical
problems for different materials which are summarized in Section 3.2. From then
on, a great amount of works have been presented applying this criterion to very
diverse problems of crack initiation in different materials. Additionally to this
discussion, in is fair to highlight two very relevant contributions by Cornetti
et al. (2006) and Taylor et al. (2005) which are key for the subsequent evolution
of the coupled criterion. A proof of the extension of the use of this criterion
is the evolution of papers in JCR journals citing this article from 2003, see
Figure 3.1. The number of cites is approximately multiplied by an approximate
factor going from 2 to 4 in other less restrictive database as Scopus or Google
Scholar. Moreover, the increasing tendency is confirmed if the number of papers
is normalized with the average rate of annual growth on scientific publication
which is 2.8 % according to National Science Board (2014), this correction is
represented as a solid line in Figure 3.1.

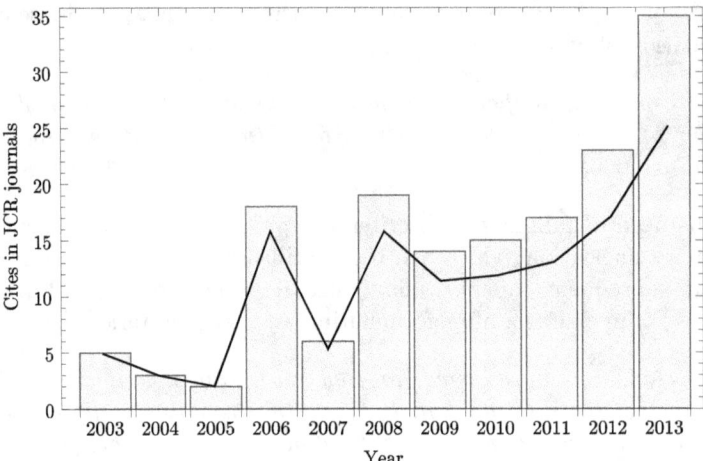

Figure 3.1: Bars: Number of cites to Leguillon (2002) per year in articles published in JCR
according to the Web of science database. Solid line: Number of cites corrected with the annual
growth of total scientific publication according to National Science Board (2014)

One of the problems for which the coupled criterion has been widely employed
is the prediction of crack initiation at V-notches in homogeneous materials. Assuming a sharp notch, a stress singularity is predicted at the notch tip. Since

this singularity is softer than the singularity due to a crack, the linear elastic fracture mechanics is not able to study the failure at the notch. On the contrary, the coupled criterion does enable to predict it, as was demonstrated by Leguillon (2001, 2002) for symmetric loading. Carpinteri et al. (2008) studied a similar problem, using a different stress criteria and compared the expression obtained with experimental results for bending specimens with notches. In general, they observed a good agreement between the model and the experiments. In addition, Henninger et al. (2007) showed that the model agrees well with the predictions obtained using a Dugdale cohesive model, see Section 2.3.3 and Figure 2.5 for a brief description of this cohesive model. The hypothesis of totally sharp notch is reviewed in other works by Leguillon and Yosibash (2003), see also Leguillon et al. (2007), who evaluated the effect of the notch tip radius on the prediction given by the coupled criterion. Yosibash et al. (2006) generalized these models to take into account the cases with asymmetric loading. The comparison with experiments showed a good agreement for both the critical load leading to the failure and the angle of crack initiation. In fact, a very interesting result was found for the crack angle: the observed crack after the onset is straight, keeping the angle predicted from the notch tip, see pictures in Yosibash et al. (2006). This is not an obvious result since the singularity due to the symmetric mode is stronger than the singularity due to the antisymmetric mode. That means that, except in the case of purely antisymmetric loading, a region exists in the vicinity of the notch tip where the antisymmetric mode is irrelevant in comparison with the symmetric mode, see e.g. Hills and Dini (2011). Thus, if the crack initiation was continuum, the angle of the crack should keep the symmetry, being deviated during its growth. However, a straight crack is observed which could be an evidence of a finite crack length at onset, or at least, of the influence of stresses along a finite length. In the case of mixed-mode loading, the hypothesis of sharp notch was also revisited by Priel et al. (2008) who verified that the agreement with experiments improve when the notch radius is considered in the analysis for mixed-mode loading. In the context of this thesis, a model for the prediction of failure at V-notches with an interface between similar materials is developed in Chapter 7, see also García and Leguillon (2012). Tran et al. (2012) also investigate a similar problem applying the model to a bending specimen obtaining a good agreement with experiments. Sapora et al. (2014) analyzed the case under pure mode II using approximative tools to obtain an almost analytical failure criterion. The related problem of failure at U-notches has been also studied: Picard et al. (2006) proposed a model for U-notches in bending specimens to evaluate how the experimental measured value for the fracture toughness is affected by the notch radius.

The coupled criterion has also been utilized to investigate the failure due to other soft singularities as a crack approaching an interface between dissimilar materials. Martin et al. (2008) applied it to study the competition between the deflection of the crack and the penetration at an interface. As an application of this model, Leguillon and Martin (2013b,c) explained the strengthening effect caused by the presence of layers of dissimilar materials. A model has been also developed to predict a crack kinking out an interface by Leguillon and Murer

(2008b). The related phenomenon of crack deflection in an homogeneous material has been also studied by applying the coupled criterion by Leguillon and Murer (2008a) obtaining a good agreement with experiments.

Crack initiation in composites has been also studied using the coupled criterion at different scales, from nano to macro scale. In the majority of cases with available experiments, a good agreement have been found with the predictions obtained. In what follows the most relevant works are briefly reviewed sorted by the scale of the analysis.

At the nanoscale, Salviato et al. (2013) presented, to the author's knowledge, the first work implementing the coupled criterion to this scale. They utilized it combined with surface elasticity to study the debonding strength of a spherical nanoparticle subjected to a hydrostatic tension. The implementation enables a relatively simple evaluation of the influence of the interface properties on the process of debonding.

At the microscale, the coupled criterion has been implemented to study the initiation of the those failure mechanisms which have the origin at this scale. This is very common in long-fiber reinforced composites due to the typical fiber radius in common composites. Regarding to the well-known matrix or interfiber failure, Mantič (2009) developed a model to predict the onset of a debond at the fiber-matrix interface when the composite is subjected to tension transverse to the fiber direction. This enables to predict a strong effect of the fiber radius on the critical tension leading to the interface debond. This model has been generalized to take into account a more general loading state. First, in the context of this thesis it was generalized to take into account a biaxial load inside the plane perpendicular to the fiber axis, see Section 4.3. Second, Carraro and Quaresimin (2014) evaluated the influence of a shear stress out of the plane transverse to the fiber axis. The effect of a pure transverse compression was also investigated by Quesada et al. (2009). This work actually focuses on stiff inclusions in soils but the model can be directly applied to the failure of fibers.

At the mesoscale, transverse cracking has been briefly revisited by Leguillon (2002) and a general theoretical model has been proposed in this thesis, see Chapter 6, showing a good agreement with experiments. An interesting characteristic of the application of the coupled criterion for this problem is that the stress and energy criteria act decoupled. The related phenomenon of free-edge delamination has also analyzed by several authors, see e.g. Martin et al. (2010), by means of the coupled criterion.

At the macroscale, the key problem of failure at an open hole has been widely investigated, see e.g. Zhang and Li (2008); Leite et al. (2010); Camanho et al. (2012); Martin et al. (2012). These models predict a size effect of the hole radius on the global strength, which agrees with experiments in some materials. A similar model was proposed by Catalanotti and Camanho (2013) to predict the net-tension failure of mechanically fastened joints. The coupled criterion enables to obtain a semianalytical expression which is very useful from an engineering point of view. Andersons et al. (2010) proposed a more general model to predict the failure due to the presence of internal U-notches. Other key problem at the macroscale is the failure of adhesive joints, several models have been proposed enabling the prediction of their failure, see e.g. Weissgraeber and Becker (2011, 2013).

Although this revision focuses mainly on the applications for composites, the range of applications for which the coupled criterion has been shown to be a useful tool is very wide. This can be applied to diverse problems as snow mechanics, see Cardu et al. (2008), or the generation of crack pattern due to thermal effects, see Leguillon (2013).

3.2 Leguillon's hypothesis

Leguillon's hypothesis is the key concept behind the coupled criterion. It postulates that the simultaneous fulfillment of both a stress and an energy condition is necessary for the crack initiation. This hypothesis combines two types of criteria commonly utilized separately in brittle and quasibrittle materials. Stress criteria are commonly used to predict failure in the absence of stress singularities, whereas the energy criterion is employed in the presence of cracks. Thus, this hypothesis enables to unify the two different visions of the failure process claiming that these cases classically associated to a single type of criterion, either on stresses or on energy, are extreme cases of the general condition of fulfilling both criteria.

Experimental and conceptual evidences supporting Leguillon's hypothesis have been presented by Leguillon (2002); Taylor et al. (2005); Cornetti et al. (2006). These evidences are far from being a closed and general proof justifying the coupled criterion. The rigorous physical justification of the criterion is therefore pending. In what follows these evidences are briefly discussed.

The experiments carried out by Parvizi et al. (1978) in cross-ply specimens shows that the first transverse crack onset occurs for a certain critical strain which depends on the inner-ply thickness. In fact, a threshold thickness is observed, the behavior being very different above or below this threshold. Above this threshold thickness the critical strain corresponding to the first transverse crack initiation does not vary with the thickness. On the contrary, below this threshold thickness the critical strain increases strongly when the inner-ply thickness decreases. The presence of two differentiated scenarios motivates the idea that two different criteria are being superimposed. Leguillon (2002) showed that the two behaviors could correspond to the two stress and energy criteria. Subsequent experiments showed a similar behavior, see Chapter 6 for a full analysis of this problem.

The requirement of fulfilling a stress criterion for crack initiation is evidenced by Leguillon (2002) employing the example of a bar subjected to a uniform strain. If only an energy condition is utilized to predict the failure, the predicted critical strain decreases with the bar length. The reason is that the elastic potential energy stored in the bar is directly proportional to the bar length and the square of the strain, whereas the dissipated energy during an abrupt fracture of the bar is given only by the fracture energy and the bar-section area. After the abrupt failure, all the potential elastic energy is released, thus, if the bar length increases, the strain level necessary to fulfill the energy balance decreases. This implies that for sufficiently large lengths, the critical strain vanishes, which is not reasonable. The addition of the stress condition proposed by Leguillon's hypothesis solves this problem because above a certain length, the stress criterion prevails, fixing the critical load to the value of the tensile strength. From a rigorous point of view, not all the elastic energy stored can be actually employed on breaking the bar if dynamics is taken into account given the finite velocity of the elastic waves. The reason is that it is unrealistic to assume that the energy density at points far from the breaking surface can be taken into account for the energetic balance given the high speed of the crack onset process. Thus, the stress criterion can be partially interpreted as a condition to take into account this limitation. At the other extreme, for short bars below a certain length, the energy criterion prevails and an increase of the critical strain is expected for shorter bars.

The transition between the criteria governing the failure is evidenced by the problem of the crack initiation at a reentrant corner. By varying the corner angle, the problem goes from a sharp crack, whose failure is governed by the energy criterion, to a straight edge without stress concentration, whose failure is ruled by the stress criterion. Thus, it is reasonable to think that the cases with intermediate angles have to be governed by a combination of both criteria, the crack and the straight edge being the extreme cases of a unifying theory. The combination of Leguillon's hypothesis and the assumption of finite length of the crack at onset enables to explain this transition and agrees well with experiments as demonstrated in Leguillon (2002) and many other subsequent works, see a review of the analysis of the failure at reentrant corners in Section 3.1.

It is interesting to remark the behavior predicted for short cracks. It is well known that the linear elastic fracture mechanics is not valid for short cracks below a certain length since the nonlinear region in the vicinity of the crack becomes of the same order of the crack. Experimental methods, see e.g. Taylor et al. (2005) for a review of several experiments, shows that the critical remote stress predicted by the coupled criterion goes from a LEFM-like variation for large cracks to a constant value for short cracks. This well-known phenomenon can be interpreted as an evidence of the transition between two scenarios governed by the stress and the energy criteria as was proposed by Taylor et al. (2005); Cornetti et al. (2006). They found that the transition observed in experiments agrees with the prediction of the coupled criterion.

In what follows, the two types of criteria are analyzed and the different formulations proposed for them are discussed. In addition, the practical evaluation of each one is detailed and the tools used are described.

3.3   Stress criterion

In brittle and quasibrittle materials, a stress criterion is commonly employed to predict failure in the absence of stress singularities as cracks or damaged zones at the macro scale. The stress criteria are normally based on predicting phenomena which occur at the lowest scales through a macroscopic magnitude as the stress tensor. Since the processes behind the stress criteria depend strongly

on the material microstructure, the different stress criteria are associated to the
materials or type of materials. Many of the stress criteria are based on phenomenological laws, often founded on processes which have not been understood
yet. From a practical point of view, these criteria are formulated by defining a
critical value for a certain combination of the components of the stress tensor, see
e.g. the Mohr-Coulomb criterion for soil mechanics or the recent failure criteria
for composites, see also París (2001); Hinton et al. (2002) for reviews.

In the context of the finite fracture mechanics, the stress criterion is evaluated
before the crack onset at the entire surface ΔS_c where the crack will initiate, see
Figure 3.2. In general, the stress criteria can be written by defining a functional
$f\left(\sigma_{ij}\left(P\right)\right)$ which depends on the stress tensor σ_{ij} for $i, j = 1, 2, 3$ at the points P
contained in the future crack surface ΔS_c and a critical value for this functional
f_c,

$$f\left(\sigma_{ij}\left(P\right), \Delta S_c\right) \geq f_c,\ P \in \Delta S_c. \tag{3.1}$$

Note that neither $f\left(\sigma_{ij}\left(P\right), \Delta S_c\right)$ nor f_c have to be necessarily scalars since the
stress criterion can be expressed as a combination of several scalar conditions.

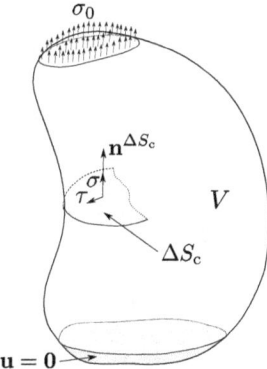

Figure 3.2: Arbitrary solid employed for the formulation of the stress criterion.

Several stress criteria have been used for the coupled criterion since the publication of the first works employing it. In the majority of cases, they are adap-
tations of the classical stress criteria commonly used out of the context of the
finite fracture mechanics. In the context of the finite fracture mechanics, the key
condition for defining a stress criterion from a classic one is that the prediction
given for an extreme problem governed exclusively by the stress criterion has to
agree with the classical prediction. A classical problem governed by the stress
criterion within a certain range of length is a cylindrical bar under tension. This
requirement for the stress criteria enables the resulting coupled criterion to be
claimed as a generalization of the stress and energy criteria.

The first stress criteria were based on a critical value of the normal component
of the traction vector associated to the surface ΔS_c normal. The critical value
can be identified with the tensile strength of the material given by a tensile test.

Thus, in what follows these stress criteria will be denoted tensile stress criteria. The adaptation of these criteria to the context of the finite fracture mechanics for static problems was carried out by Novozhilov (1969) and previously by others for fatigue problems. They were used as stress criteria with a fixed critical length, see Section 2.3.4.1 for a review.

In the context of the coupled criterion, Leguillon (2002) proposed to use the "pointwise stress criterion", see Section 2.3.4.1, in combination with the energy criterion. This criterion postulates that the normal tractions σ at all the points P in ΔS_c have to exceed or equal the tensile strength,

$$\sigma(P) = n_i(P)\sigma_{ij}(P)n_j(P) \geq \sigma_c, \ \forall P \in \Delta S_c \tag{3.2}$$

where $n_i(P)$, $i = 1, 2, 3$ are the components of the unit vector normal to ΔS_c in P. In the terms expressed in (3.1), f_c can be identified with σ_c and the function $f(\sigma_{ij})$ can be expressed as,

$$f(\sigma_{ij}(P)) = \min_{P \in \Delta S_c} (n_i(P)\sigma_{ij}(P)n_j(P)) \tag{3.3}$$

The condition for σ of equaling or exceeding σ_c at all the points of the future crack surface can be physically understood as a trigger condition for each point. Thus, a point fulfilling the stress criterion would be ready to be broken if the energy criterion is also fulfilled. At the microstructure, the fulfillment of this condition can be associated to several states depending on the material. In general, it can be interpreted to be associated to a damage level which is high enough at a lower scale to enable the onset of a macroscopic crack.

An alternative form of the tensile criterion is the "averaged stress criterion", see Section 2.3.4.1. Cornetti et al. (2006) proposed to compare the averaged value of σ at ΔS_c with the tensile strength. Note that in the case of the tensile test of a bar, this criterion also agrees with the previous one and predicts the expected value for this extreme problem governed by the stress criterion. According to this idea, (3.1) can be written as

$$\bar{\sigma}(P) \geq \sigma_c, \ P \in \Delta S_c \tag{3.4}$$

where $\bar{\ }$ represents the average value. In this case, the stress criterion can be interpreted as the competition between two forces: The driven force for the crack onset and the strength force against the onset. The driven force would be the sum of the normal tractions which would act as a single force. The strength force would be the sum of the tensile strength of each point at the surface. The condition would impose that the driven force equals or exceed the strength one to allow the crack onset. An alternative interpretation, not necessarily incompatible with the previous one, is the following: at micro scale, it can be considered that the stresses are redistributed and their value become uniform. This criterion avoids the question arising at the pointwise criterion about how the normal tractions at some points can exceed considerably the tensile strength without the presence of macroscopic irreversible effects. In this case, the uniforming process of the averaged stress criterion avoids the existence of points with a stress level which exceeds the tensile strength significantly if a redistribution is assumed. However, the fact that the stress exceeds considerably the tensile strength at a point does not have to be rejected necessarily since a post tensile strength behavior is possible although it cannot be observed in standard tensile tests. On the other hand, the average criterion presents a conceptual difficulty when it is applied to the onset of cracks which are not plane: the concept of stress redistribution is not as obvious as in the plane case because the direction of the traction vector depends on the point. Thus, the combination in a single total force would require additional assumptions.

Both stress criteria have been employed within the context of the coupled criterion to predict the crack initiation in a wide variety of problems. Several works used both criteria in order to compare their predictions, see e.g. Cornetti et al. (2012). The difference between the critical loads predicted by the two criteria is often low, at least for the problems studied. Moreover, when the critical load is compared with experimental results, the difference between the two predictions is lower that the typical dispersion of the experiments.

The stress criteria described previously are based exclusively on the normal tractions to ΔS_c, i.e. neglecting the influence of the shear component of the traction vector. This is acceptable for the majority of problems dealing with homogeneous bodies subjected mainly to tension and formed with materials with a tensile strength higher than the shear strength. In these cases, the angle of crack initiation observed usually agrees very approximately with the plane without shear component of the traction vector, see e.g. Yosibash et al. (2006). However, the shear stresses can become relevant in the presence of an interface weaker than the bulk material. A weak interface is a preferential surface for crack initiation, avoiding an optimum alignment of the angle of initiation as observed in the absence of interfaces by Yosibash et al. (2006). As a consequence, the influence of the shear stresses can become relevant if the loads promote their presence at the interface. On the other hand, even for homogeneous bodies, the role of the shear stresses should be taken into account for materials with a lower shear than tensile strength or subjected to compression. This is the typical scenario in soil mechanics.

In the context of the application of the coupled criterion to predict interface crack initiation, Mantič (2009) proposed to use a Mohr-Coulomb criterion, typically employed in soil mechanics, as a future development of his model, which was initially based on a tensile stress criterion. A Mohr-Coulomb criterion can be expressed in similar terms to (3.2) by defining a equivalent stress $\sigma_{\text{eq}}^{\text{MC}}(P)$ as,

$$\sigma_{\text{eq}}^{\text{MC}}(P) = \sigma(P) + \frac{|\tau(P)|}{\mu} \qquad (3.5)$$

where τ is the shear stress at the interface and $\mu = \tau_c/\sigma_c$ is the ratio of shear τ_c to tensile σ_c strength. Note that, according to this definition, the value of σ_{eq} tends to σ when τ vanishes or if τ_c is much higher than σ_c. In terms of this σ_{eq}, the pointwise stress criterion (3.2) can be rewritten as,

$$\sigma_{\text{eq}}^{\text{MC}}(P) \geq \sigma_c, \ \forall P \in \Delta S_c. \qquad (3.6)$$

whose treatment in the context of the coupled criterion is totally equivalent to the employment of the original tensile criterion. An averaged Mohr-Coulomb criterion can also be defined by modifying (3.4) similarly,

$$\bar{\sigma}_{eq}^{MC}(P) \geq \sigma_c, \ P \in \Delta S_c. \tag{3.7}$$

For the problem studied by Mantič (2009) subjected to a mixed-mode loading, Carraro and Quaresimin (2011, 2014) proposed a quadratic stress criterion to take into account the influence of the shear stresses. In the terms employed here, the quadratic stress criterion can be expressed defining a equivalent stress σ_{eq}^{q} as

$$\sigma_{eq}^{q}(P) = \sqrt{\sigma^2(P) + \left(\frac{\tau(P)}{\mu}\right)^2} \tag{3.8}$$

Similarly to the Mohr-Coulomb criterion, this expression can be used either as a pointwise stress criterion, see e.g. Chapter 7 or an averaged one, see e.g. Carraro and Quaresimin (2014).

At a conceptual level, the quadratic criterion solves one of the main problems of the Mohr-Coulomb criterion when it is used with dominant tension: the non-soft transition between the critical curve σ, τ from $\tau > 0$ to $\tau < 0$, see the knee at $\sigma = \sigma_c$ in Figure 3.3. In fact, through a simple analysis with Mohr circles, which is in the basis of the original Mohr-Coulomb theory, it can be demonstrated that this criterion, in the form defined in (3.5), is not able to predict a fracture plane perpendicular to the load direction under uniaxial loads in homogeneous brittle materials. In order to predict the expected fracture plane, the peak has to be rounded using the Mohr circle corresponding to a uniaxial state. This adds a complexity in the form of the stress criterion and the derived analyses. Thus, in addition to using the quadratic criterion for the materials or interfaces presenting this fracture behavior, it can be employed as a good approximation for materials which show a Mohr-Coulomb-like fracture behavior. In the presence of weak interfaces, the previous analysis with Mohr circles is not longer valid since the weak interface avoids the prediction of a unexpected angle for the fracture plane. In addition, the analysis with Mohr-circles of the stresses at the interface has not sense since not all the components of the stress tensor have to be continuous through the interface.

Tran et al. (2012); García and Leguillon (2012) proposed to employ a polynomial expression as a generalization of the previous criteria, which can be expressed by defining an equivalent stress σ_{eq}^{pol} as,

$$\sigma_{eq}^{pol}(P) = \sqrt[m]{\sigma^m(P) + \left(\frac{\tau(P)}{\mu}\right)^m} \tag{3.9}$$

where the exponent m is considered a material or interface property. Note that the values $m = 1$ and $m = 2$ correspond to the Mohr-Coulomb and the quadratic criteria, respectively.

The previous polynomial criterion for even values of the exponent m presents an unrealistic behavior for $\sigma < 0$. If m is even, this expression prescribes that a compression promotes the failure decreasing the critical value of τ. On the contrary, if m is odd, a compression increases the required value of τ for failure. In general, it is more realistic that a compression makes more difficult the failure, therefore, the required value for τ should increase for $\sigma < 0$. Carraro and Quaresimin (2011, 2014) avoided this problem assuming that failure was impossible in the presence of compressions. Actually, it is, taken to extremes, an application of the idea of that compressions difficult the failure. A softer correction of this behavior can be obtained if the expression (3.9) is slightly modified as,

$$\sigma_{eq}^{pol}(P) = \sqrt[m]{\left\langle \text{sgn}\left(\sigma(P)\right) |\sigma(P)|^m + \left(\frac{|\tau(P)|}{\mu}\right)^m \right\rangle_+} \qquad (3.10)$$

where $\langle \cdot \rangle_+$ denotes the positive part of a real number.

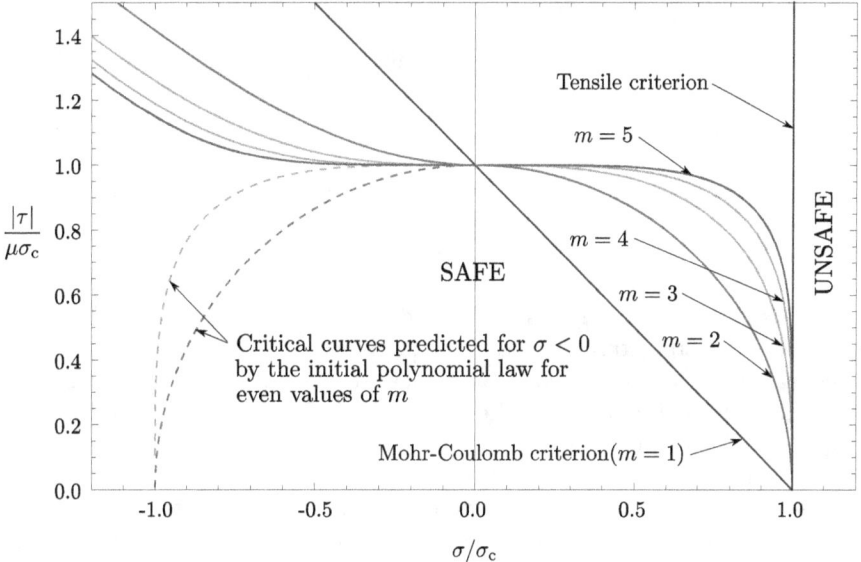

Figure 3.3: Critical curves predicted by the different stress criteria proposed to include the influence of the shear stresses at ΔS_c before the crack onset.

The stress criterion can be expressed in a general form including the different visions described previously. A monotonic and proportional increase of the external loads is assumed independently of the nature of the load: either prescribed displacement, prescribed traction or a combination of them. In the context of this thesis, a linear relation between the external loads and the stresses generated is assumed. For the sake of simplicity, a certain tension σ_0 is considered as a reference of the external loads. Thus, the solution of stresses, strains and displacements is linear with σ_0. The stress criterion can be written in a general and dimensionless form as,

$$\frac{\sigma_0}{\sigma_c} \geq s(\sigma_{eq}, \Delta \hat{S}_c) \tag{3.11}$$

where

$$s(\sigma_{eq}, \Delta S_c) = \frac{1}{f(\sigma_{eq}/\sigma_0, \Delta \hat{S}_c)}, \tag{3.12}$$

$\Delta \hat{S}_c$ being the surface of the crack after the onset ΔS_c expressed in dimensionless coordinates. Since the function s is necessarily dimensionless in (3.11), expressing ΔS_c in dimensionless terms is the most natural form. Observe that, in the definition of the function f in (3.1), the dependency on the stress tensor σ_{ij} has been substituted by the equivalent stress σ_{eq} because it summarizes the different forms of combining the components of the stress tensor σ_{ij} described previously. On the other hand, the form of the function f differs among the pointwise criterion,

$$f^{\mathrm{PW}}(\sigma_{eq}, \Delta \hat{S}_c) = \min_{\hat{P} \in \Delta \hat{S}_c} \left(\sigma_{eq} \left(\hat{P} \right) \right) \tag{3.13}$$

and the averaged criterion,

$$f^{\mathrm{AV}}(\sigma_{eq}, \Delta \hat{S}_c) = \frac{\int_{\Delta \hat{S}_c} \sigma_{eq} \left(\hat{P} \right) d\hat{S}}{\int_{\Delta \hat{S}_c} d\hat{S}_c}. \tag{3.14}$$

where $\hat{P}$ is the point P expressed in the dimensionless coordinates. In view of the previous general expression, the application of the stress criterion requires: on the one hand the knowledge of the typical strength properties of the material or interface and on the other hand the elastic solution of the problem before the crack onset. The first can be obtained for the majority of cases by well established experimental procedures because the required properties are well known. The form used to obtain the elastic solution depends strongly on the complexity of the problem under study. In some cases, the problem can be simplified to enable the employment of analytical solutions, see e.g. Sections 4.3.1 and 5.1.1. In other cases, with the crack onset sufficiently enclosed in the vicinity of a certain point, an approximate solution can be obtained by employing a matched asymptotic analysis, see e.g. Section 7.2. If neither analytical nor approximated solutions are available, the solution can be extracted from a computational analysis by means of, e.g., the finite element method, see Section 6.2.2.

3.4   Energy criterion

The energy criterion employed in the coupled criterion is an adaptation to the context of the finite fracture mechanics of the differential energetic balance proposed by Griffith. Since a crack onset of a finite extension is assumed in FFM, the energy criterion is based on an incremental energetic balance between the states before and after the crack onset. In the context of the FFM, this has been used previously to the formal proposal of the coupled criterion by some authors using a fixed crack length at the onset, see Section 2.3.4.1 for a review. The incremental energetic balance can be expressed as

$$\Delta\Pi + \Delta E_k + \Delta\Gamma = 0 \qquad (3.15)$$

where $\Delta\Pi$ and ΔE_k are, respectively, the changes in potential elastic and kinetic energy between the states before and after the crack onset. $\Delta\Gamma$ is the energy dissipated at the irreversible processes associated to the crack onset. Figure 3.4 represents the two states between which the previous changes are evaluated.

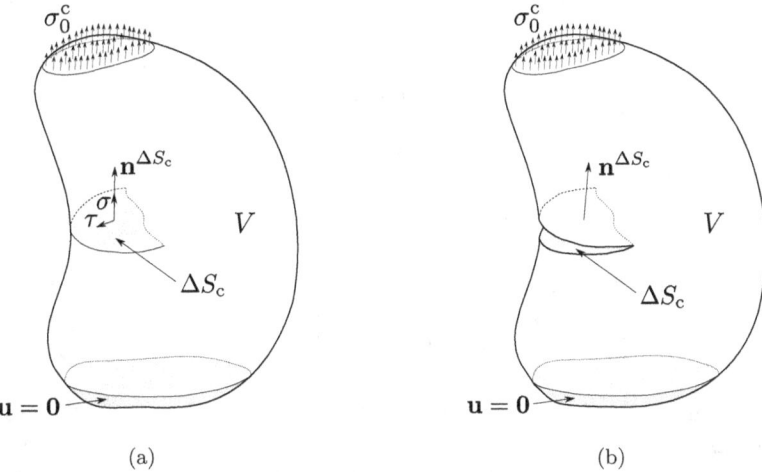

(a)   (b)

Figure 3.4: Arbitrary solid employed for the formulation of the energy criterion at the situations immediately (a) before and (b) after the crack onset.

In the context of this thesis, the initial state is considered quasistatic, implying,

$$\Delta E_k \geq 0. \qquad (3.16)$$

To the author's knowledge, this is a common assumption to all the works employing the coupled criterion presented in the literature. Under this assumption, the balance (3.15) can be written as,

$$-\Delta\Pi \geq \Delta\Gamma. \qquad (3.17)$$

This condition can be physically interpreted as the requirement of releasing enough elastic energy to fulfill the required energy which has to be dissipated during the crack onset. In what follows, the two terms are analyzed describing the different manners proposed to calculate or approximate their values.

3.4.1   Released energy

In general, when a new crack is initiated at a solid, a certain amount of elastic energy $-\Delta\Pi$ is released as a consequence of the stress redistribution in the solid. This process occurs more intensely in the vicinity of the new crack. Since the elastic potential energy is a conservative energy, the value of $-\Delta\Pi$ only depends on the elastic solutions before and after the crack onset. Since linear elasticity is assumed for both states, the unicity on the value for $-\Delta\Pi$ is assured.

The value of $\Delta\Pi$ depends in general on the value of the external loads, here summarized in a certain tension σ_0 under the same assumptions described for the stress criterion. In addition, $\Delta\Pi$ varies with the geometric properties of the crack onset given by ΔS_c, the solid geometry given by a set of lengths $L, L_1, L_2, \ldots$ and the elastic properties of the material or materials which compose the solid. In particular, under the assumption of linear elasticity, $\Delta\Pi$ depends on a finite set of properties with dimensions of elasticity modulus $E, E_1, E_2, \ldots$ and a finite set of dimensionless properties as $\nu, \nu_1, \nu_2, \ldots$.

$$\Delta\Pi(\sigma_0, \Delta S_c, \underbrace{E, E_1, E_2, \ldots, \nu, \nu_1, \nu_2, \ldots,}_{\text{Elastic properties.}} \underbrace{L, L_1, L_2, \ldots}_{\text{Geometric properties}}) \qquad (3.18)$$

The amount of arguments of $\Delta\Pi$ can be reduced taking into account that the elastic energy stored in a solid varies quadratically with the external load σ_0 under the assumption of linear elasticity. Since the states before and after the onset are considered linear elastic problems and subjected to the same σ_0, the change in the potential elastic energy varies quadratically with σ_0 as well. Moreover, the dimensional analysis defines that for a static mechanical problem, the dependency can be reduced by extracting two dimensionally independent properties. In this case, E and L are taken as references. Thus, a dimensionless released energy $\Delta\hat{\Pi}$ can be defined as,

$$\Delta\hat{\Pi}\left(\Delta\hat{S}_c, \frac{E_1}{E}, \ldots, \nu, \nu_1, \ldots, \frac{L_1}{L}, \ldots\right) =$$
$$= \frac{\Delta\Pi(\sigma_0, \Delta S_c, E, E_1, E_2, \ldots, \nu, \nu_1, \nu_2, \ldots, L, L_1, L_2, \ldots)}{\frac{\sigma_0^2 L^3}{E}} \qquad (3.19)$$

where $\Delta\hat{S}_c$ is the surface of the crack onset ΔS_c expressed in the new coordinates normalized with the length of reference L. It means that, if the crack surface is given by a set of lengths $l_c, l_{c,1}, l_{c,2}, \ldots$, in $\Delta\hat{\Pi}$, they have to be normalized with L.

The released energy during the crack onset can be obtained by several procedures. In what follows, the three main procedures which can be extrapolated to a wide variety of problems are described, analyzing their main advantages and disadvantages.

3.4.1.1 Method based on integrating the energy release rate

This method is based on the definition of the energy release rate G (ERR) given by the linear elastic fracture mechanics (LEFM) as,

$$G = -\frac{\mathrm{d}\Pi}{\mathrm{d}A} \tag{3.20}$$

where Π is the potential elastic energy and A is the crack area. The released energy by a crack onset can be calculated by integrating this expression,

$$-\Delta\Pi = \int_{\Delta S_c} G(A)\mathrm{d}A. \tag{3.21}$$

where $G(A)$ is the ERR for a crack growing from the state previous to the crack onset to the state immediately after the onset with a crack with surface ΔS_c, see Figure 3.5. Note that the value of $G(a)$ depends on the path assumed between the two states. However, the value of $\Delta\Pi$ obtained by this method is independent of the path since the elastic energy derives from a potential.

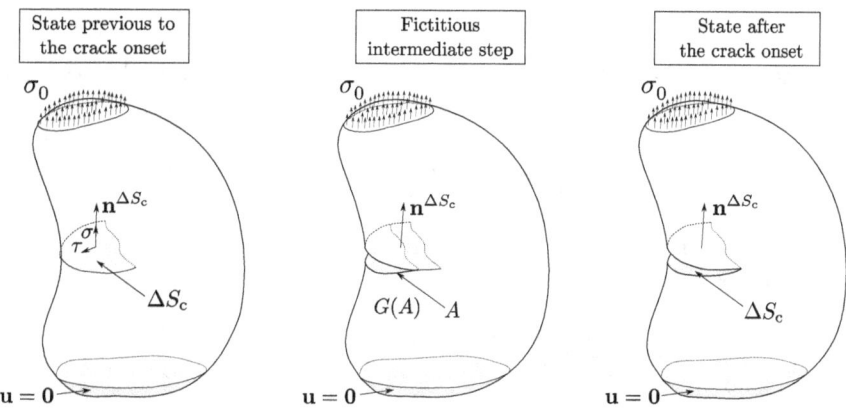

Figure 3.5: Schematic of the method to obtain the released energy $\Delta\Pi$ by integrating the energy release rate G.

The main advantage of this method is being based on a well studied magnitude as the ERR. Given its relevance in the LEFM, the methods to obtain the ERR have been very studied for decades. As a consequence, analytical expressions for a wide variety of practical problems are available in the literature. In addition, a set of mature methodologies have been developed to extract the value of ERR from computational models. Moreover, the majority of commercial FEM codes have native implementations of methodologies for the estimation of ERR. This method is particularly suitable for 2D problems since the integral in (3.21) becomes an integral along a line.

The main disadvantage is the requirement of knowing the value of ERR for the intermediate cracks between the states before and after the crack onset. This is not a relevant problem if an analytical solution is available. Nevertheless, if it is not, this method is far from being optimum because it requires the computation of a sufficiently large set of intermediate models between the states before and after the crack onset in spite of that $\Delta\Pi$ depends only on the two extreme states. Thus, the requirement of computing the intermediate models is considered superfluous, therefore this method is advised against when computing is necessary unless the intermediate states have to be computed for other purposes, see e.g. the problem studied in Section 4.1. This method is also advised against for 3D problems since the integral in (3.21) and the approximation of G become much more complex and computationally expensive.

3.4.1.2   Method based on the external-work increment

The change in potential elastic energy $\Delta\Pi$, as a consequence of a crack onset, can be obtained by the change in tractions and displacements at the parts of the boundary subjected to displacements and tractions boundary conditions respectively if the body forces are neglected. It is well known from the LEFM that $\Delta\Pi$ can be expressed as,

$$\Delta\Pi = \Delta U - \Delta W \tag{3.22}$$

where ΔU and ΔW are the changes in internal energy and external work respectively. ΔU can be expressed in terms of the stress σ and strain ε tensors in the volume V,

$$\Delta U = \frac{1}{2}\Delta\left(\int_V \sigma\cdot\varepsilon dV\right) \tag{3.23}$$

Assuming traction-free crack surfaces after the onset, the application of the divergence theorem gives,

$$\Delta U = \frac{1}{2}\Delta\left(\int_{S_t} t^0\cdot u dS\right) + \frac{1}{2}\Delta\left(\int_{S_u} t\cdot u^0 dS\right) \tag{3.24}$$

where t^0 and u^0 are the prescribed tractions and displacements respectively at the surfaces subjected to traction and displacement boundary conditions, S_t and S_u, respectively. On the other hand, ΔW can be expressed as

$$\Delta W = \Delta\left(\int_{S_t} t^0\cdot u dS\right). \tag{3.25}$$

Substituting (3.24) and (3.25) into (3.22) gives,

$$\Delta\Pi = \frac{1}{2}\Delta\left(\int_{S_u} t\cdot u^0 dS\right) - \frac{1}{2}\Delta\left(\int_{S_t} t^0\cdot u dS\right) \tag{3.26}$$

Taking into account the linearity of the Δ and integral operator and that $\Delta t^0 = 0$ and $\Delta u^0 = 0$, the previous expression can be rewritten as,

$$\Delta\Pi = \frac{1}{2}\int_{S_u}\Delta t\cdot u^0 dS - \frac{1}{2}\int_{S_t} t^0\cdot\Delta u dS. \tag{3.27}$$

Thus, $\Delta\Pi$ can be calculated directly if Δt and Δu are obtained, see Figure 3.6 for a schematic of the method. In the presence of boundary conditions with tractions or displacements prescribed depending on the direction, the previous expression can be applied similarly by separating the integrals for different directions. This separation is equivalent to the separation in S_t and S_u.

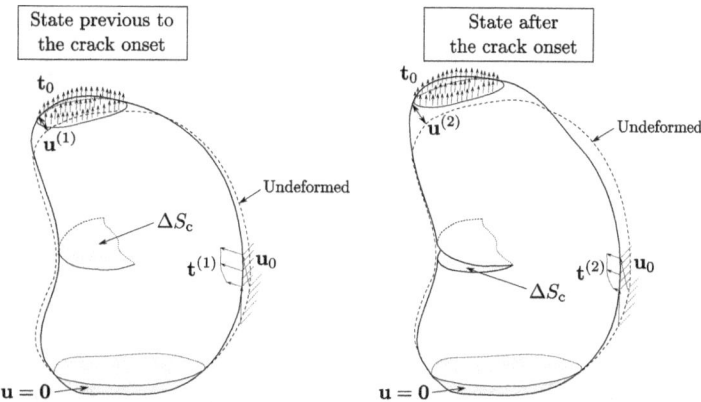

Figure 3.6: Schematic of the method to obtain the released energy $\Delta\Pi$ by the external-work increment.

The main advantage of this method is that only the elastic solutions before and after the crack onset are required. In addition, the variation of u and t at the surfaces where, respectively, either tractions or displacements have been prescribed are necessary to obtain $\Delta\Pi$. Thus, this method is suitable to approximate $\Delta\Pi$ in the case of problems without analytical solution for G since Δu and Δt can be easily extracted from the results given for a computational method as FEM or BEM for the models corresponding to the states before and after the crack onset.

This method can present problems when the crack onset is very small with respect to the solid size since the influence of the presence of the crack on the values of u and t at the external boundaries can become too low. The values of Δu and Δt required in (3.27) have to be calculated by the values of u and t before and after the crack onset. If the influence of the presence of the crack is low, the values of u and t are very similar and the difference can be several orders of magnitude lower that their values. As a consequence, this method is strongly affected by the truncation errors for small cracks. Thus, this method is advised against when cracks in very large or infinite domains are studied, see e.g. the problems in Chapters 4 and 5.

3.4.1.3 Incremental crack closure technique

The virtual crack closure technique (VCCT) has been intensely employed to obtain G for decades, mainly in combination with numerical results, see Krueger

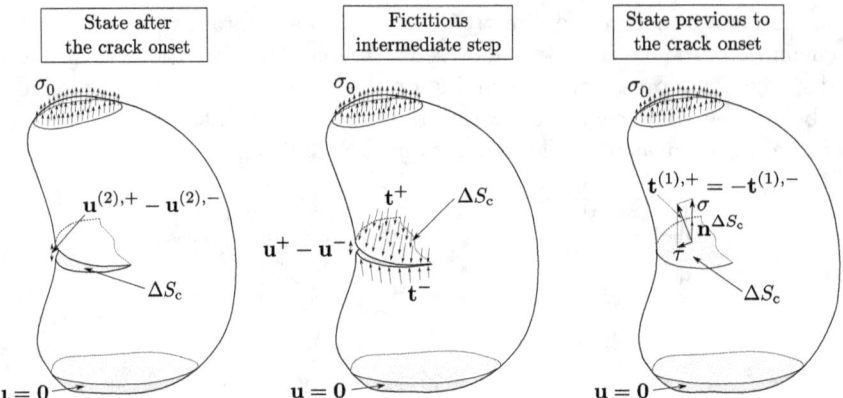

Figure 3.7: Schematic of the method to obtain the released energy $\Delta\Pi$ denoted incremental crack closure technique.

(2004) for a review. This is actually an approximation of the crack closure integral under the assumption of that the virtual infinitesimal extension of the crack does not alter significantly the elastic solution near the crack tip. Both methods are based on the idea that the change in potential elastic energy due to a certain crack growth is identical to the work necessary to close the crack an equivalent extension. In the context of FFM, this idea can be applied from an incremental perspective. Thus, $\Delta\Pi$ can be calculated as the work necessary to close the crack from the state after the onset to the state before the onset, see (3.7) for a schematic of this idea. Actually, unlike the classical VCCT, the use of this method in FFM is only valid if all the boundary conditions are displacement-based. The reason is that the work of the external forces during the crack closure cannot be neglected. Assuming that the closure is carried out in a quasistatic evolution, the work can be calculated if the tractions applied to close the crack and the displacements at points where they are applied are known,

- The tractions necessary to close the crack goes from 0 (assuming traction-free crack surfaces) to the tractions $\boldsymbol{t}^{(1)}$ found for the state before the crack onset and associated to the plane of the future crack.

- The displacements of the crack surfaces goes from $\boldsymbol{u}^{(2),+}$ and $\boldsymbol{u}^{(2),-}$ for the crack after the crack onset, superscripts $+$ and $-$ referring to the two crack surfaces, to $\boldsymbol{u}^{(1),+} = \boldsymbol{u}^{(1),-}$ before the crack onset.

Thus, assuming linear elasticity, $\Delta\Pi$ can be obtained as,

$$\Delta\Pi = -\frac{1}{2}\int_{\Delta S_c} \boldsymbol{t}^{(1),+} \cdot \left(\boldsymbol{u}^{(2),+} - \boldsymbol{u}^{(1),+}\right) \mathrm{d}S - \frac{1}{2}\int_{\Delta S_c} \boldsymbol{t}^{(1),-} \cdot \left(\boldsymbol{u}^{(2),-} - \boldsymbol{u}^{(1),-}\right) \mathrm{d}S$$

(3.28)

where $\boldsymbol{t}^{(1),+}$ and $\boldsymbol{t}^{(1),-}$ are the tractions before the crack onset associated to the normals to the two crack surfaces $+$ and $-$. This expression can be simplified taking into account that $t^{(1),-} = -t^{(1),+}$ and $u^{(1),-} = u^{(1),+}$ respectively due to equilibrium and compatibility conditions before the onset,

$$\Delta\Pi = -\frac{1}{2} \int_{\Delta S_c} t^{(1),+} \cdot \left(u^{(2),+} - u^{(2),-} \right) dS \qquad (3.29)$$

Note that, similarly to the method based on the variation of the prescribed tractions and displacements at the external boundaries, this method is only based on the states before and after the crack onset. Thus, it is avoided the requirement of approximating the intermediate steps of the crack onset discussed for the method based on integrating G. Hence, this method is suitable for its combination with numerical methods to obtain the elastic solutions before and after the crack onset. In addition, the problem of truncation errors described for the method developed in previous section is also avoided since this method is based on the variation of magnitudes in the vicinity of the crack. Thus, this method is also suitable for cracks in infinite or very large domains without loss of accuracy. Actually, a problem can arise if the variation on displacements is very small in comparison with the displacements in the state before the crack onset. However, this can be easily avoided by modifying the reference of displacements, which is given normally by the form in which boundary conditions are implemented. This method is particularly advised for 2D problems requiring to solve numerical models and for 3D models, see e.g. Section 6.2.

3.4.2 Dissipated energy

The LEFM is based on the idea introduced by Griffith (1921, 1924) that the growth of a crack requires the dissipation of a certain amount of energy on different irreversible processes associated to the separation of the two surfaces. For brittle materials as the glass studied by Griffith, the energy dissipated can be identified with the energy inversed in the rupture of the chemical bonds between atomic planes. In general, the internal mechanisms associated to the energy dissipation during the crack growth are more complex. A macroscopic property, denoted fracture toughness G_c, is used in LEFM as a measure of the dissipated energy per unit area of crack growth. In the context of the FFM, the most simple generalization of the energy dissipated during a crack onset of finite extension ΔS_c is,

$$\Delta\Gamma = G_c\Delta S_c, \qquad (3.30)$$

This generalization was already applied in the works implementing the incremental energy criterion with a fixed critical length, see e.g. Parvizi et al. (1978); Hashin (1996). In addition, it was employed by Leguillon (2002) in his formal discussion of the coupled criterion. Actually this generalization is conceptually correct only under the assumption that the dissipated energy depends only on the crack area and the material, and is independent of other factors as crack growth speed or stress mixity at the crack tip. This assumption can be considered correct under the Griffith's perspective which associates totally the energy dissipation to the cleavage of chemical bonds between atomic planes, since the work of separating atomic planes derives from a potential. However, in engineering materials as metals or composites, the contribution of other irreversible mechanisms to the total energy dissipated is much more relevant per unit area than the contribution of the cleavage of chemical bonds. In fact, it is well known that the value experimentally obtained for G_c in these materials is actually the sum of the contribution of all these phenomena to the dissipation energy per unit area of crack growth. Depending on the material, some of the irreversible processes contributing to G_c could be irrelevant when a crack onset of a finite extension is assumed. Similarly, other processes can be relevant in FFM which are neglected by a value of G_c obtained for a quasistatic crack growth. Due to these reasons, the usage in the context of FFM of a G_c obtained for LEFM has to be considered as an approximation of the actual energy dissipated per unit area of crack onset.

In the context of LEFM, it is normally assumed that G_c varies with the fracture mode mixity ψ. This variation is due to the dependence on ψ of the irreversible processes contributing to G_c. For instance, in metals, the yielding around the crack tip varies with the fracture mode mixity, therefore the contribution of this process to G_c depends on. Since these irreversible processes and the relevance of their contribution to G_c are very linked to the microstructure, the variation of G_c depends strongly on the material nature. As a consequence, the phenomenological laws for G_c are very particular of the material for which they are obtained. In homogeneous and isotropic materials, the cracks tends to propagate near pure mode I according to the local symmetry criterion by Barenblatt and Cherepanov (1961); Erdogan and Sih (1963), or following other criteria leading to very similar results, see e.g. experiments for PMMA by Maccagno and Knott (1989). In the case of ductile materials, the behavior differs slightly, observing a more relevant contribution of the mode II, see e.g. experiments for aluminum by Hallbäck and Nilsson (1994). For interface cracks, the presence of an interface with weaker properties promotes the propagation along the interface even in highly mixed fracture modes, avoiding the crack deviation following the local symmetry criterion. Due to this, the variation of G_c with ψ is particularly relevant for the propagation of interface cracks. This fact has motivated efforts to design mixed-mode fracture specimens with interface cracks to measure the variation of G_c. Experiments show a strong variation of G_c with ψ, see e.g. Hutchinson and Suo (1992).

In the context of the FFM, the influence of the fracture mixity on $\Delta\Gamma$ is controversial, in particular for interface cracks. On the one hand, the extrapolation of the microstructural causes of the variation in LEFM to FFM is not direct. On the other hand, the definition of fracture mode mixity is associated to a continuous crack growth, therefore additional assumptions are required to generalize this concept to FFM. In what follows, these controversial problems are discussed and the main contributions in the literature are presented.

A part of the mechanicians claim that, in the context of the FFM, G_c is a surface energy which does not depend on the fracture-mode mixity. The main arguments supporting this idea were detailed by Carraro and Quaresimin (2014). According to their discussion, the crack onset in FFM is supposed to occur suddenly, therefore, the concept of fracture-mode mixity at the crack tip is not valid since the first crack tip appears after the crack onset. According to Carraro and Quaresimin (2014) and in the context of the problem of fiber-matrix debonding, in the absence of a crack tip, none of the phenomena contributing to the variation of the dissipated energy with the fracture-mode mixity are expected to occur. An example of them is the plastic deformation of the matrix which requires the presence of the singularity at the crack tip, therefore, in the absence of a crack tip, the dependence on the mixity due to its influence on the plastic deformation cannot be extrapolated to the context of FFM.

Although some of physical reasons for the influence of the fracture-mode mixity do not have a clear sense in FFM, other physical phenomena which do contribute may remain valid. For instance, the microvoid mechanics, a common phenomenon in the basis of crack initiation, varies with ψ as reported by Shi et al. (1994). Additionally, fractographies show that the morphology of the surface roughness varies with ψ, see e.g. Lane (2003). As a consequence, the effective surface changes and therefore the dissipated energy per unit of macroscopic area G_c varies. Moreover, even in the absence of a crack tip, the effect of friction is expected to occur. This effect is strongly affected by ψ, thus, its contribution to the dissipated energy is an additional source of variation of G_c with ψ. The influence of ψ due to these phenomena and others, see Evans et al. (1990), motivates Mantič (2009) to propose an expression to approximate the influence of ψ on $\Delta\Gamma$. This expression is inspired by the formula employed to obtain $\Delta\Pi$ trough the integration of ERR, see Section 3.4.1.1. Thus, according to Mantič (2009), the dissipated energy can be approximated by,

$$\Delta\Gamma = \int_{\Delta S_c} G_c(\psi) \mathrm{d}S \tag{3.31}$$

where ψ is the fracture-mode mixity of a crack advancing from a void surface to ΔS_c following a certain growth sequence. Unlike the integral for the ERR, in this case the value obtained from the dissipated energy does depend on the assumption about the growth sequence. In the context of the plane strain study of fiber-matrix debonding when subjected to uniaxial transverse tension, Mantič (2009) assumed a "natural" growth sequence: the crack propagates symmetrically with respect to the tension direction and begins at the stress concentration. However, for other problems, the hypothesis of growth can be difficult to be established. This is particularly problematic for 3D problems.

In the framework of his problem, Mantič (2009) proposed to use the phenomenological Hutchinson-Suo law to approximate G_c,

$$G_c(\psi) = G_{1c} \left(1 + \tan^2\left[(1-\lambda)\psi\right]\right) \tag{3.32}$$

where G_{1c} is the fracture toughness in pure mode 1 and λ is a mode-mixity sensitivity parameter.

In order to avoid the incongruence between the assumption of sudden crack onset in the FFM and the need of a crack tip to measure the fracture mode mixity and its influence on the dissipated energy, García and Leguillon (2012) proposed to base the approximation of the dissipated energy on the stress mixity before the crack onset. Thus, $\Delta\Gamma$ is approximated by the expression (3.31), taking ψ as,

$$\psi(P) = \arctan\frac{\tau(P)}{\sigma(P)} \tag{3.33}$$

where σ and τ are the normal and shear tractions respectively at points P in ΔS_c before the crack onset. This approximation avoids to assume a slow crack growth. In addition, the value of $\Delta\Gamma$ obtained following this idea does not depend on an arbitrary crack sequence. As a consequence, this can be employed for any problem including 3D problems without the difficulties described for the previous approximation. Nevertheless, even assuming that the dissipated energy depends on the stress mixity before the crack onset, the fact of employing phenomenological laws for G_c obtained experimentally as a function of the fracture-mode mixity for quasistatic crack growth is a controversial hypothesis which is difficult to physically justify.

In general, the expression in (3.30) can be generalized to take into account any possible variation of the value of G_c in ΔS_c and not only due to the discussed dependence on ψ. Thus, the next expression can be used to account for the variation of G_c on any property depending on the point $P \in \Delta S_\mathrm{c}$,

$$\Delta\Gamma = \int_{\Delta S_\mathrm{c}} G_\mathrm{c}(P)\mathrm{d}S. \tag{3.34}$$

This expression can be normalized, defining a dimensionless dissipated energy $\Delta\hat{\Gamma}$ as,

$$\Delta\hat{\Gamma} = \frac{\Delta\Gamma}{G_{1\mathrm{c}}L^2} = \int_{\Delta\hat{S}_\mathrm{c}} \hat{G}_\mathrm{c}(\hat{P})\mathrm{d}\hat{S}. \tag{3.35}$$

where $\hat{G}_\mathrm{c} = G_\mathrm{c}/G_{1\mathrm{c}}$ and $G_{1\mathrm{c}}$ is a value of the fracture toughness taken as reference as, e.g., the fracture toughness for a quasistatic growth in pure opening fracture mode. L is the length of the solid geometry taken as reference in the dimensional analysis for $\Delta\Pi$, see (3.19). $\Delta\hat{S}_\mathrm{c}$, $\hat{S}$ and $\hat{P}$ denote the geometric entities ΔS_c, S and P expressed in the normalized coordinates with L.

Previous approximations of the dissipated energy in FFM are based on extrapolating the values of the fracture toughness obtained for a crack growth in LEFM. It is a very practical approximation because it enables to use properties of LEFM which can be measured by employing standard tests. However, the nature of the irreversible processes contributing to the energy dissipation can be very different in FFM and LEFM. An attempt to take into account this difference was carried out by Zhang and Li (2008) through the definition of a FFM fracture toughness $G_\mathrm{c}^{\mathrm{FFM}}$. They assume that the dissipated energy per unit area in the presence of a crack tip is different from its value under uniform tensile stress. The first correspond to the LEFM fracture toughness G_c, whereas the second is an additional material parameter, being denoted G_c^u. The basis of this difference is that energy dissipation occurs at a band of a certain width h in the vicinity of the fracture surface. Thus, it is reasonable to think that h differs in the presence of a crack tip from its value for uniform tensile stress. In an intermediate case, as a non-singular stress concentration, Zhang and Li (2008) proposed a FFM fracture toughness G_c^{FFM} as an interpolation between the two values,

$$G_c^{FFM} = \alpha G_c + (1 - \alpha) G_c^u \qquad (3.36)$$

where α is a parameter depending on the level of the stress concentration. The authors proposed to obtain it as,

$$\alpha = \frac{h_{crack}}{h} \qquad (3.37)$$

where h_{crack} is the width of the dissipating band for a crack and h is the width of the equivalent zone for the stress concentration. Once the value of G_c^{FFM} is obtained, it can be employed directly in (3.30). From a conceptual perspective, this idea is, *a priori*, a more congruent approximation to the FFM assumptions. However, its general application presents practical problems. The main cause is that, to the author's knowledge, no experimental method has been proposed to measure the fracture toughness under tensile stress G_c^u. In addition, the concept behind the value of h is complex enough and its approximation is not straightforward.

3.4.3   General form of the energy criterion

The expression of the energy criterion (3.17) can be rewritten by substituting the dimensionless forms of the released energy (3.19) and the dissipated energy (3.35), giving

$$-\frac{\sigma_0^2 L^3}{E} \Delta \hat{\Pi} \left(\Delta \hat{S}_c \right) \geq G_{1c} L^2 \Delta \hat{\Gamma} \left(\Delta \hat{S}_c \right) . \qquad (3.38)$$

Although it is not detailed in previous expression for the sake of brevity, $\Delta \hat{\Pi}$ and $\Delta \hat{\Gamma}$ depend also on the dimensionless elastic and geometric parameters of the solid before the crack onset.

The energy criterion in (3.38) can be rewritten as a condition over the reference of external loads σ_0 as,

$$\sigma_0 \geq \sqrt{\frac{G_{1c} E}{L}} \, g \left(\Delta \hat{S}_c \right) \qquad (3.39)$$

where $g(\cdot)$ is a dimensionless function defined as the ratio of the dimensionless dissipated energy to the released energy,

$$g \left(\Delta \hat{S}_c \right) = \frac{\Delta \hat{\Gamma} \left(\Delta \hat{S}_c \right)}{-\Delta \hat{\Pi} \left(\Delta \hat{S}_c \right)} . \qquad (3.40)$$

The function $g(\cdot)$ characterizes the resistance against the crack for a certain $\Delta \hat{S}_c$ from the perspective of the energy criterion. Note that $g(\cdot)$ is scale-independent since the geometric parameters on which this function depends are dimensionless. On the contrary, the energy criterion does depends on the scale since L appears explicitly in (3.39). Thus, for a certain solid and $\Delta \hat{S}_\mathrm{c}$, the energy criterion prescribes a critical value for σ_0, which is inversely proportional to the square root of the length of reference L. This is the typical size effect predicted by other energy criteria as Griffith's criterion. The reason for this size effect is the same in both cases: the dissimilarity between the scales of releasing energy (L^3) and dissipating energy (L^2).

3.5   Coupled criterion

In previous sections, the two stress and energy criteria have been discussed separately and the main methodologies to be formulated and approximated have been described. This section describes the form in which both criteria are combined according to Leguillon's hypothesis, described in Section 3.2. In the first part, the combination of both criteria is studied from the practical perspective of their application. In the second part, the typical failure behavior predicted by the coupled criterion is discussed.

3.5.1   General form of the coupled criterion

According to Leguillon's hypothesis, described in Section 3.2, the critical load σ_0^c leading to the onset of a crack of finite extension is given by the minimum value of σ_0 fulfilling both the stress and energy criteria. It means that it is necessary to minimize the combination of the conditions obtained for the stress (3.11) and energy (3.39) criteria. In order to express both criteria in the same terms, the expression corresponding to the energy criterion (3.39) is normalized with the tensile strength σ_c, which was taken in the stress criterion as a reference of strength. Thus, the energy criterion is expressed as,

$$\frac{\sigma_0}{\sigma_\mathrm{c}} \geq \gamma \sqrt{g\left(\Delta \hat{S}_\mathrm{c}\right)}, \tag{3.41}$$

where γ is a dimensionless brittleness number defined in the context of the FFM by Mantič (2009) as,

$$\gamma = \frac{1}{\sigma_\mathrm{c}} \sqrt{\frac{G_{1\mathrm{c}}E}{L}}. \tag{3.42}$$

This brittleness number is analogous to others being proposed in the literature in other contexts, see Mantič (2009) for a comparative review.

Once both criteria have been expressed in the same terms, the coupled criterion can be formulated as an optimization problem given by the next expression,

$$\frac{\sigma_0^\mathrm{c}}{\sigma_\mathrm{c}} = \min_{\Delta \hat{S}_\mathrm{c}} \left[\max \left(s\left(\Delta \hat{S}_\mathrm{c}\right), \gamma \sqrt{g\left(\Delta \hat{S}_\mathrm{c}\right)} \right) \right]. \tag{3.43}$$

Thus, the coupled criterion is expressed as a minimization problem, the objective function being the maximum of the dimensionless expressions of the two stress and energy criteria. The apparition of the function $\max(\cdot, \cdot)$ is the mathematical expression of the requirement of fulfilling both conditions since the critical load is given by the most limiting condition, i.e. the condition requiring the maximum value for σ_0^c/σ_c.

The minimization is carried out over the dimensionless crack surface $\Delta\hat{S}_c$. Thus, the optimum value of $\Delta\hat{S}_c$ in the minimization process corresponds to the crack geometry immediately after the onset. The morphology given by $\Delta\hat{S}_c$ does not correspond necessarily to the crack observed after the onset since an unstable growth could occur.

In general, the minimization of (3.43) over the dimensionless morphology of the crack $\Delta\hat{S}_c$ is a problem of calculus of variations. The rigorous solution of problems formulated following the expression (3.43) presents a very high complexity. This fact is incongruent with the claimed advantage of the coupled criterion of being able to obtain analytical or semianalytical expressions by simple procedures to predict the crack initiation in problems with practical interest. Motivated by this reasoning, all the works applying the coupled criterion found in the literature assume additional hypothesis on the crack geometry over which the minimization is carried out. The most common assumptions are:

- The crack onset is contained in a certain surface. This is typical in symmetric problems where the crack is assumed to be contained by the plane of symmetry, see e.g. the corner studied by Leguillon (2002). In the presence of anisotropy on the fracture and strength properties, the weaker plane is also supposed to contain the crack onset, see e.g. the problem in Section 6.2. In the presence of interfaces with weaker properties, the crack can be supposed to appear at the interface. In general, this type of assumption reduces drastically the complexity of the problem because the dimensions of the domain for $\Delta\hat{S}_c$ are reduced.

- The problem can be studied as a plane problem under the typical conditions given by the elasticity theory. The assumption of plane problem in the context of the FFM is also an assumption over the morphology of the crack since actually it implies that the crack morphology after the onset is invariable through the direction normal to the plane of study. This hypothesis also simplifies greatly the formulation of the coupled criterion. Moreover, the combination of this hypothesis with the previous one, reduces the minimization to find the optimum segment contained in a line. In fact, the majority of the problems found in the literature employing the coupled criterion are plane problems.

- A certain point of the initial geometry is necessarily contained in $\Delta\hat{S}_c$. This is a very common hypothesis in the presence of stress concentrations. The stress concentration point is normally supposed to be contained in $\Delta\hat{S}_c$ given that the stress level is maximum there.

- The crack onset is symmetric with respect to the initial planes of symmetry. Note that this is a more restrictive assumption than the hypothesis of crack contained in the plane of symmetry. For instance, in the problem of a crack onset at a holed plate under tensile stress, see e.g. Li and Zhang (2006); Zhang and Li (2008); Leite et al. (2010); Camanho et al. (2012), the assumption of crack contained by the plane of symmetry implies that the crack appears at either only one or the two symmetric stress concentration points at the hole perimeter, whereas the assumption of symmetric crack onset prescribes that the crack onset occurs simultaneously at the two stress concentration points. Although this hypothesis has been intensely employed, it is controversial since it has been demonstrated to be incorrect for some problems, see e.g. Section 4.1.

The assumption of some of the previous hypotheses enables to reduce the optimization problem presented in (3.43) to an optimization over a finite set of scalar variables. In fact, the majority of works found in the literature employing the coupled criterion are based on the optimization over one scalar variable. Moreover, the complexity is often reduced since the variation of stress and energy expressions in (3.43) with the scalar variable is monotonic, with opposed tendency for the stress and energy criterion for many of the problems studied in the literature. Thus, the optimization becomes a problem of solving a nonlinear equation.

In some problems, the assumptions enumerated previously cannot be easily applied. The cause is that different hypotheses can be taken under reasonable evidences. In these cases, the problem can be studied using separately the different assumptions and comparing the critical load obtained for each assumption. Thus, applying (3.43), the most accurate assumption will be the one for which the critical load predicted is minimum. In some cases, this result depends on the elastic, geometric or fracture parameters of the problem. For instance, Martin et al. (2008) studied the competition between deflection and penetration of a crack approaching an interface between dissimilar materials. They studied several hypotheses about the situation of the crack onset, e.g., the crack being contained either in the interface or in the material initially without crack. The comparison of the critical loads obtained for each assumption enables to predict the preferential type of crack as a function of the problem parameters.

3.5.2   Failure behavior predicted by the coupled criterion

Previous sections have focused on characterizing the crack onset by predicting the critical load leading to the onset of a crack with finite length. In addition, the morphology of the new crack immediately after the onset is obtained as a result of the optimization process. Once the process of crack onset is characterized, it is necessary to evaluate the whole loading process predicted by the coupled criterion. Thus, this section describes the stress-strain behavior predicted for an arbitrary solid subjected to monotonically increasing loads.

Figure 3.8 present examples of the typical force-displacement curve predicted by the coupled criterion for solids subjected to a monotonic increase of the external load level. The perfectly linear elastic equilibrium path is plotted with

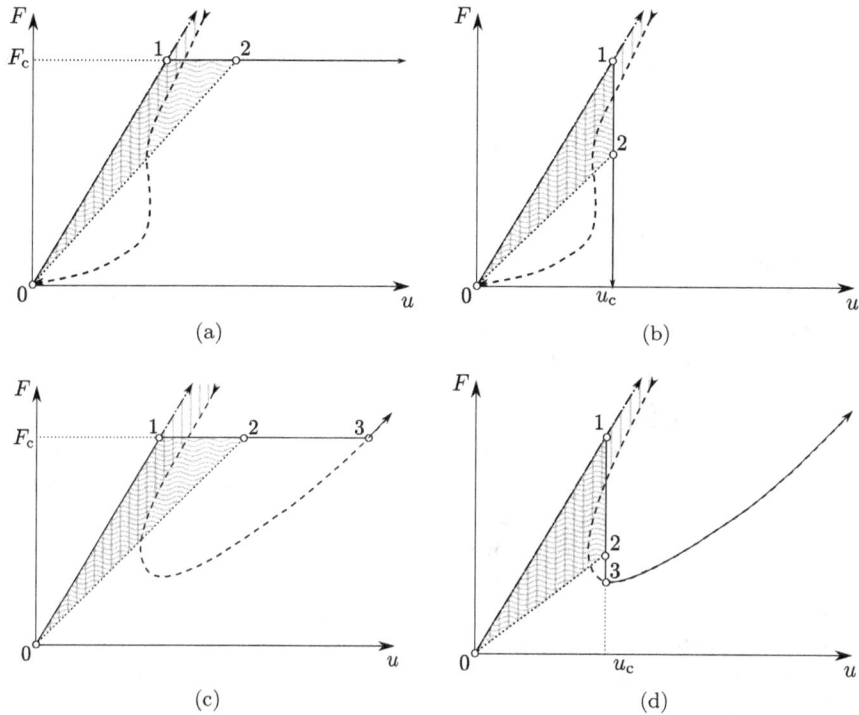

Figure 3.8: Examples of force-displacement curves predicted by the coupled criterion (solid line) for a case (a,b) with and (c,d) without global failure immediately after the onset and (a,c) force-controlled and (b,d) displacement-controlled tests. The main steps of the behavior predicted have been denoted as: (0-1) linear-elastic loading, (1-2) crack onset, (2-3) unstable crack growth after the onset, (3-) subsequent stable growth. Dashed line: LEFM equilibrium path. Dash-dot line: Undamaged linear elastic path. Areas filled with undulating lines: Released energy. Areas filled with vertical lines: Dissipated energy.

dot-dashed lines. In addition, the equilibrium paths predicted by the LEFM, given by the condition $G = G_c$, are plotted with dashed lines. Note that, in the cases plotted, the LEFM and the linear elastic equilibrium paths do not intersect. The reason is that these cases correspond to solids without preexisting cracks. As discussed previously, in the absence of cracks, the LEFM predicts an infinite critical load for crack initiation. As a consequence, the linear elastic and the LEFM equilibrium paths are asymptotically tangent for $u \to +\infty$ or $F \to +\infty$.

In general, the coupled criterion predicts initially a linear behavior following the dot-dashed line from states 0 to 1. For a certain value of either F^c (force controlled) or u^c (displacement controlled) corresponding to the critical value of the load of reference σ_0^c, both criteria are simultaneous fulfilled. Then, a crack onset occurs leading to an abrupt jump from states 1 to 2 on either u or F depending if the test is force or displacement controlled, see Figures 3.8(a) and 3.8(c) or Figures 3.8(b) and 3.8(d) respectively. Immediately after the crack onset the state is:

- For a force-controlled test, the force is $F_2 = F^c$ and the displacement corresponds to u_2.

- For a displacement-controlled test, the force is F_2 and the displacement corresponds to $u_2 = u^c$.

Note that the dissipated energy during the crack onset corresponds to the area of the triangle (0-1-2) filled with undulated lines and formed by the elastic line of loading up to the critical load (0-1), the line corresponding to the jump predicted by the coupled criterion (1-2) and a fictitious line of unloading from the situation (2) immediately after the crack onset to the origin (0). If the dissipated energy was computed by using the same G_c employed to plot the LEFM equilibrium path, the energy dissipated by the abrupt onset predicted by the coupled criterion is equivalent to the energy dissipated in the fictitious process predicted for the crack initiation by the LEFM. This process corresponds to a path which follows the linear elastic equilibrium up to an infinite load, for which the LEFM predicts crack initiation, and returns following the LEFM equilibrium path up to its intersection with the fictitious elastic unloading path from the state 2. Thus, the area enclosed by these paths, which is filled with bars in Figure 3.8, is the same that the area of the triangle (0-1-2).

The subsequent growth after the crack onset can be predicted by the LEFM. Thus, the stability of the growth depends strongly on the evolution of the LEFM equilibrium path. In some cases, see e.g. Figures 3.8(a) and 3.8(b), no mechanisms exists avoiding the unstable growth of the crack up to the global failure. This is the typical behavior of plates under tensile stress with stress concentrators as corners or holes. In these cases, the crack grows unstably after the crack onset up to reach the free edge, leading to the global failure of the specimen without any additional increase of the load. The equilibrium paths predicted by the LEFM for these problems normally end at the origin. The cause is that, when the crack approaches the free edge, the force or displacement necessary for the crack propagation vanishes. On the contrary, in other problems, the presence of certain phenomena avoids the sequence of unstable growth and global failure. For instance, when a crack approaches an interface between two dissimilar materials, the material containing initially the crack being softer, the value of $G \to 0$ when the crack approaches the interface. As a consequence, the growth after the crack onset becomes stable after (or not) a first unstable growth. In fact, according to the LEFM, the equilibrium path tends to a new straight line corresponding to the stiffness of a specimen with a crack totally arrested, see Section 4.1 for more details of this behavior for an actual problem. It is not reasonable to think that this stable growth will go on *ad infinitum* due to the same idea employed to reject an unlimited evolution along the linear elastic path. For a certain load level, a new failure mechanism will appear leading to a new crack onset, e.g., in the case of the crack approaching an interface, the new mechanisms are the onset of a crack at the interface or the penetration through the interface, see Martin et al. (2008).

CHAPTER 4

Failure initiation in long-fiber reinfonced composites under transverse loads

Composites reinforced by long fibers are commonly used as a structural material in lightweight structures at present. In aerospace applications, where lightweight is a key aspect of the design, their level of structural responsibility has significantly increased as they are massively used in primary structures. However, our understanding of the damage mechanisms occurring in these composites on different scales is still insufficient. Thus, it is necessary to generate more knowledge about these mechanisms in order to avoid the present high level of uncertainty in the failure loads predicted in the design.

In particular, the failure criteria still do not predict satisfactorily failures of unidirectional laminae under combined transverse loads (with loads perpendicular to the fiber-axis), as was highlighted in a series of coordinated studies (known as the world-wide failure exercises), e.g. Hinton et al. (2004); Hinton and Kaddour (2013). A reason for this is that the current failure criteria are still not sufficiently physically based from the microscopic point of view.

One of the most complex failure mechanisms on micro scale in these composites is associated to the matrix failure, also called inter-fiber failure. In particular, the tension dominated mechanism follows a well described sequence of stages, see Hull and Clyne (1996) and París et al. (2007): *i)* failure is initiated at the fiber-matrix interface as small debonds, *ii)* the interface cracks grow along the interface until a certain arrest angle and then *iii)* kink out the interface towards the matrix, *iv)* coalescence of growing matrix cracks generates a macrocrack which may cause the failure of the unidirectional lamina.

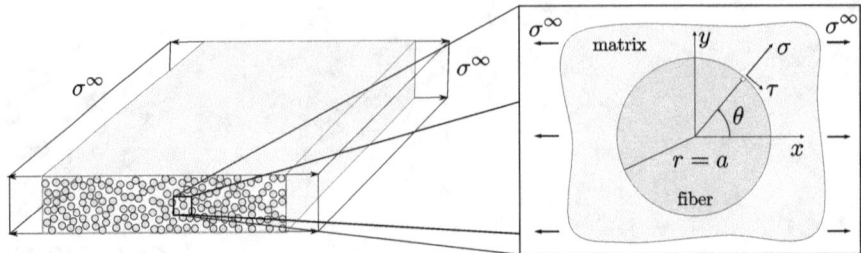

Figure 4.1: Schematic of the simplification assumed in this chapter: the plane strain problem of a single long fiber surrounded by an infinite matrix.

The present work is focused on the two first steps of this failure mechanism: crack initiation and growth at the fiber-matrix interface. A simplified model given by a single long fiber surrounded by an infinite matrix is often used as a reasonable first approximation to study some failure mechanisms at the micromechanical level in fiber-reinforced composites, at least for dilute packing. A few experimental studies of debond onset and growth in single fiber or inclusion specimens subjected to transverse loads can be found in Zhang et al. (1997); Gamstedt and Sjögren (1999); Contreras (2000); Martyniuk et al. (2013). The study of the micromechanical fiber-matrix behavior has demonstrated to be very useful for explaining some macroscopic phenomena in Fiber Reinforced Composites (FRC). A few examples are: the effect of the reinforcement size on the superplastic deformation properties observed in fiber reinforced metal matrix composites Carpinteri et al. (2005); Zahid et al. (1998), the macroscopic effect of the interface properties Yang et al. (1991), the features of failure under transverse tension Mantič (2009); París et al. (2007); Távara et al. (2011) and compression Correa et al. (2008) and the influence of a transverse compression on the failure under dominant transverse tension París et al. (2003) in FRC laminates. In the case of *matrix failure* under remote transverse tension, the single fiber model leads to a plane strain problem of a circular inclusion embedded in an infinite matrix. Problem of a single fiber embedded in a matrix has been intensively studied by many authors for a long time. The stress solution of such a problem for perfect and linear-elastic inclusion-matrix-interface, respectively, was deduced by Goodier (1933) and Gao (1995). Assuming the open model of interface cracks at perfect interfaces, England (1966) and Toya (1974) obtained analytic solutions for stresses, displacements and energy release rate (ERR) in the presence of a debond at the inclusion-matrix interface, whereas considering contact at the crack tip of interface cracks, París et al. (1996); Mantič et al. (2006) presented accurate numerical solutions for this problem. Levy (1991) proposed to model interface debonding by prescribing a cohesive interface law and presented an integral equation which governs the solution, see París et al. (2007) and Mantič (2009) for comprehensive reviews. Nevertheless, the crack initiation has not attracted a sufficient attention up to the last decade. Results presented in bibliography have usually been obtained by computational methods as cohesive zone or weak interface models, see for instance Carpinteri et al. (2005), Xie and Levy (2007), and Távara et al. (2011)).

In the context of the coupled criterion of the finite fracture mechanics, see Chapter 3, Mantič (2009) proposed a theoretical model to predict the crack initiation along the fiber-matrix interface under a uniaxial remote tension. This model is based only on analytical solutions and is able to predict the critical transverse load for which the interface crack onset occurs. In addition, the angle of debond immediately after the onset is predicted. Both values are expressed as a function of the elastic parameters of fiber and matrix, fiber radius and strength and fracture properties of the interface. This model predicts some interesting results as a strong size effect on the critical load and a brittle-to-ductile transition on the failure behavior.

In view of the successful application by Mantič (2009), this chapter aims to extend this model to the case of biaxial transverse load in order to evaluate the influence of a secondary transverse load on the main critical load leading to the crack onset. This secondary transverse load is neglected by the majority of failure criteria for composites in spite of the existence of some preliminary experiments evidencing its influence, see París et al. (2003). This chapter is organized as follows: First, a controversial hypothesis employed by Mantič (2009) is revisited. Mantič (2009) assumed that the crack onset breaks the initial symmetry of the problem because a single debond is predicted at one of the poles where the normal stress to the interface is maximum. This was motivated by some experimental observations. However, the two results can be found in the literature predicted by models based on prescribing cohesive laws at the interface. Thus, prior to the extension of the model to take into account biaxial transverse model, this hypothesis is reviewed in Section 4.1 by comparing the critical load predicted following this hypothesis with the critical load obtained if two symmetric debonds are assumed. Second, one of the main results found by Mantič (2009), the size effect, is compared with the predictions obtained by a cohesive zone model in Section 4.2. Finally, an extended theoretical model to take into account a secondary transverse load is presented in Section 4.3 along with an analysis of the influence of this secondary load as a function of the main problem parameters.

4.1 Symmetry on the crack initiation

A few works can be found in the literature studying the crack initiation at the fiber-matrix interface under transverse uniaxial loading by using different interface laws, e.g. Carpinteri et al. (2005); Han et al. (2006); Kushch et al. (2011); Távara et al. (2011). Several of these works agree on some relevant results such as a size effect on the critical load, see Section 4.2 for an analysis of this result. However, there is a disagreement on a key issue of the *post-failure configuration* (or *post debond-onset configuration*). For instance, works by Carpinteri et al. (2005); Kushch et al. (2011) predict a symmetric debond onset whereas Levy

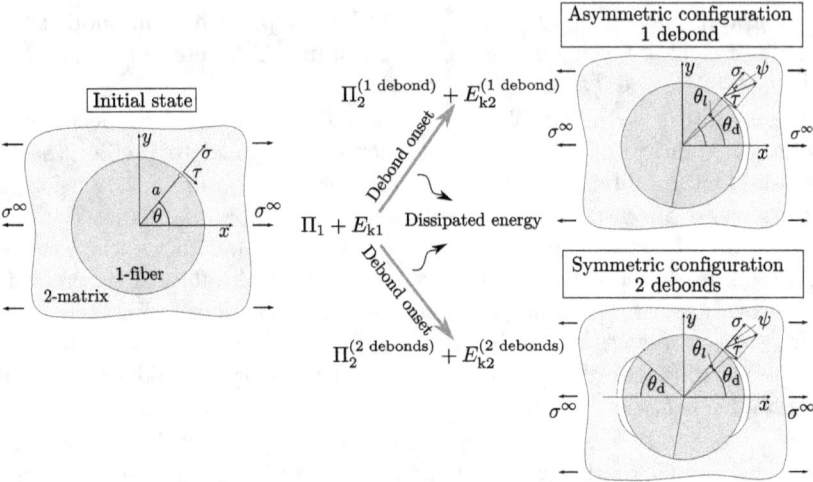

Figure 4.2: Schema of the problem under study.

and Hardikar (1999); Han et al. (2006); Távara et al. (2011) predict breaking the symmetry of the original configuration due to the onset of a single debond.

This section aims to analyze these two post-failure configurations observed in previous studies, an asymmetric debond and two symmetric debonds, by means of the coupled criterion of the FFM. The objective is to assess which of them is to be expected to appear in experiments and clarify why. It is assumed that the lower critical remote tension is predicted for the preferential post-failure configuration.

The plane strain problem under study is shown in Figure 4.2. In the initial state, a single fiber is perfectly bonded to an infinite matrix. The matrix is loaded with a uniaxial remote tension σ^∞ in the x-direction. For *a priori* two different critical values of σ^∞, one or two debonds appear as showed in Figure 4.2. As both initial state and post-failure configurations are symmetric with respect to the x-axis, only the upper-half ($y \geq 0$) of the geometry is considered. Hence, the polar angle θ and also other angles are defined as ≥ 0. Glass/epoxy composite is taken as a reference bimaterial in the present work, see Table 4.1 for its linear elastic properties, although, a summary of key results is also shown for carbon/epoxy composite (Table 4.1) and for a few virtual materials with extreme values of the main parameters governing the problem.

First, stress and energy criteria are studied and applied separately in Sections 4.1.1 and 4.1.2, respectively. Both criteria are applied in a parallel manner for both post-failure configurations. The combination of both criteria is described in Section 4.1.3. Finally, results about the loss of symmetry are discussed in Section 4.1.4 for several bimaterials, and a representation of crack onset by the stress-strain curves predicted for both post-failure configurations is introduced and interpreted.

Bimaterial	E_1(GPa)	ν_1	E_2(GPa)	ν_2	α	β	k	m
glass/epoxy	70.8	0.22	2.79	0.33	0.919	0.229	1.44	1.56
carbon/epoxy	13.0	0.20	2.79	0.33	0.624	0.136	1.32	1.43

Table 4.1: Elastic properties of the composites used (1, fiber; 2, matrix)

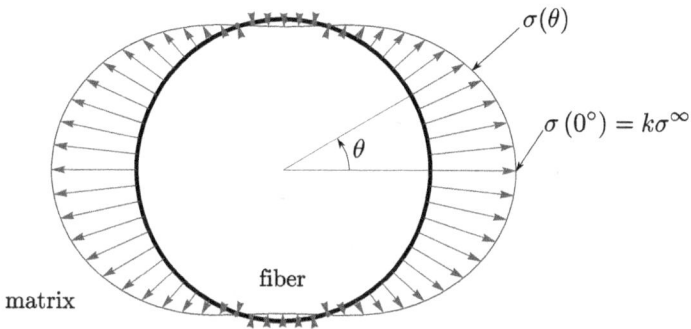

Figure 4.3: Normal tractions σ along the fiber-matrix interface at the initial state for glass/epoxy.

4.1.1 Stress criterion

A stress criterion is usually invoked for brittle or quasi-brittle materials when no pre-existing damage exists. In the framework of the FFM, the stress criterion defines a condition on stresses along the assumed future crack surface in the elastic state prior to the crack onset, see Section 3.3 for a detailed analysis.

The tensile criterion based on a pointwise analysis is used in the present study for the sake of simplicity. Assuming a critical value $\sigma_c > 0$ for normal stresses σ along the interface, the condition for the onset given by this criterion can be expressed as,

$$\sigma(\theta) \geq \sigma_c, \qquad \forall \theta \in [0, \Delta\theta], \tag{4.1}$$

where θ is the polar angle, see Figure 4.2, and $\Delta\theta$ is the debond semiangle. Analytical expression of the normal tractions σ along the fiber-matrix interface in the undamaged state can be extracted from the classical solution by Goodier (1933), cf. Mantič (2009),

$$\frac{\sigma(\theta)}{\sigma^\infty} = k - m \sin^2(\theta), \tag{4.2}$$

where k and m are dimensionless elastic bimaterial properties defined by Mantič (2009) in terms of the Dundurs (1969) parameters α and β,

$$k(\alpha, \beta) = \frac{1}{2} \frac{1+\alpha}{1+\beta} \frac{2+\alpha-\beta}{1+\alpha-2\beta}, \qquad m(\alpha, \beta) = \frac{1+\alpha}{1+\beta}. \tag{4.3}$$

Two identical maximums of $\sigma(\theta)$, see Figure 4.3, are found at $\theta = 0°$ and $180°$. These two points are the preferential points for the debond initiation which

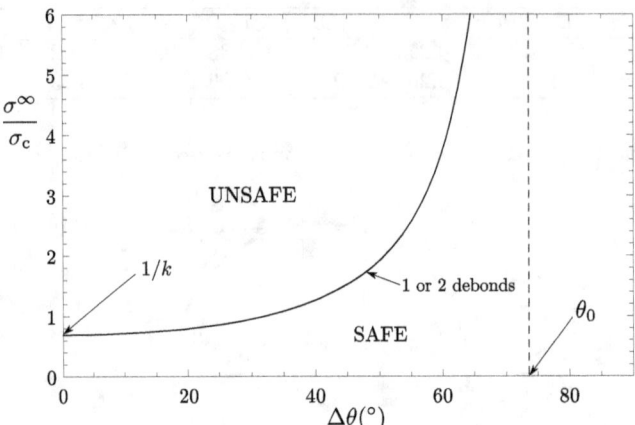

Figure 4.4: Graphical representation of the stress criterion for both post-failure configurations and glass/epoxy.

justifies the two post-failure configurations assumed here. In the following, due to the symmetry of the initial state with respect to the y-axis, only the half part for $x \geq 0$ is studied in order to unify the analysis for both post-failure configurations. In the case of the symmetric configuration, both debond onsets are obviously allowed when the stress condition is fulfilled on one of the sides. Taking into account that $\sigma(\theta)$ is a decreasing function for $0° \leq \theta \leq 90°$, see (4.2), and assuming that $\Delta\theta \leq 90°$, the condition in (4.1) is verified if $\sigma(\Delta\theta) \geq \sigma_c$. Then, introducing (4.2) into (4.1), the expression of the stress condition for the debond onset is obtained

$$\frac{\sigma^\infty}{\sigma_c} \geq s(\Delta\theta) = \frac{1}{k - m\sin^2 \Delta\theta}. \tag{4.4}$$

This gives a minimum remote tension σ^∞ necessary to originate the debond as a function of the debond semiangle $\Delta\theta$. Figure 4.4 shows how the stress criterion splits the $(\Delta\theta, \sigma^\infty/\sigma_c)$ plane into unsafe and safe regions where the debond onset is or not allowed, respectively. The stress criterion curve for glass/epoxy in Figure 4.4 has a vertical asymptote at $\theta_0 = 73.63°$, thus no debond onset with $\Delta\theta \geq \theta_0$ is possible for any remote tension. The reason is that $\sigma(\theta) \leq 0$ for $\theta \geq \theta_0$, see Figure 4.3. A discussion of this issue for any bimaterial can be found in Section 4.3.

As the stress criterion is based exclusively on the initial (undamaged) elastic state, it is quite obvious that both post-failure configurations are equivalent with respect to this criterion.

4.1.2   Energy criterion

An incremental energy criterion, in contrast to the classical (infinitesimal) Griffith criterion, is considered in FFM by evaluating the energetic balance between the elastic states before and after the debond onset for both post-failure configurations. The first law of Thermodynamics, neglecting heat transfer, gives

$$\Delta\Pi + \Delta E_k + \Delta\Gamma = 0, \tag{4.5}$$

where $\Delta\Pi$ and ΔE_k, respectively, are variations of the elastic potential and kinetic energy and $\Delta\Gamma$ is the energy dissipated in the irreversible processes associated to the debond onset. Assuming an initial static state implies $\Delta E_k \geq 0$, hence (4.5) can be rewritten as

$$-\Delta\Pi \geq \Delta\Gamma, \tag{4.6}$$

which means that the released energy has to exceed the dissipated energy during the debond onset. In the next subsections $\Delta\Pi$ and $\Delta\Gamma$ will be obtained as functions of the debond semiangle $\Delta\theta$, the bimaterial and interface properties and the post-failure configuration.

4.1.2.1   Released energy

The variation of the elastic potential energy $\Delta\Pi$ can efficiently and accurately be evaluated by using the energy release rate G (ERR) in the classical expression of the linear elastic fracture mechanics (LEFM), see Mantič (2009), as $G = -(d\Pi/d\theta_d)/(an)$ where n is the number of debonds, $n = 1$ and 2, corresponding to the asymmetric and symmetric post-failure configurations, respectively. Thus, for a debond onset of a semiangle $\Delta\theta$,

$$\Delta\Pi = -2n \int_0^{\Delta\theta} G(\theta_d, n) a\, d\theta_d, \tag{4.7}$$

where G gives the ERR of n debonds growing from $\theta = 0°$ and symmetrically with respect to the x-axis, and θ_d is the semiangle of an intermediate debond employed only to compute $\Delta\Pi$.

To evaluate $\Delta\Pi$ by (4.7) it is necessary to know $G(\theta_d, n)$ for any $0 \leq \theta_d \leq \Delta\theta$. Toya (1974) deduced an analytical solution for the case $n = 1$ using the open model of interface cracks. However, to the best of our knowledge no analytic solution is available for the case $n = 2$, though accurate numerical results, assuming the open model of interface cracks, can be found in Chen and Nakamichi (1997); Murakami (2001). Nevertheless, following París et al. (1996); Mantič et al. (2006), the values of G computed assuming the open model of interface cracks are valid only for sufficiently small debond lengths. Therefore, frictionless contact possible at the interface crack tips is considered here, which requires a computational code able to solve the pertinent nonlinear receding contact problems and accurately compute G, valid for any debond length.

For given bimaterial elastic properties, G depends, in addition to n and θ_d, on the remote tension σ^∞ and the geometry, completely defined by the fiber radius a. In view of the large amount of physical variables and in order to reduce the number of numerically solved cases required to obtain G, the Vaschy-Buckingham Π theorem, Vaschy (1892); Buckingham (1914), is applied. Thus, as is demonstrated in Appendix 4.A, a dimensionless ERR $\hat{G}$ can be defined as,

$$\hat{G}\left(\theta_{\mathrm{d}}, n, \alpha, \beta\right) = \frac{E^*}{a\left(\sigma^\infty\right)^2} G\left(\theta_{\mathrm{d}}, n, \sigma^\infty, a, E^*, \alpha, \beta\right), \tag{4.8}$$

where E^* is the harmonic mean of the effective elasticity moduli

$$\frac{1}{E^*} = \frac{1}{2}\left(\frac{1-\nu_1^2}{E_1} + \frac{1-\nu_2^2}{E_2}\right). \tag{4.9}$$

As the first part of the present study is carried out for glass/epoxy, the dependence on α and β will be omitted for the sake of brevity.

The elastic problem for each value of θ_{d} and n is solved by the Boundary Element Method (BEM) code developed by Graciani et al. (2005). This code uses continuous linear boundary elements and is optimized for the solution of frictionless contact problems of interface cracks. The BEM model is defined by a square cell of matrix significantly larger than the circle at its center representing the fiber. Boundary conditions at interface nodes are either "perfectly bonded" or "in contact", depending on the values of θ_{d} and n. A mesh suitably refined at the crack tips is used, with element lengths at the interface varying from $2°$, far from the crack tips, to $0.00004°$, at the crack tips. Normal traction $\sigma^\infty > 0$ is imposed at two parallel external edges of the square cell, whereas the other two edges are free.

The values of interface tractions and relative displacements computed by BEM in the vicinity of the crack tips are used to estimate $\hat{G}$ by the Virtual Crack Closure Technique (VCCT) with the aid of a Chebyshev-Gauss quadrature, see Graciani et al. (2010) for details.

Figure 4.5 shows the computed values of $\hat{G}$ as a function of θ_{d} and n, and includes the analytical solution by Toya (1974) for checking purposes. The values of $\hat{G}$ computed for $\theta_{\mathrm{d}} \lesssim 65°$ and $n = 1$ are very similar to the analytical values. However, for larger values of θ_{d}, a significant contact zone appears at the debond tip, that is not captured by the analytic solution assuming the open model of interface cracks, leading to an overestimation of $\hat{G}$ by Toya's solution.

A comparison of $\hat{G}$ values for $n = 1$ and 2 shows that they are larger for $n = 1$ and any value of $\theta_{\mathrm{d}} > 0$. The difference is quite relevant for larger values of θ_{d} due to the shielding effect between the two debonds for $n = 2$. Moreover, $\hat{G}$ for $n = 2$ tends to zero for $\theta_{\mathrm{d}} = 90°$, since it corresponds to a symmetric situation of two shear cracks approaching each other along the same line. The previous qualitative arguments are independent of the bimaterial properties, so the differences between the cases $n = 1$ and 2 observed in Figure 4.5 can be extrapolated to other bimaterials.

The values of $\hat{G}_{\mathrm{I}}$ and $\hat{G}_{\mathrm{II}}$, corresponding to a virtual crack increment angle of $\delta\theta = 0.5°$, are also plotted in Figure 4.5. The difference between both post-failure configuration is much larger in mode II than in mode I. The reason is that the shielding effect in the case $n = 2$ is not sufficiently strong for the small debonds where the mode I is dominant.

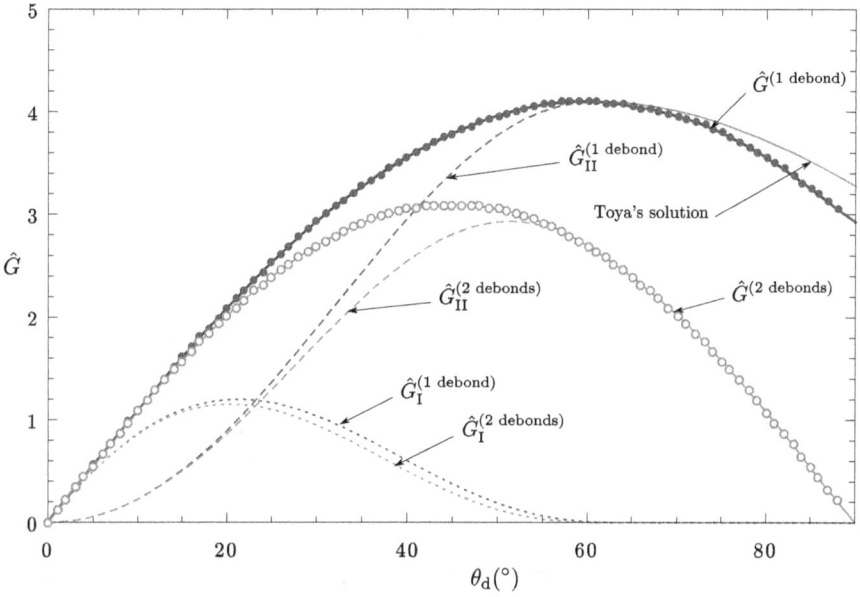

Figure 4.5: $\hat{G}$ as a function of θ_{d} for glass/epoxy, obtained by the VCCT applied with a virtual crack increment angle in VCCT of $\delta\theta = 0.5°$ to the BEM results for each value of θ_{d} (points) and $n = 1$ and 2, compared with the Toya solution for $n = 1$.

4.1.2.2   Dissipated energy

The energy dissipated during the debond onset due to the associated irreversible processes can be estimated by integrating the interface fracture toughness corresponding to an intermediate debond semiangle θ_{d} as proposed by Mantič (2009),

$$\Delta\Gamma = 2n \int_0^{\Delta\theta} G_{\mathrm{c}}(\psi(\theta_{\mathrm{d}}, n))a\mathrm{d}\theta_{\mathrm{d}}, \tag{4.10}$$

where the angle ψ is a stress-based measure of fracture mode mixity at the debond tip for θ_{d}. This expression is analogous to that for the released energy (4.7), but in (4.10), in opposite to (4.7), the value of the integral does depend on the sequence of growth chosen for the intermediate debond. In the present case, the debond growth sequence described above for the computation of $\Delta\Pi$ is assumed in accordance with Mantič (2009), see Section 3.4.2 for a discussion on this issue.

The dependence of G_{c} on ψ can be approximated e.g. by the phenomenological law due to Hutchinson and Suo (1992), which provides a good agreement with available experiments for interface cracks,

$$G_{\mathrm{c}}(\psi) = G_{1\mathrm{c}}\hat{G}_{\mathrm{c}} = G_{1\mathrm{c}}(1 + \tan^2[(1 - \lambda)\psi]), \tag{4.11}$$

where $G_{1\mathrm{c}}$ is the interface fracture toughness in pure mode I, λ is a fracture

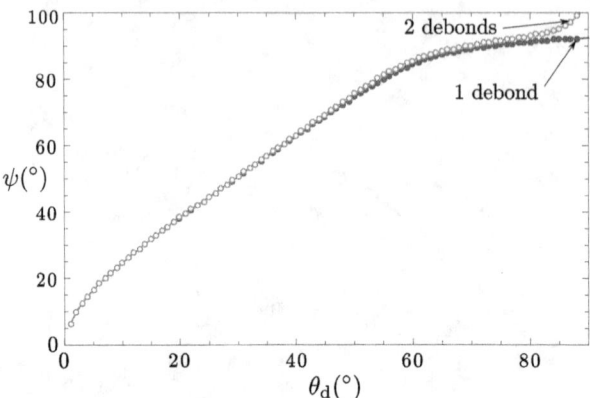

Figure 4.6: Fracture-mode-mixity angle ψ as a function of the debond semiangle θ_d and both post-failure configurations for $\lambda = 0.3$, $\theta_l = 0.1°$ and glass/epoxy.

mode-sensitivity parameter, with a typical range $0.2 \leq \lambda \leq 0.35$, and $\hat{G}_\mathrm{c}$ is a dimensionless interface fracture toughness.

The fracture-mode-mixity angle ψ can be calculated by using the interface stresses obtained from the computational model

$$\tan \psi(\theta_\mathrm{d}, \theta_l) = \frac{\tau(\theta_\mathrm{d} + \theta_l)}{\sigma(\theta_\mathrm{d} + \theta_l)}, \tag{4.12}$$

where θ_l is the distance from the debond tip where the fracture mode mixity is measured. Following Mantič (2009), $\theta_l = 0.1°$ is taken.

Figure 4.6 shows the value of ψ as a function of θ_d for the two post-failure configurations. In general the angle ψ is very similar for both configurations, though slightly larger for $n = 2$ in general. The difference becomes larger for the extreme where $\theta_\mathrm{d} \to 90°$ since a closed crack with $\psi \gtrsim 90°$ is expected for $n = 1$, whereas $\psi \to 180°$ for $\nu_2 > 0$ and $n = 2$ due to the symmetry of the limit configuration and the Poisson's effect of the matrix.

$\hat{G}_\mathrm{c}$ is plotted in Figure 4.7(a) as a function of θ_d, for the above values of ψ and $\lambda = 0.3$. As could be expected, in view of Figure 4.6 and (4.11), $\hat{G}_\mathrm{c}$ is slightly greater for $n = 2$ than for $n = 1$ and this difference increases with θ_d. Figure 4.7(b) also shows the percentage difference in $\hat{G}_\mathrm{c}$ between $n = 1$ and $n = 2$, confirming that for typical values of λ, the above observations hold. Anyway, the relative difference in $\hat{G}_\mathrm{c}$ appears to be in general smaller than in $\hat{G}$, the latter providing the main contribution to the difference in the energetic balance between both configurations.

4.1.2.3   Final expression of the energy criterion

Both terms in the energetic balance (4.6) have been analyzed and computed above. This subsection aims to collect the main results in order to achieve an expression analogous to that obtained for the stress criterion in (4.4).

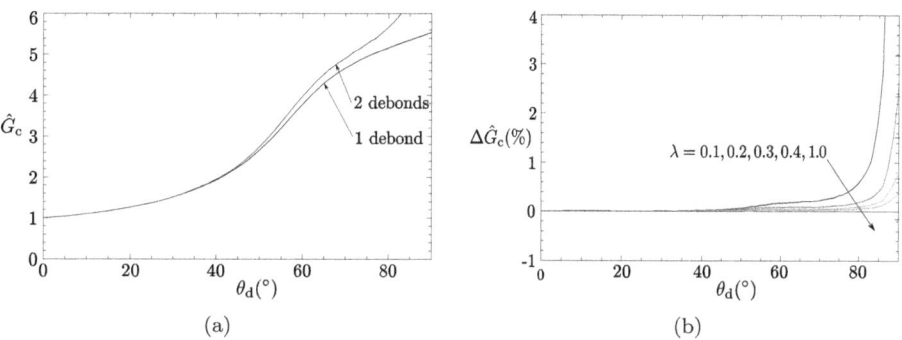

(a)   (b)

Figure 4.7: (a) dimensionless interface fracture toughness $\hat{G}_{\mathrm{c}}$ and (b) percentage difference $\Delta\hat{G}_{\mathrm{c}} = 100 \cdot \left(\hat{G}_{\mathrm{c}}^{(2\ \mathrm{debonds})} - \hat{G}_{\mathrm{c}}^{(1\ \mathrm{debonds})}\right)/\hat{G}_{\mathrm{c}}^{(1\ \mathrm{debond})}$ between the dimensionless interface fracture toughness for the symmetric $\hat{G}_{\mathrm{c}}^{(2\ \mathrm{debonds})}$ and asymmetric $\hat{G}_{\mathrm{c}}^{(1\ \mathrm{debond})}$ configurations as a function of the debond semiangle θ_{d} for $\lambda = 0.3$, $\theta_l = 0.1°$ and glass/epoxy.

Introducing the expression of the released energy (4.7) and the dissipated energy (4.10) in the energetic balance (4.6) and taking into account the definitions of $\hat{G}$ in (4.8) and $\hat{G}_{\mathrm{c}}$ in (4.11), this balance leads to

$$2n\frac{(\sigma^{\infty})^2\,a}{E^*}\int_0^{\Delta\theta}\hat{G}(\theta_{\mathrm{d}},n)\mathrm{d}\theta_{\mathrm{d}} \geq 2nG_{1\mathrm{c}}\int_0^{\Delta\theta}\hat{G}_{\mathrm{c}}(\psi(\theta_{\mathrm{d}},n))\mathrm{d}\theta_{\mathrm{d}}. \qquad (4.13)$$

This inequality can be rewritten to the form of a condition for the normalized remote tension $\sigma^{\infty}/\sigma_{\mathrm{c}}$ necessary to originate a debond of semiangle $\Delta\theta$, giving a minimum remote tension semiangle

$$\frac{\sigma^{\infty}}{\sigma_{\mathrm{c}}} \geq \gamma\sqrt{g(\Delta\theta,n)}, \qquad (4.14)$$

where

$$\gamma = \frac{1}{\sigma_{\mathrm{c}}}\sqrt{\frac{G_{1\mathrm{c}}E^*}{a}} \qquad (4.15)$$

is a brittleness number, cf. Mantič (2009), and

$$g(\Delta\theta,n) = \frac{\int_0^{\Delta\theta}\hat{G}_{\mathrm{c}}(\psi(\theta_{\mathrm{d}},n))\mathrm{d}\theta_{\mathrm{d}}}{\int_0^{\Delta\theta}\hat{G}(\theta_{\mathrm{d}},n)\mathrm{d}\theta_{\mathrm{d}}} \qquad (4.16)$$

is a dimensionless function which measures the ratio of the dimensionless dissipated to released energies of the debond and depends on the parameters that determines the post-failure situation, i.e. $\Delta\theta$ and n, and also on the dimensionless material parameters α, β and λ. According to Figure 4.8, which shows g computed by numerical integrations, $g(\Delta,1) < g(\Delta,2)$ with both the absolute and relative difference of both values increasing with $\Delta\theta$ at least for $\Delta\theta \geq 1°$.

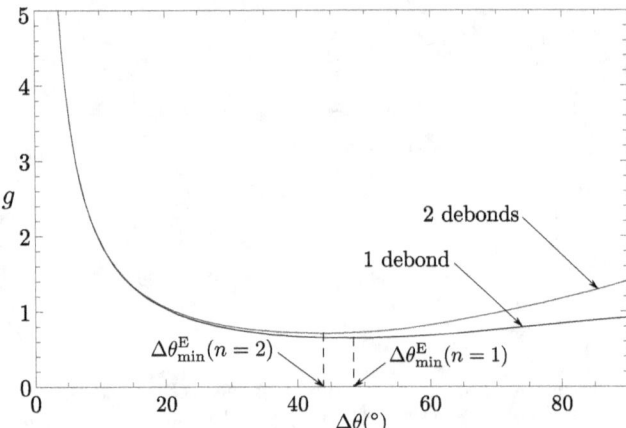

Figure 4.8: Ratio g of the dimensionless dissipated to released energy as a function of the debond semiangle $\Delta\theta$, for $\lambda = 0.3$, $\theta_l = 0.1°$ and glass/epoxy.

It is interesting to observe that g has a minimum for $\Delta\theta = \Delta\theta_{\min}^{E} < 90°$. In the limit $\Delta\theta \to 0°$, $g \to +\infty$ in accordance with the classical Griffith criterion for an infinitesimal advance of a crack.

4.1.3 Coupled criterion

Both stress and energy criteria have been analyzed previously in an independent manner leading to two conditions on the remote tension σ^{∞} required for a debond onset. Both criteria are combined in this section assuming the hypothesis by Leguillon (2002): the critical remote tension σ_c^{∞} originating a debond onset is given by the minimum remote tension σ^{∞} for which both criteria are fulfilled simultaneously for a critical value of $\Delta\theta$.

An analysis of the monotonicity of the functions on the right-hand sides of (4.4) and (4.14), corresponding to the stress and energy criteria, respectively, is required to implement Leguillon's hypothesis in a semianalytical procedure. For both post-failure configurations, the function $s(\Delta\theta)$ in (4.4) is increasing with $\Delta\theta$, whereas $g(\Delta\theta, n)$ in (4.16) is decreasing with $\Delta\theta$ up to a minimum $\Delta\theta_{\min}^{E}$. At the plane $(\Delta\theta, \sigma^{\infty}/\sigma_c)$ for $\Delta\theta \leq \Delta\theta_{\min}^{E}$, the two curves given by these criteria have either one or none at all intersection points, $s(\Delta\theta) = \gamma\sqrt{g(\Delta\theta, n)}$, depending on the value of γ, which leads to the following two scenarios:

- Scenario A: If both curves have one intersection point for $\Delta\theta \leq \Delta\theta_{\min}^{E}$, the minimum remote tension fulfilling both criteria for the debond onset σ_c^{∞} is given by this intersection point $(\Delta\theta_c, \gamma\sqrt{g(\Delta\theta_c, n)})$, $\Delta\theta_c$ being the critical semiangle of debond and $\sigma_c^{\infty}/\sigma = \gamma\sqrt{g(\Delta\theta_c, n)}$ the critical remote tension for the debond onset.

- Scenario B: If both curves have none intersection point for $\Delta\theta \leq \Delta\theta_{\min}^{E}$, the remote tension fulfilling both criteria for the debond onset σ_{c}^{∞} is given by the minimum value of g, the critical semiangle of debond being given by $\Delta\theta_{\min}^{E}$, and the critical remote tension by $\sigma_{c}^{\infty}/\sigma_{c} = \gamma\sqrt{g(\Delta\theta_{\min}^{E}, n)}$.

The scenario is determinated by the value of γ since the energy criterion curve is proportional to this value. A threshold value of γ can be defined as

$$\gamma_{\text{th}} = \frac{s(\Delta\theta_{\min}^{E})}{g(\Delta\theta_{\min}^{E}(n), n)}, \tag{4.17}$$

which separates the scenario A ($\gamma \leq \gamma_{\text{th}}$) from B ($\gamma > \gamma_{\text{th}}$). Thus, for low values of γ, the debond onset (θ_{c} and $\sigma_{c}^{\infty}/\sigma_{c}$) is governed by the combination of both criteria, the critical remote tension being essentially given by the stress criterion whereas the debond semiangle by the energy criterion. On the contrary, for large values of γ, the debond onset is governed by the energy criterion only.

An unstable growth of the debond is possible after its onset. This is studied here by means of the classical LEFM. Thus, the condition for a further growth of the debond is given by the Griffith criterion as

$$G(\sigma_{c}^{\infty}, \theta_{d}, n) \geq G_{c}(\psi(\theta_{d}, n)), \quad \text{for} \quad \Delta\theta_{c} \leq \theta_{d} \leq \theta_{a}, \tag{4.18}$$

where this unstable growth will be arrested for an arrest semiangle θ_{a} verifying still

$$G(\theta_{a}, n) = G_{c}(\psi(\theta_{a}, n)), \tag{4.19}$$

but

$$\left.\frac{dG(\theta_{d}, n)}{d\theta_{d}}\right|_{\theta_{d}=\theta_{a}} < \left.\frac{dG_{c}(\psi(\theta_{d}, n))}{d\theta_{d}}\right|_{\theta_{d}=\theta_{a}}. \tag{4.20}$$

The critical values characterizing the debond as $\sigma_{c}^{\infty}/\sigma_{c}$, $\Delta\theta_{c}$ and θ_{a} are computed by algorithms developed in Mantič (2009).

The critical and arrest semiangles are plotted in Figure 4.9 as functions of γ and the post-failure configuration. Note that $\Delta\theta_{c}$ is very similar for both post-failure configurations in scenario A whereas for scenario B, both corresponds to the value of $\Delta\theta_{\min}^{E}$, which is quite different for both post-failure configurations. On the contrary, the arrest angle is very different in both cases due to the very different values of $\hat{G}$ for large θ_{d}, see Figure 4.5.

Figure 4.10 shows $\sigma_{c}^{\infty}/\sigma_{c}$ as a function of γ for both post-failure configurations. σ_{c}^{∞} is quite constant for small γ, but it increases strongly for $\gamma \gtrsim 1$. In particular, it is a linear function of γ in scenario B, as shown above. Qualitatively the behavior of σ_{c}^{∞} is similar for both post-failure configuration in both scenarios, although a larger absolute difference is found in scenario B, as will be pointed out in Section 4.1.4.

For given material properties, the dependence of the results on γ can be interpreted as a size effect by defining a reference radius

$$a_{0} = \frac{G_{1c}E^{*}}{\sigma_{c}^{2}}, \tag{4.21}$$

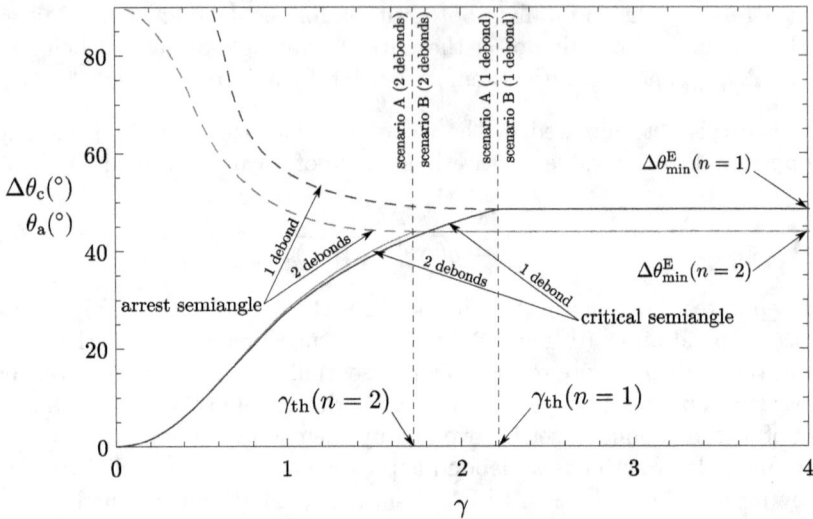

Figure 4.9: Critical and arrest semiangles, $\Delta\theta_\text{c}$ and θ_a, respectively, as functions of the brittleness number γ, $\lambda = 0.3$, $\theta_l = 0.1°$ and glass/epoxy.

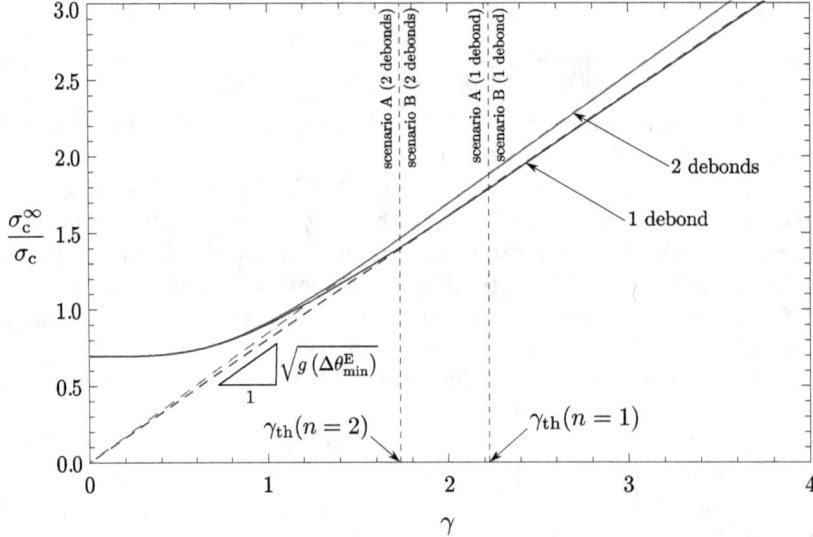

Figure 4.10: Critical remote tension σ_c^∞ as a function of the brittleness number γ for both post-failure configurations, $\lambda = 0.3$, $\theta_l = 0.1°$ and glass/epoxy.

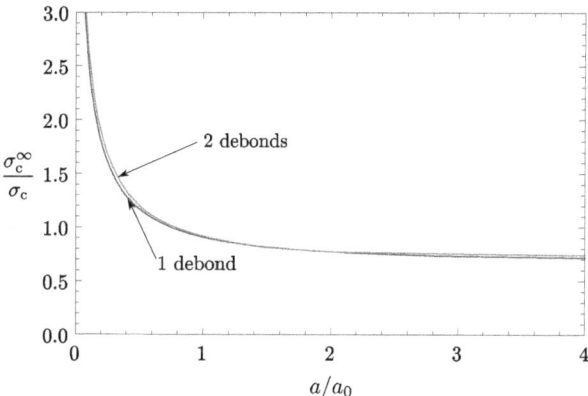

Figure 4.11: Critical remote tension σ_c^∞ as a function of the fiber radius a for both post-failure configurations, $\lambda = 0.3$, $\theta_l = 0.1°$ and glass/epoxy.

and relating a dimensionless radius a/a_0 and γ by

$$\frac{a}{a_0} = \frac{1}{\gamma^2}. \tag{4.22}$$

A strong size effect can be observed in Figure 4.11, showing σ_c^∞/σ_c as a function of a/a_0, where σ_c^∞/σ_c increases strongly for small values of a/a_0, see Mantič (2009) or Section 4.2 for a detailed analysis of this size effect.

4.1.4   Analysis and discussion of results

This subsection focuses on the main question which motivated the present section: the loss or conservation of symmetry in the post-failure configuration in a single-fiber specimen under transverse tension. First, both configurations are compared by computing the difference of their critical tensions predicted by the coupled criterion of the FFM for glass-epoxy and also for other bimaterials. Then, the stress-strain curves predicted by this criterion for both post-failure configurations are plotted and interpreted. Finally, some experimental results found in the literature are briefly revisited.

4.1.4.1   Difference in critical loads for symmetric and asymmetric post-failure configurations

It is assumed that the preferential post-failure configuration, which can be expected to appear in experiments, corresponds to the configuration requiring a lower critical remote tension for the debond onset. Thus, in view of Figures 4.10 and 4.11, the asymmetric configuration is preferential for glass-epoxy. The percentage difference between the critical value for the tension predicted for the two

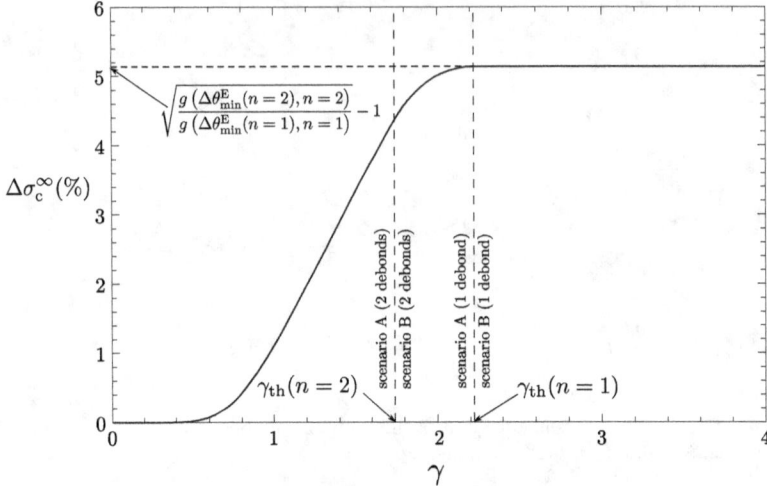

Figure 4.12: Percentage difference $\Delta\sigma_c^\infty = 100 \cdot \left(\sigma_c^{\infty(2\ \mathrm{debonds})} - \sigma_c^{\infty(1\ \mathrm{debond})}\right)/\sigma_c^{\infty(1\ \mathrm{debond})}$ between the critical remote tensions for the symmetric and asymmetric post-failure configurations, taking $\lambda = 0.3$, $\theta_l = 0.1°$ and glass/epoxy.

configurations,

$$\Delta\sigma_c^\infty(\%) = 100 \cdot \frac{\sigma_c^\infty(n=2) - \sigma_c^\infty(n=1)}{\sigma_c^\infty(n=1)}, \qquad (4.23)$$

is shown in Figure 4.12, where it is clearly seen that $\Delta\sigma_c^\infty > 0$ for $\gamma > 0$. Therefore, the asymmetric configuration is preferential independently of γ. However, $\Delta\sigma_c^\infty$ varies between two limit values. In the limit $\gamma \to 0_+$, which corresponds to scenario A, $\Delta\sigma_c^\infty = 0$. The reason is that in this limit case the critical remote tension is governed by the stress criterion, which is equivalent for both post-failure configurations, see Section 4.1.1. At the other extreme, for large values of γ, corresponding to scenario B, $\Delta\sigma_c^\infty$ is given by its maximum value

$$\max \Delta\sigma_c^\infty(\%) = \Delta\sigma_c^\infty\left(\gamma \geq \max_{n=1,2} \gamma_{\mathrm{th}}(n)\right) = 100 \cdot \left(\sqrt{\frac{g(\Delta\theta_{\mathrm{min}}^{\mathrm{E}}(n=2))}{g(\Delta\theta_{\mathrm{min}}^{\mathrm{E}}(n=1))}} - 1\right),$$

$$(4.24)$$

obtained in view of the expression $\sigma_c^\infty/\sigma_c = \gamma\sqrt{g(\Delta\theta_{\mathrm{min}}^{\mathrm{E}}, n)}$ in scenario B. $\max \Delta\sigma_c^\infty$ obviously corresponds to scenario B governed by the energy criterion, as this is the source of difference between both configurations. The main reason for this is the shielding effect between both debonds in the symmetric configuration, as discussed in Section 4.1.2.

Since $\Delta\sigma_c^\infty$, the key result of this work, is bounded by the above described two limit values, the behavior of a bimaterial is essentially characterized by the upper limit (4.24), as the lower limit is null independently of the bimaterial. To generalize the previous study to other materials, this upper limit is computed for

		$\lambda =$	0	0.1	0.2	0.3	0.4	0.5	1
Bimaterial	α	β							
glass/epoxy	0.919	0.229	2.53 %	3.16 %	4.02 %	5.13 %	6.63 %	8.78 %	15.1 %
carbon/epoxy	0.624	0.136	4.26 %	5.03 %	5.97 %	7.08 %	8.39 %	9.93 %	15.8 %
A	0	0	8.29 %	9.06 %	9.91 %	10.8 %	11.8 %	12.8 %	16.5 %
B	1	0	4.56 %	5.26 %	6.11 %	7.18 %	8.48 %	10.0 %	16.1 %
C	1	0.5	0.85 %	1.65 %	2.63 %	4.31 %	7.16 %	9.84 %	14.2 %

Table 4.2: Maximum percentage difference in critical tensions $\max \Delta \sigma_c^\infty$ (4.24) between the symmetric and asymmetric configurations for several bimaterials and values of λ. Virtual bimaterials corresponds to: A: Similar elastic properties for fiber and matrix. B: Rigid fiber ($E_1 \gg E_2$) and incompressible matrix ($\nu_2 = 0,5$). C: Rigid fiber ($E_1 \gg E_2$) and matrix with $\nu_2 = 0$.

several bimaterials and values of the fracture-mode-mixity sensitivity parameter λ. Table 4.2 shows the maximum percentage difference, $\max \Delta \sigma_c^\infty$, predicted for two usual composites, glass-epoxy and carbon-epoxy, and three virtual bimaterials corresponding to some extreme values of the Dundurs parameters α and β. The bimaterial A corresponds to the extreme case of identical elastic materials, whereas the bimaterials B and C correspond to cases with the fiber much stiffer than the matrix. The difference between them is given by the Poisson ratio of the matrix, $\nu_2 = 0$ for B and 0.5 for C. This table also shows the variations of $\max \Delta \sigma_c^\infty$ with λ to evaluate the influence of the sensitivity of the fracture toughness $\hat{G}_c$ on the fracture-mode-mixity angle ψ.

According to Table 4.2, λ has a strong influence on the values of $\max \Delta \sigma_c^\infty$. The reason for this is that a variation of λ modifies the position of the steeply increasing part of $\hat{G}_c$-curve. For large values of λ, which correspond to a lower sensitivity of $\hat{G}_c$ to variations of ψ, the steeply increasing part of $\hat{G}_c$-curve moves toward large values of θ_d. At the extreme $\lambda = 1$, $\hat{G}_c$ is a constant; this case was considered by Carraro and Quaresimin (2014). Therefore, in view of the definition of g in (4.16), if the increasing part of $\hat{G}_c$ is translated towards large values of θ_d, the value $\Delta \theta_{min}^E$ of the minimum of g is expected to increase. Thus, since the shielding effect between two symmetric debonds increases for large debonds, $\max \Delta \sigma_c^\infty$ increases with increasing λ. This argument will be repeatedly used in the following.

The influence of the elastic bimaterial properties on the value of $\max \Delta \sigma_c^\infty$ is elucidated in Table 4.2 by analyzing the extreme cases A, B and C of α and β, see Section 4.3.3.3. This influence decreases with increasing λ. In fact, the value of $\Delta \sigma_c^\infty$ for $\lambda = 1$ is very similar for all the bimaterials studied. It means that the main source of difference between bimaterials is the different evolution of the fracture mode mixity for each bimaterial. This is particularly remarkable by comparing the results for the virtual bimaterials with a rigid fiber ($\alpha = 1$): B ($\beta = 0$) and C ($\beta = 0.5$). It can be shown, following Dundurs (1967), that for a bimaterial with $\alpha = 1$, stress solution only depends on the Poisson's ratio of the matrix, which is $\nu_2 = 0.5$ and $\nu_2 = 0$ for B and C, respectively. In view of this and the values in Table 4.2, ν_2 has a strong influence on $\max \Delta \sigma_c^\infty$, at least for the extreme $\alpha = 1$. This is due to the influence of ν_2 on the evolution of ψ

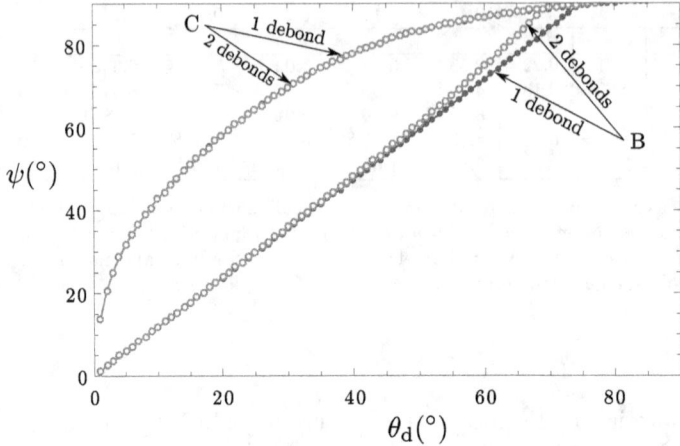

Figure 4.13: Fracture-mode-mixity angle ψ as a function of the debond semiangle θ_d and both post-failure configurations for the virtual bimaterials corresponding to rigid fiber ($E_1 \gg E_2$) with either (B) incompressible matrix ($\nu_2 = 0.5$) or (C) matrix with $\nu_2 = 0$ obtained with $\theta_l = 0.1°$.

with θ_d for small debonds, see Figure 4.13. The constraining effect of the Poisson ratio in the zones of the matrix near the crack tip countervails the effect of the variation of the angle between the normal to the interface at the crack tip and the direction of the remote tension which is the main source of the variation of the fracture mode mixity with θ_d. Consequently, ψ grows more steeply with θ_d when the constraining effect is vanishing (material C: $\beta = 0.5$) than for incompressible matrix (material B: $\beta = 0$). This implies that the increasing part of $\hat{G}_c$ moves toward small values of θ_d when reducing ν_2, thus max $\Delta\sigma_c^\infty$ decreases in a similar way as described previously in the analysis of the influence of λ.

For identical elastic materials of the fiber and matrix ($\alpha = \beta = 0$), ψ increases less steeply than for the cases with $\alpha = 1$. The reason is that a more compliant fiber is associated to the growth of ψ since the crack tip is allowed to move towards the symmetry plane. This explains that, according to the previous discussion about the relation between ψ and max $\Delta\sigma_c^\infty$, the largest max $\Delta\sigma_c^\infty$ is predicted for the extreme case when the fiber and matrix materials are identical.

The behavior of $\Delta\sigma_c^\infty$ for carbon/epoxy and glass/epoxy can be understood as an intermediate case of the extreme cases discussed previously. In this case the two effects described act combined, max $\Delta\sigma_c^\infty$ being slightly higher for carbon/e-poxy than for glass/epoxy, which could be foreseen from the above analysis.

4.1.4.2   Stress-strain curves predicted by the coupled stress and energy criterion of the FFM

In addition to different critical values of the remote tension predicted for both post-failure configurations, the coupled criterion also predicts different paths in the stress-strain plane for the two configurations. Figures 4.14 and 4.15 show the dimensionless remote tension σ^∞/σ_c versus a homogenized normal strain between two pairs of matrix points, for two arbitrarily chosen values of γ corresponding to two scenarios A and B described in Section 4.1.3. This homogenized strain is evaluated as an adequately normalized difference between the displacements u_x at these reference pairs of points. The first pair of points A'A corresponds to the two poles at the fiber-matrix interface located on the matrix side, whereas the other pair B'B corresponds to the points situated at the intersection of the symmetry axis and the external edges of matrix-cell. The pair A'A is chosen because the distance between the points A' and A leads to a stress-strain curve with the debond onset affecting significantly the represented stiffness, therefore all the steps described below are easily visible, see Figures 4.14(a) and 4.14(b). However, the areas enclosed by the curves do not provide a useful interpretation in terms of the work or energy. This is why it is also useful to plot σ^∞/σ_c versus the homogenized strain for the pair B'B, showed in Figures 4.15(a) and 4.15(b). Actually, strictly speaking $u_x^{B'}$ and u_x^{B} represent the mean displacements along the two external edges in order to be able to interpret the areas enclosed by the curves in terms of work or energy. The problem of this pair is that the influence of a debond is insignificant in this model due to the dimensions of the matrix square-cell, and the stress-strain curve would seem just as a straight line. This is solved here, following Bažant and Cedolin (1991), by subtracting the homogenized strain due to a purely linear elastic deformation $\Delta u_x^{B'B,\text{elast}}$ between the same points when the interface is perfectly bonded from the term $u_x^{B'} - u_x^{B}$. This makes visible the stages of the debond onset predicted by the model used, while keeping the meaning of the represented areas in terms of work or energy.

Dashed lines shown in Figures 4.14 and 4.15 represent the equilibrium paths given by the classical LEFM, and their asymptotes corresponding to the elastic solutions with the whole fiber-matrix interface either bonded or debonded. These asymptotes are obtained directly by the computation of the models fully debonded and perfectly bonded. The equilibrium path predicted by the LEFM is defined by those points with,

$$G(\theta_d, n, a, \sigma^\infty, E^*, \alpha, \beta) = G_c(G_{1c}, \theta_d, n, \lambda), \qquad (4.25)$$

and using the dimensionless expressions defined in (4.8)–(4.11) for G and G_c respectively, the expression of the remote tension σ^∞ leads to

$$\frac{\sigma^\infty}{\sigma_c} = \gamma \sqrt{\frac{\hat{G}_c(\theta_d, n, \lambda)}{\hat{G}(\theta_d, n, \alpha, \beta)}}. \qquad (4.26)$$

Then, the displacements at the reference points shown in the plots can be obtained by evaluating the ratio of remote tensions to displacements extracted from the computational model for each value of θ_d. Thus, the homogenized normal strain for a reference pair of points $P'P$ can be obtained as,

$$\frac{u_x^P - u_x^{P'}}{a\sigma_c/E^*} = \frac{(\tilde{u}_x^P(\theta_d) - \tilde{u}_x^{P'}(\theta_d))/a}{\tilde{\sigma}^\infty/E^*} \frac{\sigma^\infty}{\sigma_c}, \qquad (4.27)$$

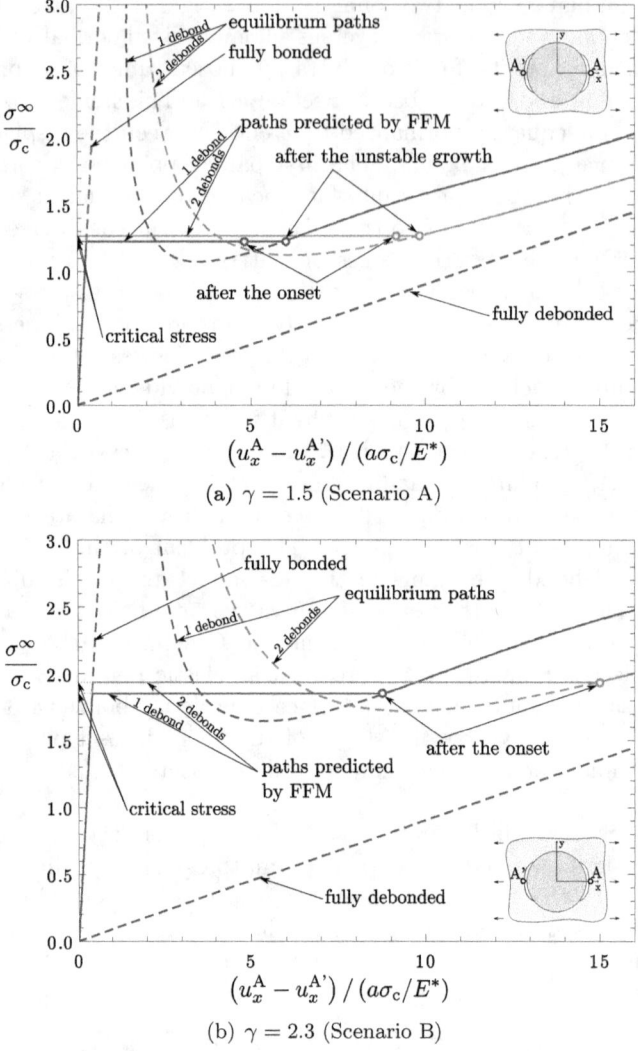

Figure 4.14: Stress versus homogenized strain curves for the reference pairs of points A'A predicted by the coupled stress and energy criterion, for $\lambda = 0.3$, $\theta_l = 0.1°$ and glass/epoxy.

where $\tilde{\sigma}^\infty$ is the remote tension applied in the computational model and $\tilde{u}_x^P$ is the corresponding value of the displacement obtained at the point P in this model for a certain value of θ_d. The parametric equation of the equilibrium path predicted by the LEFM and showed in Figure 4.14(a) is given by (4.26)–(4.27), where θ_d is the parameter. As expressions on the right-hand-side of (4.26)–(4.27) are proportional to γ, the equilibrium path expands (or shrinks) proportionally to increasing (or decreasing) γ.

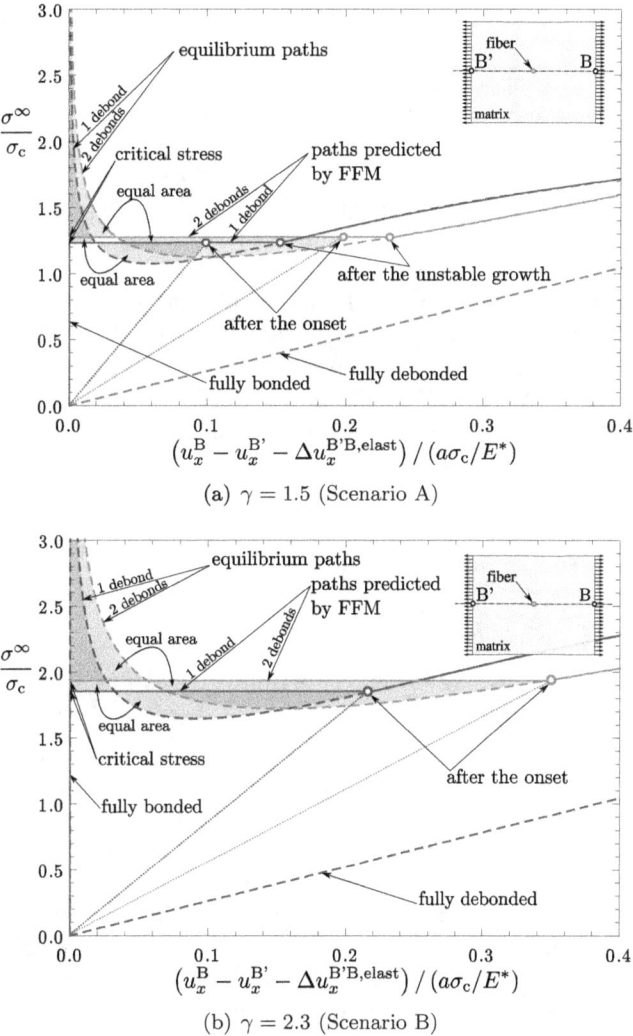

(a) $\gamma = 1.5$ (Scenario A)

(b) $\gamma = 2.3$ (Scenario B)

Figure 4.15: Stress versus homogenized strain curves for the reference pairs of points B'B predicted by the coupled stress and energy criterion, for $\lambda = 0.3$, $\theta_l = 0.1°$ and glass/epoxy.

The path predicted by the coupled criterion of the FFM is represented with solid lines in Figures 4.14 and 4.15. After following the linear elastic path without debond, when the critical tension is reached, a jump in the value of displacements is predicted. This jump ends either at the corresponding equilibrium path defined by the LEFM, in scenario B (Figures 4.14(b) and 4.15(b)), or at a lower value of the homogenized normal strain, in scenario A (Figures 4.14(a) and 4.15(a)). In the latter case, an unstable debond growth is predicted to occur immediately up to reaching the corresponding equilibrium path. Afterwards, in both scenarios, a stable growth is predicted following the equilibrium path.

The areas between the equilibrium paths defined by the LEFM and the jump-line predicted by the FFM are highlighted in Figures 4.15(a) and 4.15(b). According to the energy criterion the highlighted areas are equal as indicated in these figures. Thus the jump defined by the FFM can be understood as a kind of Maxwell line, and the equilibrium paths above this line correspond to metastable states. According to this idea, since the path predicted by FFM for the symmetric post-failure configuration is above the asymmetric one, the former is an unstable solution for the problem of fiber-matrix debonding. It can also be observed that less energy is dissipated in the asymmetric debond onset than in the onset of two symmetric debonds.

4.1.4.3   Discussion of previous experimental results

The above predicted fact that asymmetric debond is generally the preferential post-failure configuration, although for some composites (with a small brittleness number γ) this difference could be negligible, is in accordance with the experimental evidences presented in the literature for single-fiber tests under transverse loading. Gamstedt and Sjögren (1999) presented a single-fiber test under transverse cyclic loading at low frequency. The microscope observations showed a clear asymmetric initiation of the debond along the fiber-matrix interface. Contreras (2000) carried out a macroscopic test of a circular inclusion embedded in a more compliant matrix. The specimen was subjected to a monotonous loading and for a critical load showed an abrupt debond onset along one of two sides of the interface, the debond being arrested after the onset.

It should be mentioned, that, recently, Martyniuk et al. (2013) carried out related experiments for a single fiber specimen under transverse tension with two debonds appearing, in a genuine 3D situation, at a circular fiber-matrix interface edge, where a weak singularity of stresses takes place. Thus, the results of the present 2D study cannot directly be extrapolated to those experiments, although a very similar 3D study could be carried out.

4.2 Comparison of the size effect predicted by the coupled criterion with the predictions of a cohesive zone model

An effect of the fiber size for a fixed volumetric fiber content on the macroscopic properties of composites has been observed in some experiments. For instance, the superplastic behavior of metal-matrix composites, see Zahid et al. (1998), and the tensile strength, see Cho et al. (2006); Fisher and Gurland (1981); Leidner and Woodhams (1974), are found to be dependent on the inclusion size. These size effects have been explained using different cohesive zone models (CZM) for describing the fiber-matrix interface in Carpinteri et al. (2005); Távara et al. (2011), and FFM in Mantič (2009) and Sections 4.1 and 4.3. In view of this background, the aim of the present work is comparing the predictions given by the coupled criterion with the well established CZM and providing a better understanding of the mechanisms leading to this size effect. The CZM compared prescribes the Tvergaard (1990) law for the fiber-matrix interface implemented in a FEM code by Carpinteri et al. (2005). These results have shown a strong fiber-size effect on the transverse tension leading to an unstable decohesion at the fiber-matrix interface.

Both models, CZM and FFM, are governed by different parameters which rule the interface behavior. Therefore, it is necessary to find a relation between these parameters in order to select them in a coherent manner in both models. Additionally, these correlations will allow understanding the scope of applicability of these models and the differences between their predictions. Hence, this section includes three parts: first, each model is briefly presented analyzing its governing parameters from the perspective of the other model in order to clarify the relation between them and set adequately the governing parameters and laws. Second, a comparison of the predicted size effect is carried out for the materials and interface properties considered in Carpinteri et al. (2005). Third, the differences will be interpreted and analyzed. The objective of this study is therefore twofold: first, these models will provide an explanation for the size effect found in the particular problem of the fiber-matrix behavior. Second, some outcomes will contribute to the knowledge about the relations between the CZM and the FFM in general. In the same vein, an analysis of the limits of validity of both approaches is given.

The CZM is briefly presented and analyzed from a FFM perspective in Section 4.2.1. A short review of the application of the coupled criterion of the FFM to the present problem is presented in Section 4.2.2 introducing slight modifications with respect to previous works Mantič (2009) in order to be coherent with the CZM notation. A comparison and an interpretation of the predicted size effect is given in Section 4.2.3. Finally, some limits of validity of the models used are highlighted in Section 4.2.4.

4.2.1 CZM: formulation and predictions

The problem of fiber-matrix debonding caused by transverse loads is studied under the assumption of plane strain. Then, the problem geometry is reduced, in the plane perpendicular to the fiber axis, to a circular inclusion with radius a surrounded by a matrix perfectly bonded along its interface, see Figure 4.1. Both matrix and fiber are assumed to be linear-elastic materials. Mechanical parameters are chosen to represent an aluminium alloy 2124-SiC whisker-reinforced material, i.e., $E_m = 60$ GPa, $E_f = 340$ GPa, $\nu_m = 0.3$, $\nu_f = 0.18$, where subscripts (f) and (m) refer, respectively, to the fiber and matrix. Initially, it is considered the case of a regular rectangular packing with a low volumetric fiber content equal to $v = 18\%$, leading to $b/a = 4.18$, see the the geometry modeled in the FEM simulation, Figure 4.16(a), and the FEM mesh in Figure 4.16(b). Note that the geometry shown in this figure is not necessarily a representative volume element (RVE) of the material microstructure, but rather the smallest element of the composite in the case of perfect periodicity of the material microstructure.

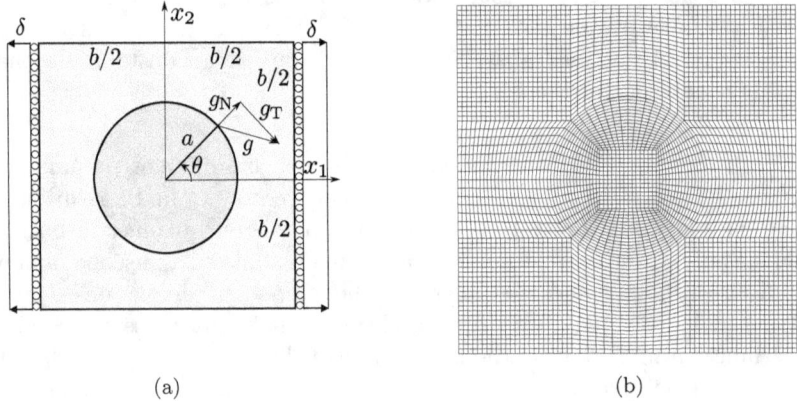

(a)   (b)

Figure 4.16: Scheme of the FEM simulation: (a) geometry and boundary conditions and (b) mesh used for $v = 18\%$

The fiber-matrix interface failure is modeled by a CZM describing a nonlinear behavior of the interface. The CZM is defined by a constitutive relation for a zero-thickness interface written in terms of cohesive tractions and relative displacements. Two pure fracture modes, Mode I (opening) and Mode II (sliding) are suitably coupled in the case of mode mixity. In the sequel, σ and τ denote the normal and tangential cohesive tractions, where the adjectives normal and tangential refer to the local axes of the curvilinear reference system along the interface. Correspondingly, a relative normal displacement (also called normal gap), g_N, and a relative tangential displacement (tangential gap), g_T, are admitted. The normal gap g_N must be non negative, since overlapping of fiber and matrix is not physically allowed.

Different CZM constitutive laws can be adopted depending on the materials under consideration, see Elices et al. (2002); Paggi and Wriggers (2011) for a critical examination. In any case, regardless of the shape of the traction-gap relation, the CZM introduces a nonlinearity into the problem. Moreover, the satisfaction of the unilateral contact constraint is another source of nonlinearity. In the present study, the FEM is employed to solve the problem numerically. The corresponding nonlinear problem is discretized using standard 2D plane strain elements with quadratic shape functions. The curvilinear interface is discretized applying the *virtual node technique*, firstly proposed in Boso et al. (2005) for contact problems and then extended to nonlinear fracture mechanics applications in Carpinteri et al. (2008). As a main advantage with respect to the classic interface element formulation Schellekens and de Borst (1993), no fixed coupling between FEM nodes along the interface has to be enforced a priori, which can be a limitation in the case of large displacements. The kinematic variables g_N and g_T are determined using a node-to-segment contact strategy Wriggers (2002) and the CZM relation is adopted for the computation of the cohesive tractions. Depending on the sign of g_N, an automatic switching procedure is adopted to choose between cohesive and contact models for the normal direction. In the case of a negative g_N, a penalty formulation is used to enforce the contact constraint Carpinteri et al. (2005, 2008); Paggi et al. (2006). For the CZM relation, the model by Tvergaard Tvergaard (1990) is selected, often used for particle-matrix debonding and similar interface problems. The normal and tangential cohesive tractions vs. normal and tangential relative displacements are computed as follows (see also Figure 4.17 for their visualization):

$$\sigma = \sigma_c \frac{g_N}{l_{Nc}} P(\lambda), \tag{4.28a}$$

$$\tau = \mu\sigma_c \frac{g_T}{l_{Tc}} P(\lambda), \tag{4.28b}$$

with $g_N \geq 0$ and

$$P(\lambda) = \begin{cases} \dfrac{27}{4}(1-\lambda)^2, & \text{for } 0 \leq \lambda \leq 1 \\ 0, & \text{otherwise} \end{cases} \tag{4.29a}$$

$$\lambda = \sqrt{\left(\frac{g_N}{l_{Nc}}\right)^2 + \left(\frac{g_T}{l_{Tc}}\right)^2}. \tag{4.29b}$$

The parameters l_{Nc} and l_{Tc} are, respectively, the critical relative opening and sliding displacements, σ_c and τ_c are the maximum interface normal and tangential tractions, $\mu = \tau_c/\sigma_c$, and λ is the dimensionless effective relative displacement.

The energy dissipated after complete separation by cohesive elements under Mode I (the interface fracture toughness in Mode I), G_{1c}, is calculated by integrating the cohesive law,

$$G_{1c} = \frac{27}{4}\sigma_c \cdot l_{Nc} \cdot \int_0^1 \lambda(1-\lambda)^2 \, d\lambda = \frac{27}{48}\sigma_c l_{Nc}, \tag{4.30}$$

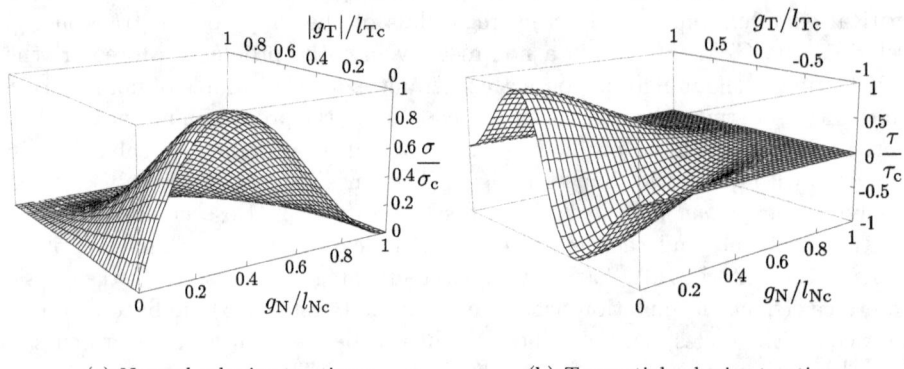

(a) Normal cohesive tractions   (b) Tangential cohesive tractions

Figure 4.17: Tvergaard CZM law used to describe the nonlinear behavior of the fiber-matrix interface.

considering that in a cohesive element monotonically loaded in pure Mode I the dimensionless effective opening displacement is $\lambda = g_N/l_{Nc}$. The energy dissipated after complete separation in pure Mode II, can be obtained analogously as G_{1c} in (4.30): $G_{2c} = (27/48)\tau_c l_{Tc}$.

The energy dissipated by the cohesive elements in mixed mode, G_c, can be evaluated as a function of the stress-based fracture mode mixity $\psi_\sigma = \tan^{-1}\frac{\tau}{\sigma}$ by the following integral,

$$G_c(\psi_\sigma) = \int_0^{l_{Nc}\sqrt{1-\left(\frac{g_T}{l_{Tc}}\right)^2}} \sigma\left(\frac{g_N}{l_{Nc}}, \frac{g_T}{l_{Tc}}\right) \mathrm{d}g_N + \int_0^{l_{Tc}\sqrt{1-\left(\frac{g_N}{l_{Nc}}\right)^2}} \tau\left(\frac{g_N}{l_{Nc}}, \frac{g_T}{l_{Tc}}\right) \mathrm{d}g_T.$$
(4.31)

This expression can be calculated using the next change of variables

$$\frac{g_N}{l_{Nc}} = \lambda \cos \psi$$
(4.32a)

$$\frac{g_T}{l_{Tc}} = \lambda \sin \psi$$
(4.32b)

where ψ is a effective displacement based fracture mode mixity which can be related to ψ_σ from (4.28) as

$$\tan \psi = \frac{g_T/l_{Tc}}{g_N/l_{Nc}} = \frac{1}{\mu} \tan \psi_\sigma.$$
(4.33)

Introducing (4.33) in (4.28), (4.29) and (4.31), the expression for $G_c(\psi)$ is obtained

$$G_c(\psi) = \frac{27}{48} \left(\sigma_c l_{Nc} \cos^2 \psi + \tau_c l_{Tc} \sin^2 \psi\right)$$
(4.34)

and using trigonometric identities and the relation between ψ and ψ_σ (4.33), the expression of the equivalent interface fracture toughness in the Tvergaard cohesive law as a function of the stress based fracture mode mixity ψ_σ is obtained

Assuming a fixed stress-based fracture mode mixity $\psi_\sigma = \arctan(\tau/\sigma)$ during the loading, the energy dissipated by a cohesive element can be expressed as a function of ψ_σ,

$$G_c(\psi_\sigma) = G_{1c} \left(\frac{\mu^2 + \frac{G_{2c}}{G_{1c}} \tan^2 \psi_\sigma}{\mu^2 + \tan^2 \psi_\sigma} \right) \tag{4.35}$$

where $G_{2c} = (27/48)\tau_c l_{Tc}$, the interface fracture toughness in Mode II, can be obtained analogously as G_{1c} in (4.30). It is interesting to compare this expression to the Hutchinson and Suo (1992) law commonly used in applications (e.g. Mantič (2009); Távara et al. (2011); Correa et al. (2008)). Figure 4.18 shows the above expression of $G_c(\psi_\sigma)/G_{1c}$ for several values of μ and G_{2c}/G_{1c} compared to the Hutchinson and Suo law. Note that whereas $G_c(\psi_\sigma)$ from the Tvergaard's CZM depends on μ, the Hutchinson and Suo law does not. As a consequence the difference between both depends on the value of μ. In general, two main differences can be observed: Hutchinson and Suo law is convex in the whole domain whereas Tvergaard CZM law has an inflection point between convex and concave parts because $G'_c(\psi_\sigma = 0°) = G'_c(\psi_\sigma = 90°) = 0$.

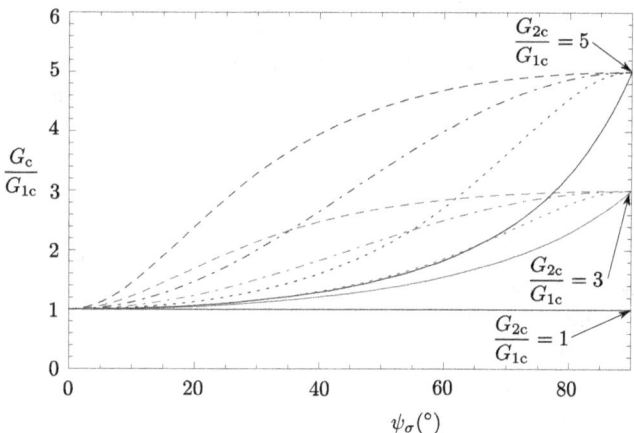

Figure 4.18: Interface fracture toughness G_c as a function of the stress based fracture mode mixity ψ_σ extracted from the Tvergaard CZM for several values of G_{2c}/G_{1c} and $\mu = 0.5$ (dashed lines), $\mu = 1$ (dotted-dashed lines) and $\mu = 2$ (dotted lines) in comparison to the Hutchinson and Suo phenomenological law Hutchinson and Suo (1992) (solid line).

In the present work the same values as in Carpinteri et al. (2005) are used: $\mu = 1$, $l_{Tc} = l_{Nc} = 0.03\,\mu m$ and $\sigma_c = \tau_c = 300$ MPa. In this case, since $l_{Nc} = l_{Tc}$ and $\sigma_c = \tau_c$ it is possible to demonstrate that the energy dissipated after complete separation G_c is independent of the stress path with value,

$$G_c = G_{1c} = G_{2c} = 5.0625 \text{ J/m}^2. \tag{4.36}$$

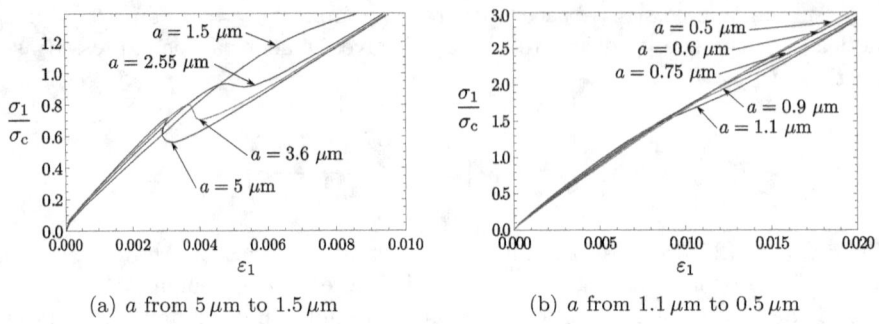

(a) a from $5\,\mu$m to $1.5\,\mu$m   (b) a from $1.1\,\mu$m to $0.5\,\mu$m

Figure 4.19: Homogenized stress-strain curves by changing the fiber size for $\nu = 18\%$. Results courtesy of Professor M. Paggi published in García et al. (2014).

Imposing horizontal (direction 1) displacements on the vertical boundaries of the unit square cell as shown in Figure 4.16(a), the homogenized stress-strain response can numerically be computed. As in Paggi and Wriggers (2011), the average stress σ_1 is determined as the sum of the horizontal reactions divided by the length of the vertical cell side b. Similarly, the average strain ε_1 is evaluated as the imposed relative displacement 2δ divided by the original length of the horizontal side of the cell b. For the same volumetric fiber content, different fiber radii a are considered, obtaining the family of curves shown in Figure 4.19. A strong nonlinear behavior can be observed. This is associated to the progressive decohesion of the interface leading to two separated debonds symmetric with respect to the x_2-axis as shown in Figure 4.20. Reducing the fiber radius from 5 μm to 2.55 μm, a transition from snap-back to snap-through instability is observed in the post-peak response, see Figure 4.19(a). A further reduction of the fiber radius leads to hardening behavior, see Figure 4.19(b). At the other extreme when fibers are large, see Figure 4.19(a), all curves approach to the same asymptotic behavior before the peak. To capture the post-peak branch in the case of large fibers, a simple displacement control is not sufficient and the position of the real interface crack tip has to be used as the control parameter monotonically increasing with time Carpinteri et al. (2005). These curves show a pronounced fiber-size effect that will be discussed in Section 4.2.3.

The effect of the fiber volumetric content v can be analyzed by changing it and repeating the above virtual uniaxial tensile tests. The homogenized stress-strain curves are shown in Figure 4.21 for two different fiber radii $a = 1.5\,\mu$m and $a = 3.6\,\mu$m. The progress of interface decohesion leads to a nonlinear stress-strain response. For small fiber radii, the post-peak response substantially depends on the volumetric content and may present a hardening ($v = 17.8\%$ and 35\% in Figure 4.21(a)) or a snap-through branch ($v = 50\%$ in Figure 4.21(a)) by increasing v. On the other hand, for larger fiber sizes, the post-peak branch does not change type (always snap-through as in Figure 4.21(b) or snap-back for larger radius).

From the numerical results shown in Figure 4.19 it is possible to define the

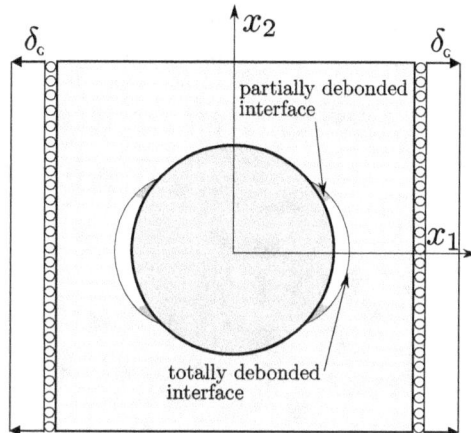

Figure 4.20: Fiber-matrix interface decohesion predicted by the CZM-FEM simulation.

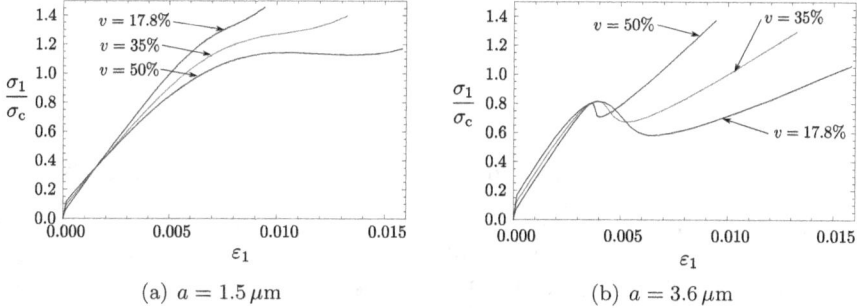

(a) $a = 1.5\,\mu\mathrm{m}$ (b) $a = 3.6\,\mu\mathrm{m}$

Figure 4.21: Homogenized stress-strain curves by changing the fiber volumetric content v. Results courtesy of Professor M. Paggi published in García et al. (2014).

critical stress for the onset of fiber-matrix debonding, σ_{1c}, as the value of the homogenized stress corresponding to a local maximum in the stress-strain curve. This maximum is, however, only present at curves for $a \geq 2.55\,\mu\mathrm{m}$. For smaller fiber radii (curves for $a \leq 1.5\mu\mathrm{m}$), this local maximum is not present and the critical stress is estimated in correspondence of the point where the curvature of the longitudinal stress-strain curve changes sign, which corresponds to a similar mechanism.

Examining the critical stress for the onset of debonding, the results are summarized in Figure 4.22 for the two fiber radii and the three volumetric contents considered in Figure 4.21. For a small radius ($a = 1.5\,\mu\mathrm{m}$), the ratio σ_{1c}/σ_c is a decreasing function of v. For larger fibers ($a = 3.6\,\mu\mathrm{m}$ and so forth), the critical stress ratio σ_{1c}/σ_c becomes almost independent of the volumetric content. This result is important in practical applications, since the fiber size is often in the micrometer scale in existing composite materials Chawla (1998). On the other hand, these results suggest that the volumetric content may be relevant for the critical stress in nano-scale fiber-reinforced composites.

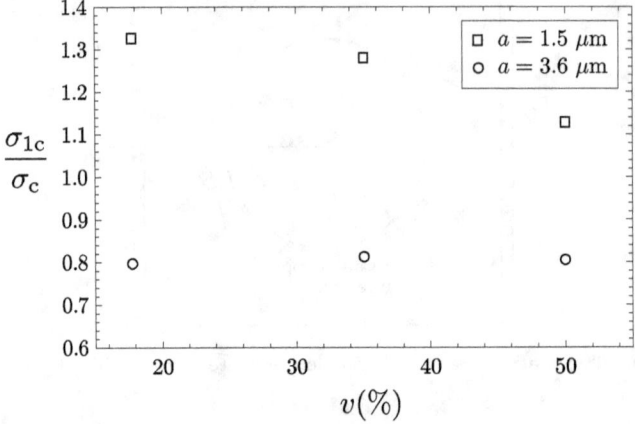

Figure 4.22: Effect of the fiber volumetric content on the dimensionless critical stress. Results courtesy of Professor M. Paggi published by García et al. (2014).

4.2.2   FFM coupled criterion: formulation and predictions

This subsection summarizes the FFM analysis of this problem carried out by Mantič (2009) and highlight the modifications made to be coherent with the cohesive law studied previously. The FFM model is based on the next hypotheses assumed in Mantič (2009):

- Dilute fiber packing with a low fiber volumetric content. This allows using Goodier's Goodier (1933) and Toya's Toya (1974) analytical solutions for an unbounded matrix as reasonable approximations. Small variations of the critical stress for small volumetric contents in the FEM results for CZM shown in Figure 4.22 and large enough fibers supports this hypothesis.

- Only one debond is supposed to appear at the interface symmetrically situated with respect to the loading direction (x_1-axis) (see Figure 4.23) as discussed in Section 4.1. Thus, the problem configuration symmetry with respect to the x_2-axis before the crack onset is broken by the one crack onset. Nevertheless, this is the situation usually observed in experiments Zhang et al. (1997). Finally, due to the symmetry with respect to the x_1-axis, only the upper half domain will be studied, all the angles being measured from the positive x_1-axis, see Figure 4.23.

Stress criterion is invoked assuming no pre-existing cracks. It states that an interface crack onset of semiangle $\Delta\theta$ cannot occur if an equivalent stress σ_{eq}

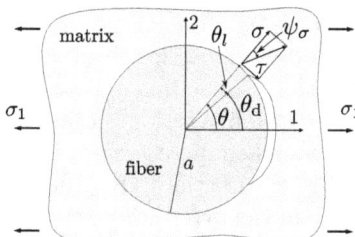

Figure 4.23: The elastic problem after the interface crack onset.

(function of the interface traction vector) at some point with semiangle θ at the expected crack path is below a critical value σ_c. Hence, the crack onset requires

$$\sigma_{\text{eq}}(\theta) \geq \sigma_c, \quad \forall \theta \in [0, \Delta\theta], \tag{4.37}$$

where $\sigma_{\text{eq}}(\theta)$ depends on the particular stress criterion used. In the case of the tensile stress criterion used in Mantič (2009) and Section 4.1, $\sigma_{\text{eq}} = \sigma$, where σ is the normal traction at the interface. In the present work, in view of the experimental evidences in Ogihara and Koyanagi (2010) and in order to take into account the influence of the tangential tractions τ prescribed by the cohesive law, the Mohr-Coulomb criterion is employed,

$$\sigma_{\text{eq}} = \sigma + \frac{|\tau|}{\mu} \quad \text{with } \mu > 0. \tag{4.38}$$

Note that, the limit cases $\mu \to \infty$ and $\mu \to 0^+$, respectively, correspond to the tensile and shear criteria. The value $\mu = 1$ is adopted here in accordance with setting of the CZM law used, which corresponds to $\tau_c = \sigma_c$. Introducing the analytical Goodier's solution, rewritten following Mantič (2009) in terms of the Dundurs bimaterial parameters α and β, into (4.37) leads to

$$\frac{\sigma_1}{\sigma_c} \geq s(\theta, \mu, \alpha, \beta) = \frac{1}{k - m \sin^2 \theta + \frac{m \sin\theta \cos\theta}{\mu}}, \quad \forall \theta \in [0, \Delta\theta], \tag{4.39}$$

where k and m are elastic bimaterial parameters and $s(\theta, \mu, \alpha, \beta)$ is a function giving the minimum value of σ_1/σ_c required to fulfill the Mohr-Coulomb criterion at a certain interface point θ,

$$k(\alpha, \beta) = \frac{1}{2} \frac{1+\alpha}{1+\beta} \frac{2+\alpha-\beta}{1+\alpha-2\beta} \quad \text{and} \quad m(\alpha, \beta) = \frac{1+\alpha}{1+\beta}. \tag{4.40}$$

It is easy to check that $\partial s(\theta, \mu, \alpha, \beta)/\partial\theta < 0$ for $\theta = 0°$ and $\mu < \infty$, whereas $\partial s(\theta, \mu, \alpha, \beta)/\partial\theta > 0$ for $\theta > \theta_{\text{min}}^\sigma$, where

$$\theta_{\text{min}}^\sigma = \frac{1}{2} \arctan \frac{1}{\mu}. \tag{4.41}$$

Thus, $s(\theta, \mu, \alpha, \beta)$ is not a monotonic function for $\theta \in [0°, 90°]$, and consequently a debond (in fact, two debonds symmetrically situated with respect to the x_1-axis) could be predicted without including the interface point $\theta = 0°$. The reason is that according to the Mohr-Coulomb criterion the first point fulfilling this stress criterion, for $\mu < \infty$, when increasing σ_1 is not $\theta = 0°$ (see Figure 4.24 and a discussion at the end of this section). In order to use the Toya analytical solution for the energy criterion, the hypothesis of "symmetric debond onset with respect to x_1-axis" has to be assumed. Taking into account this assumption, an onset of semiangle $\Delta\theta$ cannot occur if the maximum of the required values for σ_1/σ_c for points $\theta \in [0°, \Delta\theta]$ is not exceeded. Hence, the crack onset requires

$$\frac{\sigma_1}{\sigma_c} \geq \bar{s}(\Delta\theta, \mu, \alpha, \beta) = \max_{0 \leq \theta \leq \Delta\theta} [s(\theta, \mu, \alpha, \beta)], \qquad (4.42)$$

where $\bar{s}(\Delta\theta, \mu, \alpha, \beta)$ is the minimum value of σ_1/σ_c which fulfills the stress criterion for an interface crack onset with semiangle $\Delta\theta$.

Figure 4.24 shows functions discussed above. The difference between the tensile stress criterion employed in Mantič (2009) where the tangential traction influence was neglected, and functions $s(\theta)$ and $\bar{s}(\Delta\theta)$ is remarkable for bimaterial properties used here. Note that the difference between the last two functions is reduced to a narrow range for small semiangles.

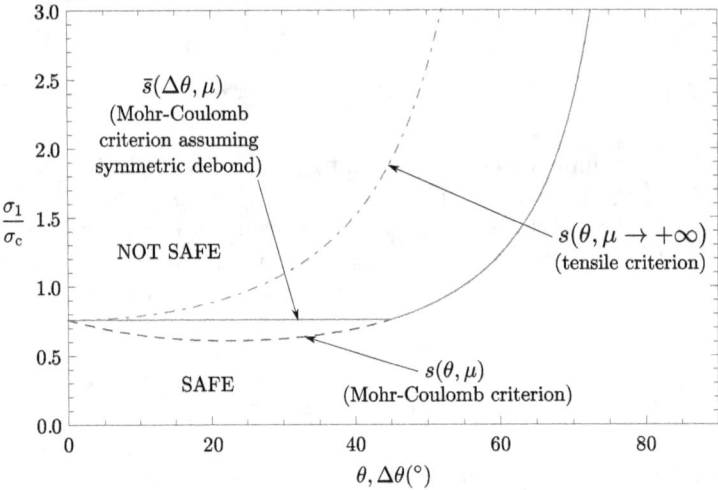

Figure 4.24: Graphical representation of the stress criteria for $\mu = 1$.

The (incremental) energy-criterion is based on an energy balance between the state prior to the debond and after the appearance of a finite debond of an angle $2\Delta\theta$,

$$\Delta\Pi(\Delta\theta) + \Delta E_k(\Delta\theta) + \Delta\Gamma(\Delta\theta) = 0, \qquad (4.43)$$

where $\Delta\Pi$ and ΔE_k, respectively, denote the change in the potential and kinetic energy, whereas $\Delta\Gamma$ is the energy dissipated during the crack onset. According

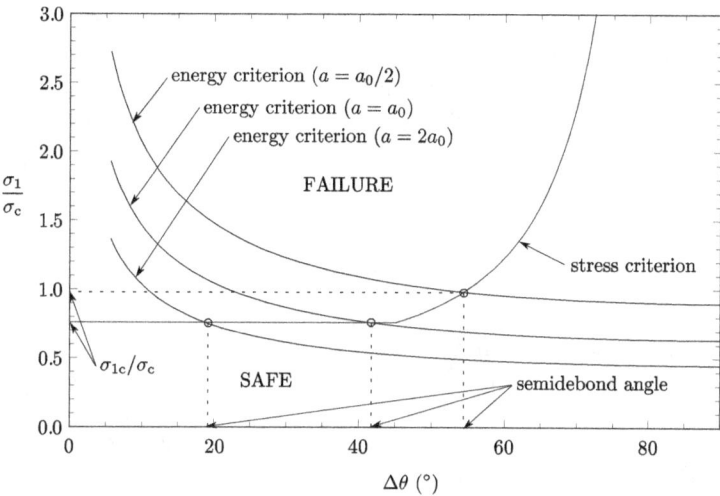

Figure 4.25: Plots of the coupled stress (stress criterion for $\mu = 1$) and (incremental) energy criterion.

to the (classical) fracture mechanics, $\Delta\Pi = -2\int_0^{\Delta\theta} G(\theta_d)a\mathrm{d}\theta_d$, where $G(\theta_d)$ is the Energy Release Rate (ERR) for a debond of semiangle θ_d. $\Delta E_k \geq 0$ since the initial state is assumed to be quasi-static. Regarding to the dissipated energy $\Delta\Gamma$, this can be estimated assuming an interface fracture toughness $G_c(\psi_\sigma)$ which, in general, depends on the fracture-mode-mixity angle ψ_σ at the interface crack tip. However, according to (4.36) and related discussion in Section 4.2.1, it is assumed that the interface fracture toughness is a constant[1] independent of ψ_σ and write $\Delta\Gamma = G_{1c}2a\Delta\theta$. Taking into account all these assumptions, the energy balance (4.43) can be expressed as a condition for the crack onset by employing Toya's explicit expression Toya (1974) of the ERR, rewritten as $G(\theta_d, \sigma_1, a, \alpha, \beta, E^*) = \sigma_1^2 a\hat{G}(\theta_d, \alpha, \beta)/E^*$ in Mantič (2009):

$$\frac{\sigma_1}{\sigma_c} \geq \gamma\sqrt{g(\Delta\theta; \alpha, \beta)}, \qquad (4.44)$$

where $g = \Delta\theta/\int_0^{\Delta\theta} \hat{G}(\theta_d; \alpha, \beta)\mathrm{d}\theta_d$ and γ is a dimensionless brittleness number introduced in Mantič (2009) as a generalization of Carpinteri's brittleness number s to interface cracks Carpinteri (1981, 1982),

$$\gamma = \frac{1}{\sigma_c}\sqrt{\frac{G_{1c}E^*}{a}} = \sqrt{\frac{a_0}{a}}, \qquad (4.45)$$

with a_0 defining an interface characteristic length. The dimensionless function g, introduced in Mantič (2009), characterizes the variations of the ratio of the

---

[1]In general the interface fracture toughness may depend strongly on the fracture mode mixity at the crack tip. This dependence can easily be introduced in the FFM model Mantič (2009) and also in the CZM.

(normalized) dissipated energy at the debond onset $\Delta\Gamma(\Delta\theta)$ to the (normalized) released potential energy at the debond onset $\Delta\Pi(\Delta\theta)$.

Stress and (incremental) energy criteria impose two conditions on the critical value of σ_1 as functions of the debond semiangle $\Delta\theta$. Coupling both criteria gives the critical stress σ_{1c} for the debond onset as the minimum value of σ_1 that fulfills both conditions, (4.42) and (4.44). Actually, this is a constrained minimization problem as stated in some works in the literature, see e.g. Hebel et al. (2010),

$$\frac{\sigma_{1c}}{\sigma_c} = \min_{0<\Delta\theta\leq 90°} \left[\max\left(\bar{s}(\Delta\theta; \mu; \alpha, \beta), \gamma\sqrt{g(\Delta\theta; \alpha, \beta)} \right) \right] \qquad (4.46)$$

Figure 4.25 shows the coupling of both criteria for several values of the fiber radius a. Note that, whereas the expression of the stress criterion (4.42) does not depend on the fiber radius a, the energy criterion does. As a consequence, a size effect is predicted as can be seen in Figure 4.26 discussed in Section 4.2.3. The algorithm used for the evaluation of the critical stress σ_{1c} is omitted here for the sake of brevity (see Mantič (2009) for details).

4.2.3   Fiber-size effect predictions by CZM and FFM

Both models described above predict a strong effect of the fiber radius on the critical stress σ_{1c}. This section aims to compare the results of both models as well as understand physically both similarities and discrepancies found. First, results focusing on the size effect will be presented and compared. Subsequently, the asymptotic behavior of each model will be studied by the means of simplified models to understand physically the previous comparison.

The critical stress σ_{1c} predicted by the CZM has been defined in Section 4.2.1. The ratio σ_{1c}/σ_c is plotted as a function of $\sqrt{a_0/a}$ for $a_0 = 6.25$ μm in Figure 4.26 with dots. The results show a remarkable fiber-size effect. In Carpinteri et al. (2005), where the range 565 m$^{-1/2}$ $< a^{-1/2} <$ 848 m$^{-1/2}$ was explored, the numerical results were fitted according to a linear expression. Actually, outside that range, a strong nonlinearity can be observed with two very different behaviors for very small and very large fiber radii a as can be clearly seen in Figure 4.26.

FFM predictions of the critical stress for the onset of debonding, based on the Mohr-Coulomb criterion, are plotted in the same figure with solid line. Also in this case it is possible to observe a strong fiber-size effect with two distinct regimes. For $\sqrt{a_0/a} \lesssim 1$ (very large fiber radii), there is a horizontal asymptote slightly above the CZM one. For $\sqrt{a_0/a} \gtrsim 1$ (very small fiber radii) the ratio σ_{1c}/σ_c increases linearly with $\sqrt{a_0/a}$. In general, a fairly good agreement between the CZM and the FFM predictions can be observed for $\sqrt{a_0/a} \lesssim 2.5$.

4.2.3.1   Asymptotic behavior for CZM and its physical interpretation

The source of the size effect in both models is the existence of at least one characteristic length which is not scaled with fiber radius. In the case of CZM,

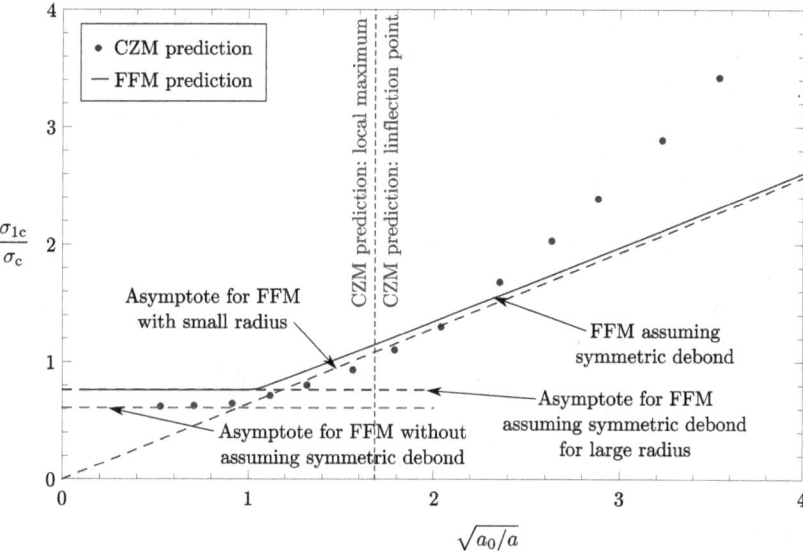

Figure 4.26: Comparison of fiber-size effect on the dimensionless critical stress predicted by both approaches (CZM with volumetric content $v = 18\%$).

these lengths are the critical relative opening and sliding displacements l_{Nc} and l_{Tc}. In fact, both lengths can be considered proportional to the reference radius a_0 defined previously as combining (4.30) and (4.45) gives $l_{\text{Nc}} = \frac{48}{27} \frac{\sigma_c}{E^*} a_0$ and $l_{\text{Tc}} = l_{\text{Nc}} \frac{G_{2c}}{G_{1c}} \frac{\sigma_c}{\tau_c}$. Since both length are related, for the sake of simplicity, subsequent discussion will be referred to the influence of the length l_{Nc} and pure opening mode.

As a consequence of the independent scaling of a and l_{Nc}, the fiber and matrix stiffness with respect to the stiffness of the cohesive elements varies with a if the cohesive law parameters are fixed, e.g. l_{Nc} and σ_c. The critical length l_{Nc} along with the interface tensile strength σ_c determines the initial stiffness of the cohesive elements k_{CE}. In fact, this can be obtained from the cohesive law (4.28)–(4.29),

$$k_{\text{CE}} = \left. \frac{\mathrm{d}\sigma}{\mathrm{d}g_{\text{N}}} \right]_{g_{\text{N}} = g_{\text{T}} = 0} = \frac{27}{4} \frac{\sigma_c}{l_{\text{Nc}}}. \tag{4.47}$$

Note that the stiffness has been defined here with dimensions of stresses over displacements. Regarding matrix and fiber, taking them as isolated linear elastic domains, an equivalent stiffness can be obtained. The matrix is taken as a linear-elastic square with length b and a circular hole in the center of radius a. If this body is loaded like in the complete simulation with CZM, the averaged transverse stress σ_1 is proportional to the averaged transverse strain ε_1, both defined in the same sense as in the CZM simulation. If an equivalent stiffness was defined as stresses over strains, stiffness would only depend on b/a. However, it is necessary to define an equivalent stiffness which can be dimensionally compared to the initial stiffness of the cohesive law. Thus, an equivalent stiffness of the matrix can be defined as the ratio of σ_1 to the averaged relative displacement between the opposite edges where the load is applied. This displacement is directly equal to $b\varepsilon_1$. Taking into account that the ratio b/a is fixed by the volumetric content, the stiffness of the matrix k_m, defined dimensionally compatible to that defined for the cohesive law, is inversely proportional to the radius a. An analogous result can be obtained for the fiber equivalent stiffness k_f selecting two lines of reference within the fiber. In view of the results, whereas the equivalent stiffness of the matrix and fiber is scaled with a, the equivalent stiffness of the cohesive elements is independent of it.

The different scaling of stiffness for matrix, fiber and cohesive elements with the radius results in two very different scenarios for very small or large fibers. In the following, the limit models for both extremes will be studied.

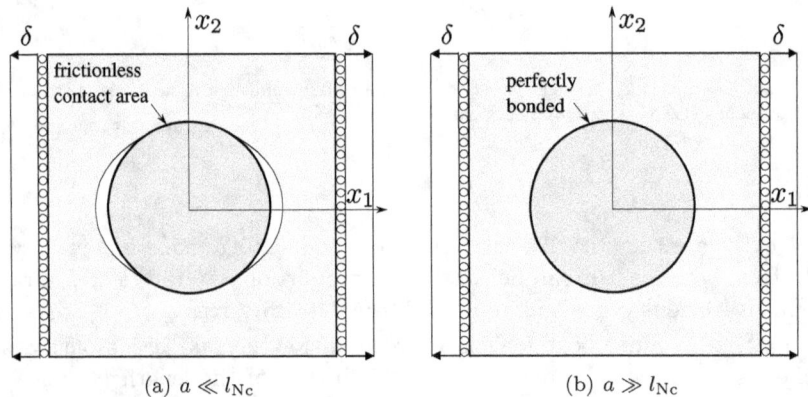

(a) $a \ll l_{Nc}$ (b) $a \gg l_{Nc}$

Figure 4.27: Asymptotic models of the problem when the CZM law used is prescribed at the interface

For very small fibers ($a \ll l_{Nc}$), the fiber and matrix are much stiffer than the cohesive elements, $k_f \gg k_{CE}$ and $k_m \gg k_{CE}$. In the limit, when the relative stiffness of the cohesive elements is vanishing, the interface behavior is similar to a fully debonded one. Thus, the problem becomes equivalent to that shown in Figure 4.27(a). This is an elastic problem with a frictionless contact zone along the interface.

The contact in the problem shown in Figure 4.27(a) is receding. It means that the contact area is either smaller than or equal to that before loading. In this case, as demonstrated, e.g., by Dundurs (1975), the contact area is load-independent and therefore stresses, strains and displacements varies linearly with load. In addition, in this limit case, stresses and strains are independent of the radius if b/a is fixed. As a consequence, displacements are linear with the radius. In particular, the relative normal and tangential displacements along the interface

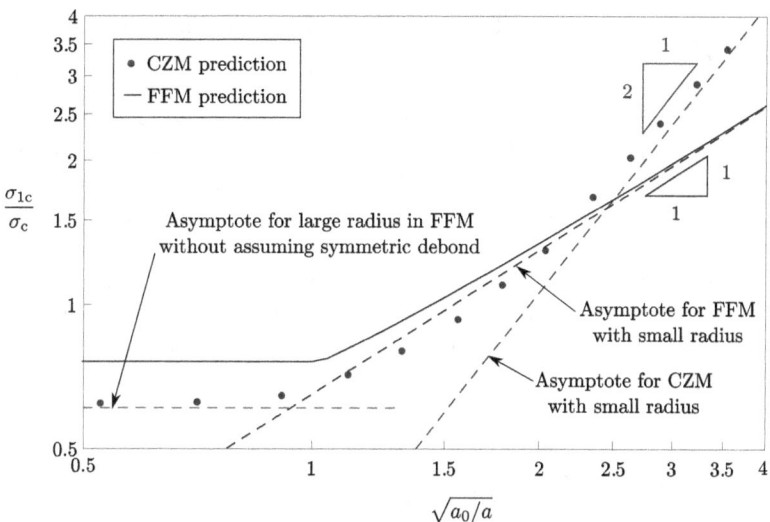

Figure 4.28: Comparison of the asymptotic behavior for small fibers predicted by both approaches for the fiber-size effect in log-log scale.

g_N and g_T can be expressed,

$$g_N = a \frac{\sigma_1}{\sigma_c} \hat{g}_N(\theta; \alpha, \beta), \tag{4.48a}$$

$$g_T = a \frac{\sigma_1}{\sigma_c} \hat{g}_T(\theta; \alpha, \beta), \tag{4.48b}$$

where $\hat{g}_N(\theta; \alpha, \beta)$ and $\hat{g}_T(\theta; \alpha, \beta)$ are dimensionless functions, which, if necessary, could be extracted from the elastic solution of the problem shown in Figure 4.27(a) and σ_c has been taken as a reference stress. Then, introducing (4.48) into the effective relative displacement λ (4.29b) reads

$$\lambda = \frac{a}{l_{Nc}} \frac{\sigma_1}{\sigma_{1c}} \hat{\lambda}(\theta; \alpha, \beta), \tag{4.49}$$

where $l_{Tc} = l_{Nc}$ has been assumed as in Section 4.2.1, and $\hat{\lambda}(\theta, \alpha, \beta)$ can be obtained from $\hat{g}_N(\theta, \alpha, \beta)$ and $\hat{g}_T(\theta, \alpha, \beta)$. Recall that λ can be interpreted as a damage parameter, $\lambda = 0$ and $\lambda = 1$ corresponding to undamaged and fully debonded interface, respectively. Thus, in view of (4.49), if a/l_{Nc} decreases, the damage level only can be maintained if σ_1/σ_c increases in the same proportion. It means that the critical stress σ_{1c} is,

$$\frac{\sigma_{1c}}{\sigma_c} \sim \frac{l_{Nc}}{a} \sim \frac{a_0}{a}. \tag{4.50}$$

and consequently in the limit for very small a, the critical remote stress σ_{1c} is inversely proportional to a. This result agrees with CZM results as can be seen in Figure 4.28, where results of σ_{1c} have been shown as a function of $\gamma = \sqrt{a_0/a}$ using log-log scale.

On the other extreme, for very large fibers ($a \gg l_{Nc}$), the cohesive elements are much stiffer than fiber and matrix, $k_{CE} \gg k_f$ and $k_{CE} \gg k_m$. In the limit, the interface behavior is similar to perfect bonding. Thus, the problem becomes equivalent to that shown in Figure 4.27(b). This is an elastic problem of a fiber surrounded by a square of matrix, both perfectly bonded along the common interface.

This linear elastic problem can be solved by computational methods and in the particular case of an infinite matrix, the elastic solution is given by Goodier Goodier (1933). In general, normal σ and tangential τ tractions along the interface are,

$$\frac{\sigma}{\sigma_1} = f_\sigma(\theta; b/a; \alpha, \beta), \qquad (4.51a)$$

$$\frac{\tau}{\sigma_1} = f_\tau(\theta; b/a; \alpha, \beta), \qquad (4.51b)$$

where $f_\sigma(\theta; b/a; \alpha, \beta)$ and $f_\tau(\theta; b/a; \alpha, \beta)$ are dimensionless functions which can be extracted from the elastic solution. For the objective of the following discussion, it is only necessary to know the arguments of these functions. For a fixed volumetric content, which corresponds to fix b/a, these expressions show that, in the limit, tractions along the interface are independent of the fiber radius a. As a result, for a fixed σ_1, tractions at an interface point tend to a constant when the fiber radius a increases. Therefore, the critical averaged transverse stress σ_{1c} tends to be independent of a. This is in agreement with results shown in Figures 4.26 and 4.28 where a horizontal asymptote can be observed for CZM results approaching $\sqrt{a_0/a} \to 0$.

Note that the shape of the CZM law has an important role on the size effect since the initial stiffness of the cohesive elements is a key parameter in the previous analysis. In addition, it has also a role in the homogenized stress-strain response. Távara et al. (2011) observed similar fiber-size effects on the homogenized peak stress using the analytic solution by Gao (1995) and the numerical solution by Boundary Element Method (BEM) considering a linear elastic-brittle interface constitutive relation. In fact, for $\sqrt{a_0/a} \gg 1$, the asymptote was $\sigma_{1c}/\sigma_c \sim a_0/a$ as in the CZM results presented here. However, in their case, since the interface constitutive law was linear, the pre-peak part of the stress-strain curve was also linear. The nonlinear CZM used in Section 4.2.1, on the other hand, leads to a nonlinear response even in the pre-peak part of the stress-strain curve. Ngo et al. (2010) have also proposed a study of fiber-matrix debonding, generalizing the Mori-Tanaka method in the case of imperfect bonding by considering an exponential softening CZM with a very stiff initial part. In that case, the pre-peak stress-strain curve was also linear but no size effects on the homogenized peak stress were observed. This is due to the fact that, until $\sigma = \sigma_c$, the model by Ngo et al. (2010) predicts the stress distribution at the interface as for a fully bonded one, which is independent of the fiber size, whereas the interface laws in the CZM of Section 4.2.1 and by Távara et al. (2011) predict a gap growth before the onset of debonding. Hence, the initial compliance of the interface CZM, often

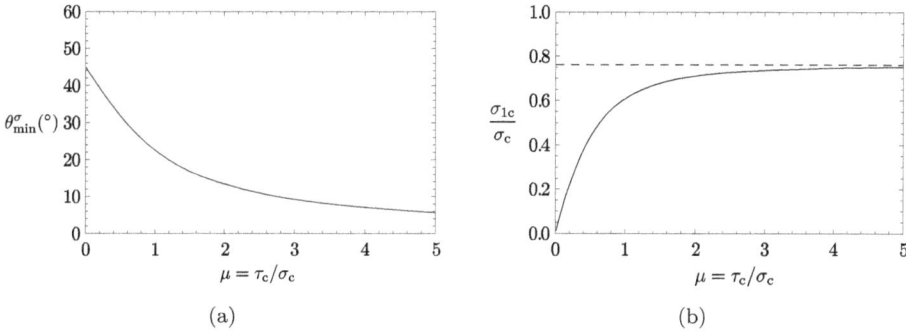

Figure 4.29: (a) Angle $\theta^{\sigma}_{\min}$ of the interface where the function $s(\theta, \mu)$ has a minimum, (b) critical stress for the limit $\sqrt{a_0/a} \to 0$

considered as an aspect of minor importance as compared to other parameters like the fracture energy and the peak cohesive traction, plays an important role for the homogenized response of the unit cell. This is also consistent with the recent theory developed by Paggi and Wriggers (2012) on the estimation of the elastic properties of heterogeneous materials with cohesive interfaces.

4.2.3.2 Asymptotic behavior for FFM and its physical interpretation

Size effect predicted by FFM is very different depending on which of the two criteria governs the solution. For tough configurations, which correspond to large values of γ (small fiber radius), the failure is governed by the energetic condition of the coupled criterion, see Mantič (2009). The released potential energy $\Delta\Pi$ at the abrupt onset is given as $\Delta\Pi \sim \sigma_1^2 a^2$ (see equations (16) and (34) in Mantič (2009)) and the dissipated energy as $\Delta\Gamma \sim a$. Therefore the critical value of debond σ_1 for the debond onset scales as $1/\sqrt{a}$. This asymptotic behavior is shown in Figure 4.28. From a mechanical point of view, in plane elasticity, the energy is released in an area (mainly in a crack neighbourhood), but it is dissipated along a line (the crack path). As a consequence of the different geometrical dimensions of energy release and dissipation, the energetic balance contains a dependence on the scale. Note that this size effect $\sigma_{1c} \sim 1/\sqrt{a}$ is analogous to the well-known size effect predicted by the Griffith criterion of LEFM, where a would correspond to the initial crack length.

On the other extreme, when $a \gg a_0$, the solution for the critical transverse stress σ_{1c} leading to the onset is mainly given by the stress criterion as discussed in Section 4.2.2. Since the stress criterion is independent of the fiber radius, when $a \gg a_0$, σ_{1c} tends to be independent of the radius a. This is in agreement with results shown in Figures 4.26 and 4.28.

Let us recall that the results shown in Figures 4.26 and 4.28 are calculated under the hypothesis of symmetric initial debond including the interface point

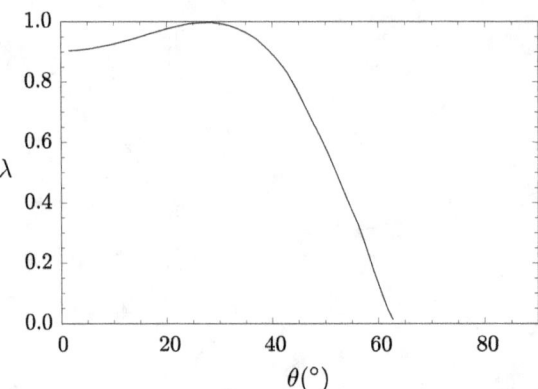

Figure 4.30: Dimensionless effective opening displacement λ at the interface from the CZM-FEM results at the load leading to the onset of debonding ($a = 5\,\mu$m). Results courtesy of Professor M. Paggi published by García et al. (2014).

with $\theta = 0°$. Thus, it seems interesting to discuss the influence of this hypothesis on the predictions of the coupled criterion. In particular this is going to be analyzed for the extreme $a \gg a_0$ since there the stress criterion governs the value of σ_{1c}. By minimizing the stress criterion function $s(\theta; \mu; \alpha, \beta)$ in (4.39) the most favorable point to failure at the fiber-matrix interface is given by $\theta^{\sigma}_{\min}$ defined in (4.41). Figure 4.29(a) shows the angle $\theta^{\sigma}_{\min}$ as a function of μ. Note that this angle is far from $\theta = 0°$ for moderate values of μ. In particular, for $\mu = 1$ used in this comparison, $\theta^{\sigma}_{\min} = 22.5°$. This reasonably agrees with the CZM predictions when looking for the effective opening displacement distribution λ corresponding to the load leading to the onset of debonding, Figure 4.30 plots an example for $a = 5\,\mu$m. As expected, the most critical point is not at $\theta \cong 0°$, but at $\theta \cong 27°$. Therefore, in FFM a debond without including the interface point $\theta = 0°$ would be possible for sufficiently small values of the debond angle. It can be shown Mantič (2009) that, in general, the debond angle at the onset tends to zero for $\sqrt{a_0/a} \to 0$, corresponding to large values of the fiber radius. Furthermore, σ_{1c} depends only on the stress criterion (4.39) as discussed previously and, thus, is given directly by its minimum value:

$$\frac{\sigma_{1c}}{\sigma_c} = \frac{2}{2k + m\left(\sqrt{1 + \frac{1}{\mu^2}} - 1\right)}. \tag{4.52}$$

Figure 4.29(b) shows this value of the critical stress and the value $\sigma_{1c}/\sigma_c = 1/k$ predicted by the Mohr-Coulomb criterion (4.42) assuming a symmetric debond with respect to x_1-axis, for $\sqrt{a_0/a} \to 0$. Note that the difference is large for small and moderate values of μ. However, no difference in the predicted critical stress is expected for small fiber radii since this case corresponds to larger debond angles which necessarily include the interface point $\theta = 0°$.

The use of the Mohr-Coulomb criterion releasing the assumption of symmetric debond with respect to x-axis in FFM improves the agreement with CZM predictions for ($\sqrt{a_0/a} \lesssim 1$), see Figures 4.26 and 4.28.

4.2.4 Limits of validity of CZM and FFM methods

The scaling laws obtained according to both approaches described in Sections 4.2.1 and 4.2.2 and compared in Section 4.2.3 present some limits of validity. This section focuses only on the limits of validity derived from some conceptual incongruences. A complete study of validity would require experiments to compare, which are not available at the moment.

Regarding the numerical predictions based on CZM, a discretization issue arises for capturing the mechanical response in the case of large fibers. In this case, the process zone size becomes relatively very small with respect to the fiber radius and a suitably refined mesh is necessary to approximate adequately the distribution of cohesive tractions obtaining mesh independent results. Similar convergence problems have been pioneeringly put into evidence by Carpinteri Carpinteri (1989a,b) in the case of cohesive crack propagation in three-point bending concrete beams. Provided that the size of the finite elements is small as compared to the process zone size, the horizontal asymptote in Figure 4.26 can be correctly captured.

In general, the FFM model has no problem for large fibers. In fact, as demonstrated in Section 4.1 for large fibers which correspond to brittle configurations, FFM predicts that the critical load leading to the failure is essentially governed by the stress criterion. This implies that critical load can directly be predicted from the stress state prior to the crack onset. As a consequence the analysis becomes simpler since the elastic state prior to the crack onset can be obtained quite easily. In some cases, as in the present work an analytical solution is available, thus an explicit expression of the critical load can be extracted. Additionally, small stresses in large part of specimen make the linear elastic approximation more accurate.

Regarding the opposite asymptote for very small fiber sizes, a deviation of the actual results from the CZM-based scaling laws can be due to the fact that the large stresses required are not compatible with the linear elastic behavior assumed for the matrix. In fact, releasing the assumption of linear elasticity for the matrix and considering an elasto-plastic behavior, the interplay between cohesive fracture and matrix plasticity could take place. As compared to the same simulations with purely linear elastic behavior (Figure 4.19), it is possible to note that for large fibers the response is not so different, since only the snap-back branches are smooth out at least for $a < 5$ μm. This implies that cohesive crack propagation prevails over plasticity. For small fibers, on the other hand, the hardening response is no longer visible due to a large amount of plasticity in the matrix. This has an important consequence on the scaling law, which can be deviated from that predicted by the CZM. In this context, the ratio between the yield strength and the interface tensile strength, σ_y/σ_c is an important quantity for the definition of the upper threshold to the CZM scaling law. Ratios larger than unity, admissible in the case of poor quality of fiber-matrix cohesion, may increase the validity of the CZM predictions that would be admissible in the range $\sigma_{1c}/\sigma_c < \sigma_y/\sigma_c$. This behavior agrees with the experimental evidences which show that finer particles limit the onset of cavities and microcracks Zahid et al. (1998) enhancing superplasticity. The explanation, following these results, would be that below a given fiber size, plasticity of the matrix predominates over fiber-matrix debond, so superplasticity properties are improved.

Regarding to the FFM model for small fibers, some problems related to those found in CZM appear as well. FFM is used here by supposing the cracks follows a determinated path. For instance, in the present work, the crack path supposed is the fiber-matrix interface. This implies that other possible preferential paths or failure points are not taken into account. In particular, for small fibers the FFM model predicts critical loads which exceed the interface strength by two or three times as can be seen in Figure 4.26. In this case as demonstrated by the FEM analysis, a plastic zone is expected to appear in the matrix. Therefore, in the case of FFM, limits of validity are very dependent on the ratio σ_y/σ_c

Additional problems can be found in CZM for very small fibers since the present model considers constant material lengths l_{Nc} and l_{Tc}. As a result of this, for very small fibers, the radius a can become similar to l_{Nc}. In this case, the numerical analysis would not provide representative results.

4.3   Crack initiation under biaxial transverse loads

In the present work, an extension of the theoretical procedure developed by Mantič (2009) to the case of tension dominated remote biaxial transverse loads is developed. This study is mainly motivated by the effect of the secondary compression in addition to a dominating tension shown in the preliminary experiments by París et al. (2003) and Correa (2008). The FFM analysis is carried out for dilute fiber packing where the influence of adjacent inclusions is almost negligible, whereas for dense fiber packing, a numerical method should be employed based on other solutions, e.g. Kushch and Mishnaevsky Jr. (2010). In addition, several models developed under the assumption of dilute fiber packing have demonstrated being useful to explain experimental results even for densely packed composites, see Correa (2008). After a revision of the solution of the elastic inclusion perfectly bonded to the matrix in Section 4.3.1, the elastic problem of an inclusion with a debond is analyzed in Section 4.3.2, where a general analytical solution is particularized for this problem. The coupled criterion is developed and applied in Section 4.3.3 obtaining results for the critical crack length and remote load at the onset. Section 4.3.4 describes the influence of the remote secondary load and the inclusion size on the results obtained. Finally a new experimental procedure for an indirect measurement of the strength and fracture interface properties is presented in Section 4.3.5.

4.3.1 Stresses in a single inclusion under a remote biaxial transverse load

Stress based failure criteria usually consider the stress state prior to the damage appearance. Hence the aim of this section is to study the stresses at the fiber-matrix interface under remote biaxial load transverse to the fiber axis. Consider a circular cylindrical inclusion of radius a embedded in an infinite matrix and perfectly bonded along its lateral interface. Let (x, y, z) and (r, θ, z) be suitably defined cartesian and cylindrical coordinate systems, the z-axis being coincident with the inclusion (longitudinal) axis. Remote uniform biaxial transverse load $(\sigma_x^\infty, \sigma_y^\infty)$ is applied parallel to the the two axes, x and y, transverse to the cylindrical inclusion axis, see Figure 4.31.

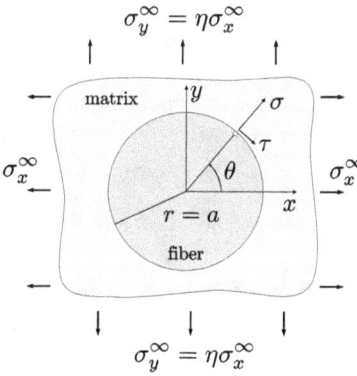

Figure 4.31: The inclusion problem configuration before the crack onset.

An analytic solution for stresses in this problem was deduced by Goodier (1933). As was shown by Hardiman (1954) the stresses inside the inclusion are constant. Following Mantič (2009), these inclusion stresses can be rewritten in terms of the Dundurs bimaterial constants α and β (Dundurs (1967, 1969)) as follows:

$$\begin{pmatrix} \sigma_x^{(1)} & \sigma_{xy}^{(1)} \\ \sigma_{xy}^{(1)} & \sigma_y^{(1)} \end{pmatrix} = \sigma_x^\infty \begin{pmatrix} k & 0 \\ 0 & k-m \end{pmatrix} + \sigma_y^\infty \begin{pmatrix} k-m & 0 \\ 0 & k \end{pmatrix}, \qquad (4.53)$$

where

$$k(\alpha, \beta) = \frac{1}{2} \frac{1+\alpha}{1+\beta} \frac{2+\alpha-\beta}{1+\alpha-2\beta} \quad \text{and} \quad m(\alpha, \beta) = \frac{1+\alpha}{1+\beta}, \qquad (4.54)$$

with the Dundurs bimaterial constants being defined as

$$\alpha = \frac{\mu_1(\kappa_2+1) - \mu_2(\kappa_1+1)}{\mu_1(\kappa_2+1) + \mu_2(\kappa_1+1)} \quad \text{and} \quad \beta = \frac{\mu_1(\kappa_2-1) - \mu_2(\kappa_1-1)}{\mu_1(\kappa_2+1) + \mu_2(\kappa_1+1)}, \qquad (4.55)$$

with $\mu_k = E_k/(2(1+\nu_k))$ and $\kappa_k = 3 - 4\nu_k$, E_k and ν_k denoting the Young's modulus and Poisson's ratio, respectively. It can be shown that $0 \leq k \leq 5/3$ and $0 \leq m \leq 2$.

Let a dimensionless load-biaxiality parameter η be defined as the ratio of remote stresses

$$\eta = \frac{\sigma_y^\infty}{\sigma_x^\infty}. \tag{4.56}$$

Then, normal and shear stresses, σ and τ, acting along the interface $(r = a)$ can be expressed as a function of the polar angle θ (see Figure 4.31) and the parameter η as

$$\frac{\sigma(\theta)}{\sigma_x^\infty} = \frac{\sigma_r(\theta)}{\sigma_x^\infty} = k + (k - m)\eta - (1 - \eta)m\sin^2\theta, \tag{4.57a}$$

$$\frac{\tau(\theta)}{\sigma_x^\infty} = \frac{-\sigma_{r\theta}(\theta)}{\sigma_x^\infty} = (1 - \eta)m\sin\theta\cos\theta. \tag{4.57b}$$

In view of the problem symmetry only angles $0° \le \theta \le 90°$ will be considered for the sake of simplicity.

In the following and without loss of generality it will be assumed that $\sigma_x^\infty > 0$ and $\sigma_x^\infty \ge \sigma_y^\infty$, i.e. $\eta \le 1$. The derivative of $\sigma(\theta)$ evaluated from (4.57a),

$$\frac{\partial\frac{\sigma(\theta)}{\sigma_x^\infty}}{\partial\theta} = (\eta - 1)m\sin(2\theta). \tag{4.58}$$

This shows that normal stress is a decreasing function of θ for $\theta \in [0°, 90°]$ and any $\eta \le 1$. According to this and the expression in (4.57a), $\sigma(\theta)$ achieves its maximum value at $\theta = 0°$:

$$\sigma_{max} = \sigma(\theta = 0°) = \sigma_x^\infty \cdot k + \sigma_y^\infty \cdot (k - m). \tag{4.59}$$

Note that k and $k - m$ represent, respectively, the relative contribution of remote stresses σ_x^∞ and σ_y^∞ to this maximum value of $\sigma(\theta)$.

Following (4.57a) and (4.59), the influence of σ_y^∞ on the normal stresses $\sigma(\theta)$ is given by the ratio k/m. In particular, tension $\sigma(\theta = 0°) > 0$ is generated by a remote compression $\sigma_y^\infty < 0$ (tension $\sigma_y^\infty > 0$) for $k/m < 1$ ($k/m > 1$), assuming small $\sigma_x^\infty \gtrsim 0$.

Recalling that $\sigma_x^\infty > 0$, the semiangle θ_0 for which the interface normal stresses vanish is given from (4.57a) by

$$\theta_0(\eta; \alpha, \beta) = \arcsin\sqrt{\frac{\frac{k}{m} + \left(\frac{k}{m} - 1\right)\eta}{1 - \eta}}. \tag{4.60}$$

According to this expression and the analysis in 4.B, the angle $\theta_0 \in [0°, 90°]$ does not exist for all the considered values of $\eta \in (-\infty, 1]$ and all admissible values of $\frac{k}{m} \in [3/4, +\infty)$, see Mantič (2009).

The condition of vanishing derivative of (4.57a) with respect to η gives the angle θ_η for which the interface normal stress is independent of η,

$$\theta_\eta(\alpha, \beta) = \arccos\sqrt{\frac{k}{m}} = 90° - \theta_0(\eta = 0). \tag{4.61}$$

Table 4.3: Examples of isotropic bimaterials constants (1, inclusion; 2, matrix)

Bimaterial	E_1(GPa)	ν_1	E_2(GPa)	ν_2
Glass/epoxy	70.8	0.22	2.79	0.33
Carbon/epoxy	13.0	0.20	2.79	0.33

	α	β	ε	E^* (GPa)
Glass/epoxy	0.919	0.229	-0.074	6.01
Carbon/epoxy	0.624	0.136	-0.044	5.09

	k	m	k/m	$\theta_\eta(°)$	η_0
Glass/epoxy	1.44	1.56	0.9205	16.3	0.086
Carbon/epoxy	1.32	1.43	0.9200	16.4	0.087

This expression makes sense only if $k \leq m$. Then, θ_η divides the interface sector $0 \leq \theta \leq 90°$ into two regions. In the first region ($\theta < \theta_\eta$) where $\frac{\partial \sigma}{\partial \eta} < 0$, σ decreases with increasing the remote secondary load σ_y^∞. On the contrary, in the second region ($\theta > \theta_\eta$) where $\frac{\partial \sigma}{\partial \eta} > 0$, σ rises with increasing the remote secondary load σ_y^∞. Hence, θ_η is a key angle to evaluate the influence of load biaxiality.

The values of the above defined and used constants characterizing the interface stress distribution for two typical fiber reinforced composites are presented in Table 4.3. The glass/epoxy bimaterial will be used as an example in the present work.

Plots of normal stress distributions along the interface obtained from (4.57a) for glass/epoxy are shown in Figure 4.32.

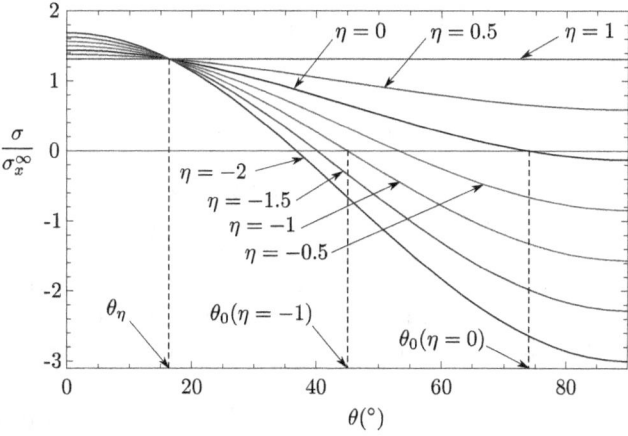

Figure 4.32: Distribution of the normal stresses along the interface for several values of η and glass/epoxy.

4.3.2   The solution for a crack at the interface of a single inclusion under a remote biaxial transverse load

Energy based fracture criteria consider a cracked configuration. Hence the aim of this section is to analyze the problem of a partial debond at the fiber-matrix interface. Under certain assumptions, a classic elastic solution particularized for the present problem provides closed form expressions for a fracture mode mixity and the energy release rate, their dependence on the key problem parameters being pointed out.

Consider the problem configuration from the previous section altered by the presence of a crack at the interface. In view of the fact that the maximum of $\sigma(\theta)$ is achieved at $\theta = 0°$, see (4.58), it will be assumed that this crack is symmetrically situated with respect to the x-axis, with a semidebond angle $\theta_d \geq 0$ and an infinite length in the z-axis direction, see Figure 4.33.

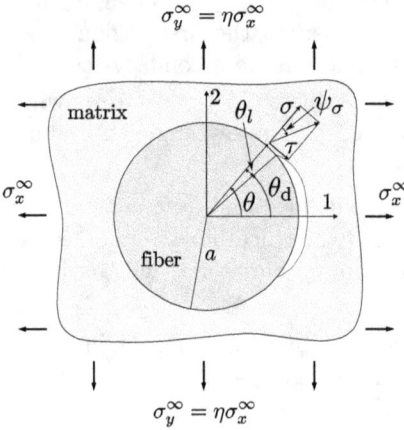

Figure 4.33: The interface crack problem configuration.

A more general problem has been analytically studied by Toya (1974) and several other authors using the open model of interface cracks. As will be seen later on, the validity of the analytic solutions based on the open model is limited and, thus, computational methods should sometimes be used employing the contact model of interface cracks, see Section 4.1 for an example on this problem and París et al. (2007) for a review.

The interface stresses at a point placed ahead of the crack tip at the polar angle $\theta = \theta_d + \theta_l$, $\theta_l > 0°$, see Figure 4.33, can be expressed by particularizing Toya's solution for stresses[2] and rewriting it in terms of the Dundurs parameters,

$$\sigma(\theta, \theta_d, \eta, \beta) - i\tau(\theta, \theta_d, \eta, \beta) = -\frac{\sigma_x^\infty}{2}\frac{1-\alpha}{1-\beta}\chi(\theta, \theta_d, \beta)p(\theta, \theta_d, \eta, \beta), \qquad (4.62)$$

where $\chi(\theta, \theta_d, \beta)$ and $p(\theta, \theta_d, \eta, \beta)$ are defined in Appendix 4.C). It should be noticed that the stresses along the interface are independent of the inclusion

---

[2]The following values of the parameters used by Toya: $\phi = 0$ and $\varepsilon_\infty = 0$ are taken.

radius a. Ratio of the interface shear and normal stresses ahead the crack tip at a small reference length (either geometry or material based) gives a measure of fracture mode mixity of an interface crack. Thus, the angle ψ at a reference angle θ_l, measured from the crack tip (see Mantič (2009) for a discussion about this reference angle for a similar problem), is defined as:

$$\tan \psi(\theta_{\mathrm{d}}; \theta_l, \eta, \varepsilon) = \frac{\tau(\theta_{\mathrm{d}} + \theta_l)}{\sigma(\theta_{\mathrm{d}} + \theta_l)}. \tag{4.63}$$

This angle will be used as a suitable measure of the fracture mode mixity. Figure 4.34 shows the evolution of $\psi(\theta_{\mathrm{d}})$ for different values of the load-biaxiality parameter η.

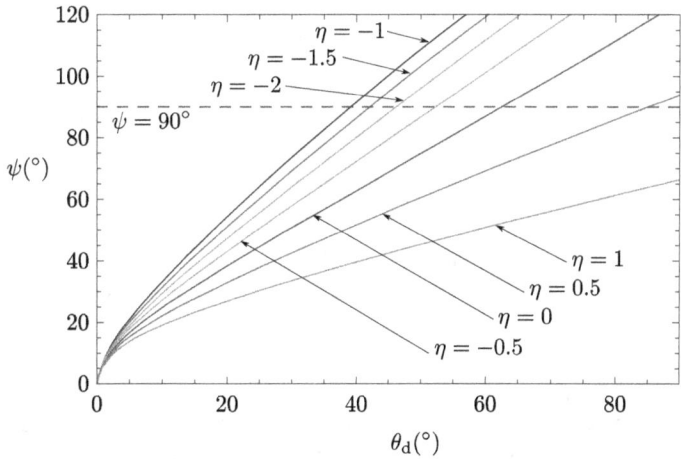

Figure 4.34: Examples of the evolution of the fracture mode mixity angle ψ (obtained from Toya's solution of the open model of interfacial cracks) taking $\theta_l = 0.1°$, for different values of η and glass/epoxy.

The ERR of the interface crack propagating at its upper crack tip at an angle θ_{d} can be expressed, rewriting Toya's expression as previously, by

$$G(\theta_{\mathrm{d}}; \sigma_x^\infty, \sigma_y^\infty; a; E^*, \alpha, \beta) = \frac{(\sigma_x^\infty)^2 a}{E^*} \hat{G}(\theta_{\mathrm{d}}; \eta; \alpha, \beta), \tag{4.64}$$

where E^* is the harmonic mean of the effective elasticity moduli

$$\frac{1}{E^*} = \frac{1}{2} \left(\frac{1 - \nu_1^2}{E_1} + \frac{1 - \nu_2^2}{E_2} \right) \tag{4.65}$$

and $\hat{G}$ is a dimensionless normalized ERR whose expression is presented in Appendix 4.D. According to expression (4.64), the ERR varies linearly with the ratio a/E^* and quadratically with the remote load σ_x^∞.

Figure 4.35 shows the evolution of the normalized ERR $\hat{G}(\theta_{\mathrm{d}})$, and also of its asymptotes for $\theta_{\mathrm{d}} \approx 0°$ given by (4.125), for different values of the load-biaxiality parameter η, see Appendix 4.E. Validity of these plots is limited by

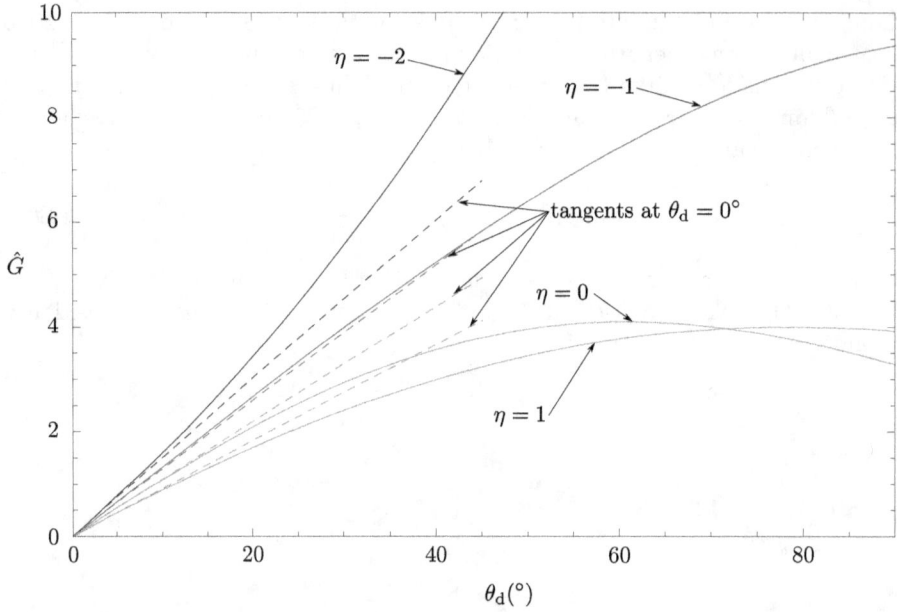

Figure 4.35: Examples of the normalized ERR (obtained from Toya's solution of the open model of interfacial cracks) and its asymptotes for different values of η and glass/epoxy.

the validity of the open model of interface cracks. Notice, in relation to Figure 4.34, that compressions ahead of the crack tip correspond to $|\psi| > 90°$ and can become relevant for $\eta < 0$. These compressions may have associated a relevant overlapping of crack faces close to the crack tip. This is not physically admissible, so it may invalidate Toya's solution for some values of θ_d and η.

Due to the above mentioned overlapping, an additional unrealistic term, corresponding in some sense to Mode I, appears in the computation of the ERR, causing some overestimation of $\hat{G}$. This overestimation can be studied by using the relation between the ERR based fracture mode mixity and the stress based fracture mode mixity according to Mantič and París (2004). This relation allows partitioning $\hat{G}$ into two components $\hat{G}(\theta_\mathrm{d}) = \hat{G}_\mathrm{I}(\theta_\mathrm{d}, \delta\theta) + \hat{G}_\mathrm{II}(\theta_\mathrm{d}, \delta\theta)$, for a given virtual-crack-step angle $\delta\theta$, as shown in Appendix 4.F, leading to,

$$\hat{G}_{I,II}(\theta_\mathrm{d}, \delta\theta) = \frac{1}{2}\hat{G}(\theta_\mathrm{d})(1 \pm F(\varepsilon)\cos(2(\psi(\theta_\mathrm{d}, \theta_l) + \psi_0(\delta\theta/\theta_l, \varepsilon)))), \qquad (4.66)$$

where θ_l is the reference angle for ψ (4.63) and the oscillation index ε is given in terms of β in (4.123).

Figure 4.36 shows the individual components of the ERR corresponding to a small virtual-crack-step angle $\delta\theta = 0.5°$. These plots allow clarifying the range of validity of Toya's expression of ERR for larger values of θ_d and different values of η. Decreasing values of η decreases the range of the values of θ_d, where Toya's expression of ERR is valid. This figure also confirms that the cause of the strong increase of $\hat{G}$ for large values of θ_d and $\eta < 0$ is associated to a fictitious contribution of $\hat{G}_I$ due to a large overlapping.

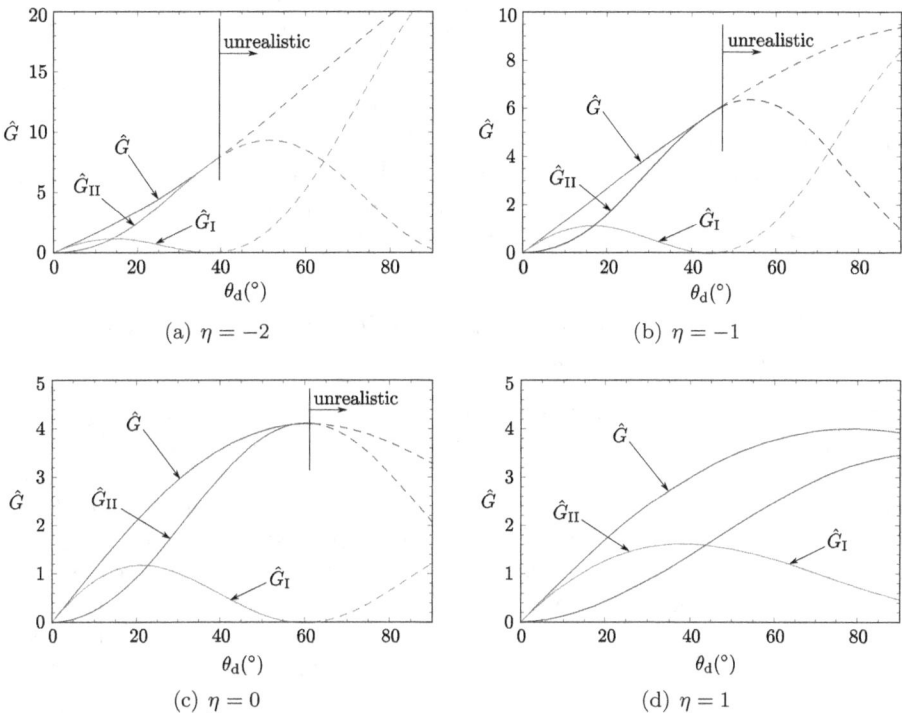

Figure 4.36: Plots of the individual components of the ERR associated to $\delta\theta = 0.5°$ for different values of the load-biaxiality parameter η and glass/epoxy. (a) $\eta = -2$. (b) $\eta = -1$. (c) $\eta = 0$. (d) $\eta = 1$.

In order to clarify the influence of the remote secondary load σ_y^∞ on the values of $\hat{G}$, it is useful to study the variation of the derivative of $\hat{G}$ (4.125) with respect to the load-biaxiality parameter η at $\theta_\mathrm{d} \cong 0°$

$$\left.\frac{\mathrm{d}^2\hat{G}}{\mathrm{d}\eta\mathrm{d}\theta_\mathrm{d}}\right|_{\theta_\mathrm{d}=0} = \frac{2\pi(k-m)\left(k+(k-m)\eta\right)\left(1+4\varepsilon^2\right)}{\cosh\left(\pi\varepsilon\right)}. \tag{4.67}$$

This expression shows again the importance of the parameter k/m. In fact, the sign of the variation of the asymptotic slope of $\hat{G}$ at $\theta_\mathrm{d} \cong 0°$ with η is directly characterized by

$$\mathrm{sign}\left.\frac{\mathrm{d}^2\hat{G}}{\mathrm{d}\eta\mathrm{d}\theta_\mathrm{d}}\right|_{\theta_\mathrm{d}=0} = \mathrm{sign}\left(\frac{k}{m}-1\right)\left(\frac{k}{m}+\left(\frac{k}{m}-1\right)\eta\right). \tag{4.68}$$

According to (4.68), a change of monotonicity of the asymptotic slope of $\hat{G}$ at $\theta_d \cong 0°$ occurs for (see Appendix 4.B)

$$\eta = \frac{\frac{k}{m}}{1 - \frac{k}{m}} = \frac{1}{\eta_0}. \tag{4.69}$$

Thus, the sign of derivative in (4.67), cf. (4.68), for $\frac{k}{m} > 1$ is positive for $1/\eta_0 < \eta \leq 1$ and negative for $\eta < 1/\eta_0$, whereas for $\frac{k}{m} < 1$ it is negative for all the values of $\eta \leq 1$. Consequently, for bimaterials with $\frac{k}{m} < 1$, a remote secondary tension $\sigma_y^\infty > 0$ will hinder the crack onset from the energetic approach point of view, while a secondary compression $\sigma_y^\infty < 0$ will facilitate it.

4.3.3 Interface crack onset at a single inclusion under a remote biaxial transverse load

This section aims to obtain the conditions derived from the stress and energy criteria and combine them to formulate the theoretical model. Thus, this section is organized as follows: First, in Section 4.3.3.1, the stress criterion is presented and applied to the stress state analyzed in Section 4.3.1. Second, a condition imposed by the incremental energy criterion is obtained in Section 4.3.3.2 with the aid of the analysis introduced in Section 4.3.2. Then, both conditions are combined in Section 4.3.3.3 leading to the prediction of the critical load and semiangle. Finally, the post-crack-onset evolution and the applicability of the open model of interface cracks in the present problem are discussed in Sections 4.3.3.4 and 4.3.3.5, respectively.

4.3.3.1 Stress criterion

A stress criterion is usually invoked if no crack exists *a priori*. The present stress criterion is based on the idea of the existence of an interface tensile strength σ_c, defined as the maximum tension that the interface can sustain. Thus, in the present problem, the inclusion-matrix interface can break at the points with a polar angle θ where,

$$\sigma(\theta) \geq \sigma_c, \tag{4.70}$$

defining a tensile criterion which is employed here in accordance to the previous model by Mantič (2009). According to Figure 4.49 and Section 4.3.1, this criterion cannot be fulfilled for $\eta \leq 1/\eta_0$ and $k/m > 1$ because the whole interface is under compression and no crack onset can be predicted following the stress criterion. Hence, in the following analysis, it will be assumed that either $\eta > 1/\eta_0$ or $k/m \leq 1$.

Then, combining (4.70) and (4.57a), the stress criterion can be expressed as

$$\frac{\sigma_x^\infty}{\sigma_c} \geq \frac{1}{k + (k - m)\eta - (1 - \eta)m \sin^2 \theta} = s(\theta, \eta, \alpha, \beta). \tag{4.71}$$

Assuming a sufficiently large remote loading, given by (4.71) for $\theta = 0°$,

$$\frac{\sigma_x^\infty}{\sigma_c} \geq \min_\theta s(\theta, \eta) = s(0°, \eta) = \frac{1}{k + (k - m)\eta} > 0, \tag{4.72}$$

an angle $\theta_c^\sigma \in [0°, 90°]$ can be defined by $\sigma(\theta_c^\sigma) = \sigma_c$. Then, due to the decreasing character of $\sigma(\theta)$ (see (4.58) and discussion in Section 4.3.1), condition (4.70) is verified for all $\theta \in [0°, \theta_c^\sigma]$

$$\theta_c^\sigma = \arcsin \sqrt{\frac{k + (k - m)\eta - \frac{\sigma_c}{\sigma_x^\infty}}{(1 - \eta)m}}, \tag{4.73}$$

According to a discussion in Section 4.3.1, for a given value of η an angle θ_0 (4.60) may exist where the normal stress is zero. Then, condition (4.71) leads to an infinite load for $\theta = \theta_0$, which is an upper limit for the values of θ_c^σ

$$\theta_c^\sigma < \theta_0(\eta; \alpha, \beta). \tag{4.74}$$

Then, combining all the conditions related to the stress criterion, the maximum angle of a debond and the function s are defined in a rigorous manner suitable for computational proposes in Appendix 4.G.

Figure 4.37 shows a representation of the stress criterion for glass/epoxy as defined in Table 4.3 for different values of η. As predicted, all the curves of the stress criterion are increasing. Thus, for a load (values of σ_x^∞/σ_c and η) two zones can be defined in this diagram: if θ_c^σ exists, a zone where a debond is possible $[0°, \theta_c^\sigma]$, and another where it is not possible $(\theta_c^\sigma, 180°]$.

Note that, if $k \leq m$, an angle θ_η (see Figure 4.37) can be defined where the stress criterion is independent of the remote secondary load σ_y^∞ as demonstrated in Section 4.3.1. This semiangle separates the interface into two regions, a region $(\theta < \theta_\eta)$ where a secondary compression $\sigma_y^\infty < 0$ facilitates a debond onset and another region $(\theta > \theta_\eta)$ where it hinders a debond.

4.3.3.2 Incremental energy criterion

An incremental Griffith criterion is used here with the aid of expressions developed in Section 4.3.2. First, an energy balance for the onset of an interface crack of a finite length is introduced and its different terms are particularized for this problem and analyzed. Finally, a condition for the minimum load originating an energetically allowed fiber-matrix debond is deduced by means of a dimensionless function of the crack length representing the ratio of the dissipated to the released energy.

Similarly as in Mantič (2009), the energy balance can be written as

$$\Delta\Pi + \Delta E_k + 2 \int_0^{\Delta\theta} G_c(\theta_d)a d\theta_d = 0, \tag{4.75}$$

where $\Delta\Pi$ is the change in the potential energy between the states prior to and after the onset of the finite length crack, ΔE_k is the change in the kinetic energy

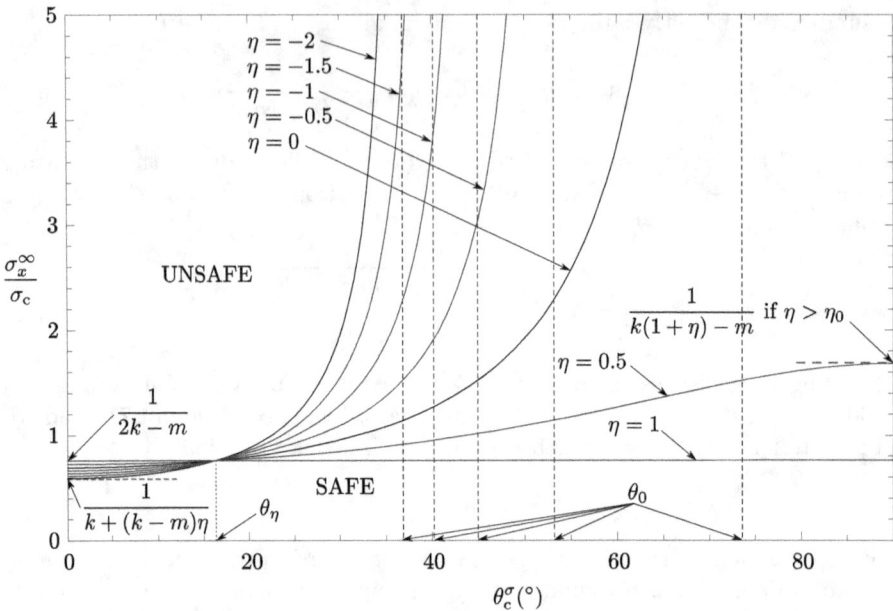

Figure 4.37: Graphical representation of the stress criterion for different values of η and glass/epoxy.

of the body, G_c is the interface fracture toughness (called also fracture energy) and $\Delta\theta$ is the semiangle of the finite crack originated at onset. Note that heat transfer and other types of energy dissipation have been neglected.

Interface fracture toughness G_c is considered to be dependent on θ_d as explained in the following. According to Hutchinson and Suo (1992), see also Mantič et al. (2006) and further references therein, the variation of fracture toughness of an interface crack depends on the fracture mode mixity. Fracture mode mixity of the crack growing along the inclusion-matrix interface can be characterized by the phase angle ψ defined in (4.63), writing $G_c(\theta_d) = G_c(\psi(\theta_d, \eta))$. The following phenomenological law proposed by Hutchinson and Suo (1992):

$$G_c(G_{1c}, \psi, \lambda) = G_{1c}\hat{G}_c(\psi, \lambda) = G_{1c}\left(1 + \tan^2\left[(1-\lambda)\psi\right]\right),\qquad(4.76)$$

will be used in the present analysis. G_{1c} is considered as the fracture Mode I toughness, λ is a fracture mode-sensitivity parameter, typical range $0.2 \leq \lambda \leq 0.35$ being characteristic of moderately strong fracture mode dependence, $\hat{G}_c$ is a dimensionless normalized fracture toughness function.

Figure 4.38 shows the evolution of the interface fracture toughness as a function of θ_d, taking $\theta_l = 0.1°$ and $\lambda = 0.3$, for different values of the load-biaxiality parameter η. Fracture toughness plots for negative values of η have vertical asymptotes at moderate values of θ_d. This is due to the effect of the secondary compression $\sigma_y^\infty < 0$ on the fracture mode mixity increasing the participation of the fracture Mode II.

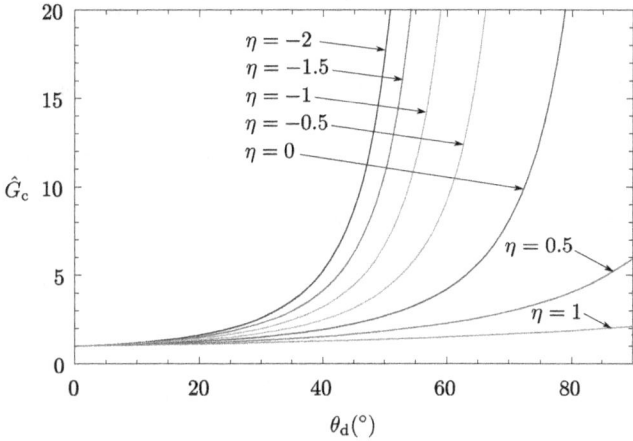

Figure 4.38: Examples of the normalized interface fracture toughness evolution $\hat{G}_c(\psi(\theta_d, \eta))$ for different values of η, taking $\lambda = 0.3$, $\theta_l = 0.1°$ and glass/epoxy.

The energetic balance (4.75) can be rewritten, in view of the above analysis and assuming the production of kinetic energy because of the quasi-static initial state $(\Delta E_k \geq 0)$, as

$$- \Delta\Pi \geq \int_0^{\Delta\theta} G_c(\psi(\theta_d, \eta)) a \, d\theta_d. \tag{4.77}$$

Employing the relation between the differential ERR and the derivative of the potential energy with respect to the crack length $G = -\frac{d\Pi}{d(2a\theta_d)}$, this inequality leads to the energetic condition

$$\int_0^{\Delta\theta} G(\theta_d; \sigma_x^\infty, \eta; a; E^*, \alpha, \beta) d\theta_d \geq \int_0^{\Delta\theta} G_c(G_{1c}, \psi(\theta_d, \eta)) d\theta_d. \tag{4.78}$$

Inasmuch as $G(0°) = 0$ and $G_c(\psi(0°, \eta)) > 0$, there is no solution of (4.78) for values of $\Delta\theta$ lower than a minimum semiangle θ. Thus, the energy criterion imposes, at least, a lower limit for the length of the originated crack.

By substituting G from (4.64) and G_c from (4.76) into (4.78), the expression of the incremental energy criterion takes the form

$$\frac{(\sigma_x^\infty)^2 a}{G_{1c} E^*} \geq g(\Delta\theta, \eta), \tag{4.79}$$

where

$$g(\Delta\theta, \eta; \alpha, \beta; \lambda, \theta_l) = \frac{\int_0^{\Delta\theta} \hat{G}_c(\psi(\theta_d, \eta)) \, d\theta_d}{\int_0^{\Delta\theta} \hat{G}(\theta_d, \eta) d\theta_d} > 0. \tag{4.80}$$

Note that, the dimensionless function g is independent of the particular values of the strength and fracture toughness parameters that characterize the interface, except for the model parameters λ and θ_l. It represents the ratio of the dimensionless forms of the incremental dissipated energy to the incremental released energy.

Figure 4.39 shows the evolution of the dimensionless function g (computed by numerical integration) and its curvilinear asymptote calculated in Appendix 4.E for different values of η and for glass/epoxy defined in Table 4.3. The function g has a minimum at an angle that will be denoted as $\Delta\theta_{\min}^{E}(\eta; \alpha, \beta; \lambda, \theta_l) > 0$. The existence of this minimum can be deduced from the behavior of the functions $\hat{G}_c$ and $\hat{G}$.

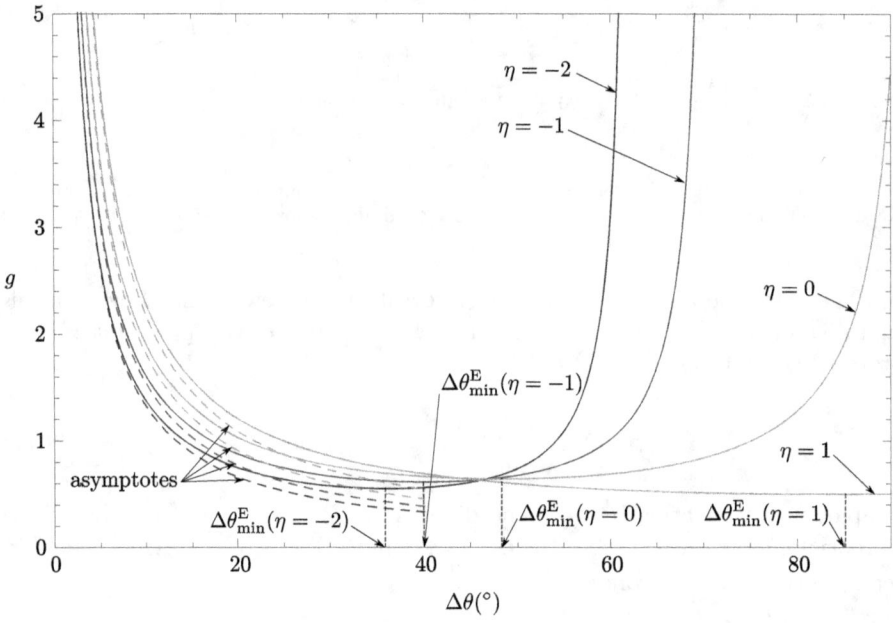

Figure 4.39: Universal dimensionless function $g(\Delta\theta, \eta)$, taking $\lambda = 0.3$, $\theta_l = 0.1°$ and glass/epoxy.

Taking into account that $g(\Delta\theta)$ is a decreasing function for $\Delta\theta < \Delta\theta_{\min}^{E}$, for a sufficiently large σ_x^∞, there exists a lower limit $\Delta\theta_c^E \le \Delta\theta_{\min}^E$ for the semiangle $\Delta\theta$ of energetically allowed debonds, defined by the equality in (4.79):

$$\Delta\theta_c^E(\sigma_x^\infty, \eta, G_{1c}; a; E^*, \alpha, \beta; \lambda, \theta_l) \le \Delta\theta. \tag{4.81}$$

In the particular case of $\Delta\theta_c^E = \Delta\theta_{\min}^E$, according to the analysis carried out in Mantič (2009),

$$\int_0^{\theta_{\min}^E} G(\theta_d, \eta)\mathrm{d}\theta_d = \int_0^{\theta_{\min}^E} G_c(\psi(\theta_d, \eta))\mathrm{d}\theta_d \quad \text{and} \quad G(\theta_{\min}^E, \eta) = G_c(\psi(\theta_{\min}^E, \eta)). \tag{4.82}$$

Additionally, as demonstrated in 4.H, it can be shown that,

$$
\left.\frac{\mathrm{d}G(\theta_\mathrm{d})}{\mathrm{d}\theta_\mathrm{d}}\right|_{\theta_\mathrm{d}=\Delta\theta_\mathrm{c}^\mathrm{E}=\Delta\theta_\mathrm{min}^\mathrm{E}} \leq \left.\frac{\mathrm{d}G_\mathrm{c}(\psi(\theta_\mathrm{d},\eta))}{\mathrm{d}\theta_\mathrm{d}}\right|_{\theta_\mathrm{d}=\Delta\theta_\mathrm{c}^\mathrm{E}=\Delta\theta_\mathrm{min}^\mathrm{E}}. \tag{4.83}
$$

Assuming that $-G$ and G_c are strictly convex functions in the range of interest, in fact, strict inequality holds in (4.83), see Mantič (2009). Notice that, according to Figures 4.35, 4.36 and 4.38, $-G$ is strictly convex for $-1 \leq \eta \leq 1$ whereas G_c is convex in all the situations studied.

The existence of a minimum of g leads to the prediction of a minimum load originating an energetically allowed debond,

$$
\sigma_{\mathrm{c}x}^{\infty,E} = \sqrt{\frac{G_{1\mathrm{c}}E^*}{a}g(\theta_\mathrm{min}^\mathrm{E},\eta)}. \tag{4.84}
$$

According to Figure 4.39, a semiangle $\Delta\theta_\eta^\mathrm{E}$ exists where the function g is roughly independent of the value of η for the bimaterial considered. For $\Delta\theta < \Delta\theta_\eta^\mathrm{E}$, the value of g reduces with decreasing σ_y^∞ with respect to a fixed σ_x^∞. However, for $\Delta\theta > \Delta\theta_\eta^\mathrm{E}$ the effect of σ_y^∞ is inverted. Nevertheless, the value of $\Delta\theta_\eta^\mathrm{E}$ is slightly large for the model assumptions. Therefore, in most cases of interest, increasing σ_y^∞ for the same σ_x^∞ increases $\Delta\theta_\mathrm{c}^\mathrm{E}$, the lower limit for $\Delta\theta$ allowed energetically.

4.3.3.3   Coupled stress and energy criterion

The above stress and energy conditions are combined here taking into account the different monotony of both criteria: stress criterion is an increasing function, whereas energy criterion is decreasing up to a minimum point. As a consequence of this, two different scenarios, brittle and ductile, are presented depending on the parameters of the problem. Next, an algorithm necessary to solve the problem is described. Then, main results are presented focusing on the influence of the load biaxiality on the failure behavior. Results presented in the following show the secondary transverse load modifies slightly the main critical transverse load for glass/epoxy. For tough configurations, an analytical expression of this dependence is obtained as a function of a bimaterial elastic parameter. A study of this influence as a function of bimaterial parameters is presented, in particular for typical composites and extreme cases.

The stress criterion (4.137) and the incremental energy criterion (4.79), respectively, essentially impose an upper limit $\Delta\theta_\mathrm{c}^\sigma$ (4.135) and a lower limit $\Delta\theta_\mathrm{c}^\mathrm{E}$ (4.81) for an initial semidebond angle considering a given remote load. Then, assuming a sufficiently large remote load, the onset of a new crack of a semiangle $\Delta\theta$ is only possible if it verifies:

$$
\Delta\theta_\mathrm{c}^\mathrm{E} \leq \Delta\theta \text{ and } \Delta\theta \leq \Delta\theta_\mathrm{c}^\sigma. \tag{4.85}
$$

Typically the two limits approach each other when decreasing the value of σ_x^∞ for a fixed η. Nevertheless, there is an exception when $\Delta\theta_\mathrm{c}^\sigma > \Delta\theta_\mathrm{min}^\mathrm{E}$ for the load

$\sigma_{cx}^{\infty,E}$ (4.84). Therefore, two scenarios are possible as described in the following. Without loss of generality, only configurations where $\tilde{\theta}_0$ (4.119) is defined will be analyzed.

In scenario A, both criteria, (4.137) and (4.79) are fulfilled as equalities. This implies that the curves of both criteria have an intersection for a semiangle of the crack after the onset $\Delta\theta = \Delta\theta_c$, called critical semidebond angle, giving the minimum value of σ_{cx}^{∞},

$$\Delta\theta_c = \Delta\theta_c^E = \Delta\theta_c^\sigma \le \Delta\theta_{\min}^E. \tag{4.86}$$

As the function $s(\theta, \eta)$ on the right-hand side of the stress criterion is increasing with θ (strictly increasing up to $\tilde{\theta}_0$), it occurs if

$$\gamma\sqrt{g(\Delta\theta_{\min}^E, \eta)} \le s(\Delta\theta_{\min}^E, \eta). \tag{4.87}$$

where γ is a dimensionless parameter defined as (see Mantič (2009))

$$\gamma = \frac{1}{\sigma_c}\sqrt{\frac{G_{1c}E^*}{a}}. \tag{4.88}$$

The value of $\Delta\theta_c$ can be computed by solving the following nonlinear equation:

$$\gamma\sqrt{g(\Delta\theta_c, \eta)} = s(\Delta\theta_c, \eta). \tag{4.89}$$

Thus, the value of the critical load originating a crack is easily calculated, using the previously obtained value of $\Delta\theta_c$, from

$$\frac{1}{\eta}\frac{\sigma_{cy}^{\infty}}{\sigma_c} = \frac{\sigma_{cx}^{\infty}}{\sigma_c} = \gamma\sqrt{g(\Delta\theta_c, \eta)}. \tag{4.90}$$

In scenario B, where condition (4.87) is not fulfilled, the minimum load originating a crack is always associated to

$$\Delta\theta_c = \Delta\theta_{\min}^E, \tag{4.91}$$

because of the increasing character of g for $\Delta\theta > \Delta\theta_{\min}^E$. Hence,

$$\frac{1}{\eta}\frac{\sigma_{cy}^{\infty}}{\sigma_c} = \frac{\sigma_{cx}^{\infty}}{\sigma_c} = \gamma\sqrt{g(\Delta\theta_{\min}^E, \eta)}, \tag{4.92}$$

the interface crack onset being essentially governed by the energy criterion. In fact, $\sigma_{cx}^{\infty} = \sigma_{cx}^{\infty,E}$.

Let a threshold value of γ be defined from the equality in (4.87) as

$$\gamma_{th}(\eta; \alpha, \beta; \lambda, \theta_l) = \frac{s(\Delta\theta_{\min}^E, \eta)}{\sqrt{g(\Delta\theta_{\min}^E, \eta)}}. \tag{4.93}$$

It is easy to see, that γ_{th} separates scenario A ($0 < \gamma \le \gamma_{th}$) from scenario B ($\gamma > \gamma_{th}$). If $\Delta\theta_{\min}^E \ge \theta_0$ then $\gamma_{th} = +\infty$ in view of (4.136), which means that only scenario A is possible.

```
If (k/m > 1 and η > 1/η₀) or k/m ≤ 1  then
```

Find $\min_{\Delta\theta} g(\Delta\theta) \overset{\text{def}}{=} g(\theta^{\text{E}}_{\min})$

If $\gamma\sqrt{g(\theta^{\text{E}}_{\min})} < s(\theta^{\text{E}}_{\min})$

Solve the next equation for $\Delta\theta < \theta^{\text{E}}_{\min}$:

$$\gamma\sqrt{g(\Delta\theta)} = s(\Delta\theta)$$

θ_{c} =the solution $\Delta\theta$ of this equation

```
Else
```

$$\theta_{\text{c}} = \theta^{\text{E}}_{\min}$$

```
Endif
```

Compute the critical load $\sigma^{\infty}_{\text{c}x}$ by

$$\frac{\sigma^{\infty}_{\text{c}x}}{\sigma_{\text{c}}} = \gamma\sqrt{g(\theta_{\text{c}})}$$

and $\sigma^{\infty}_{\text{c}y}$ by

$$\frac{\sigma^{\infty}_{\text{c}y}}{\sigma_{\text{c}}} = \eta \cdot \frac{\sigma^{\infty}_{\text{c}x}}{\sigma_{\text{c}}}$$

```
Else
```

```
No debond is possible under the present hypotheses
```

```
Endif
```

```
End
```

Figure 4.40: Computational procedure for the evaluation of θ_{c}, $\sigma^{\infty}_{\text{c}x}$ and $\sigma^{\infty}_{\text{c}y}$, assuming $\eta \leq 1$.

According to the definition of the two scenarios, the critical values of θ_{c}, $\sigma^{\infty}_{\text{c}x}$ and $\sigma^{\infty}_{\text{c}y}$ can be computed by the procedure shown in Figure 4.40.

The above defined dimensionless structural parameter γ (4.88) can be referred to as *stress oriented brittleness number* (see Mantič (2009) and further references therein). The adjective "stress oriented" corresponds to the fact that the critical load (remote stress) is in some sense proportional to this number, see (4.90) and (4.92). In fact, in scenario B (with $\gamma > \gamma_{\text{th}}$) the critical load is linearly proportional to γ. The brittleness number γ is governing brittle-to-tough transition in the fiber-matrix debond onset, small values of γ corresponding to brittle and large values to tough configurations (cf. Kushch et al. (2011)).

Whereas the values of the critical angle and load in tough configurations, usually associated to scenario B, are simply described by (4.91) and (4.92), the asymptotic behavior of these values in brittle configurations with vanishing values of γ ($\gamma \to 0^+$), associated to scenario A, requires a further analysis.

Looking at equation (4.89), defining $\Delta\theta_{\text{c}}$, for $\gamma \to 0^+$, it holds $g(\Delta\theta_{\text{c}}, \eta) \to \infty$ because $s(\Delta\theta_{\text{c}}, \eta)$ on the right hand side of (4.89) is bounded from below by its positive minimum value (4.72) for a fixed η. Taking into account condition (4.86) and definition of g in (4.80), see also Figure 4.39 and approximation (4.127), it is obtained that

$$\lim_{\gamma \to 0^+} \theta_{\text{c}} = 0°. \tag{4.94}$$

In the same manner, an asymptotic expression can be extracted as demonstrated in Appendix 4.E.

When evaluating the critical load for $\gamma \to 0^+$, which implies $\Delta\theta_c \to 0^+$, the following approximation is obtained, see (4.72):

$$\frac{\sigma_{cx}^\infty}{\sigma_c} \gtrsim s(0°, \eta) > 0 \quad \text{for} \quad \gamma \to 0^+. \tag{4.95}$$

Combining this equation with the definition of η in (4.56), the relation between σ_{cx}^∞ and σ_{cy}^∞ can be approximated as

$$(k-m) \cdot \frac{\sigma_{cy}^\infty}{\sigma_c} + k \cdot \frac{\sigma_{cx}^\infty}{\sigma_c} \gtrsim 1 \quad \text{for} \quad \gamma \to 0^+. \tag{4.96}$$

This linear relation between critical stresses is given, in fact, by the stress criterion (4.137) when considered for a small angle $\Delta\theta$.

Figure 4.41, computed using the computational procedure in Figure 4.40, presents the effect of the brittleness number γ on the critical semiangle $\Delta\theta_c$, the arrest semiangle θ_a (defined later in Section 4.3.3.4) and on the critical remote load σ_{cx}^∞. This figure is a nice illustration of the above mentioned brittle-to-tough transition in the fiber-matrix debond onset. It can be seen that the behavior of $\Delta\theta_c$ and σ_{cx}^∞ agrees with the above analytic predictions, namely (4.128) with (4.95) for small and (4.91) with (4.92) for large values of γ.

Figure 4.42 shows the safe regions and failure envelopes in $(\sigma_{cx}^\infty, \sigma_{cy}^\infty)$ plane for glass/epoxy bimaterial computed for different values of γ by applying the procedure described in Figure 4.40. According to this figure, an increase of remote secondary load σ_y^∞ increases the critical load σ_{cx}^∞ for sufficiently small values of γ (recall that $k/m < 1$ and $\eta_0 > 0$ for glass/epoxy). However, a non-monotonic boundary curve of the safe region is observed for greater values of γ in Figure 4.42. In fact, it is observed that the curve which joins the points with the maximum critical remote load σ_{cx}^∞ shows that the maximum is situated at $\eta = 1$ just for $\gamma \to 0^+$. For moderate values of γ, the maximum critical remote load σ_{cx}^∞ corresponds to $\eta < 0$. The reason for this behavior is clarified in Figure 4.43 where two situations are explained.

In the first case, the value of γ is relatively small, Figure 4.43(a), which is associated to small values of $\Delta\theta_c$, as previously demonstrated. For small values of $\Delta\theta_c$, the effect of the secondary load σ_y^∞ is the same for both the stress criterion and energy criterion curves, both curves descending when η reduces. In the second case, the value of γ is larger, Figure 4.43(b), and the values of $\Delta\theta_c$ are greater than θ_η, see (4.61) and the discussion below. Then, for values of γ originating $\theta_c > \theta_\eta$, an increase of the remote secondary load σ_y^∞ makes less restrictive the stress criterion and more restrictive the energy criterion. Thus, the monotony of the function $\sigma_{cx}^\infty(\sigma_{cy}^\infty)$ can be broken down as observed in Figure 4.42 for moderate values of γ.

The straight line defined by (4.96) represents, according to Figure 4.42, a limit of failure envelope curves for $\gamma \to 0^+$. Note that the failure envelope curves for $\gamma \to 0^+$ in Figure 4.42 show the most relevant influence of the secondary load σ_y^∞ on the value of the critical load σ_{cx}^∞.

Figure 4.42 also shows the "threshold curve" which separates scenarios A and B. It is interesting to remark that greater values of γ correspond to a larger

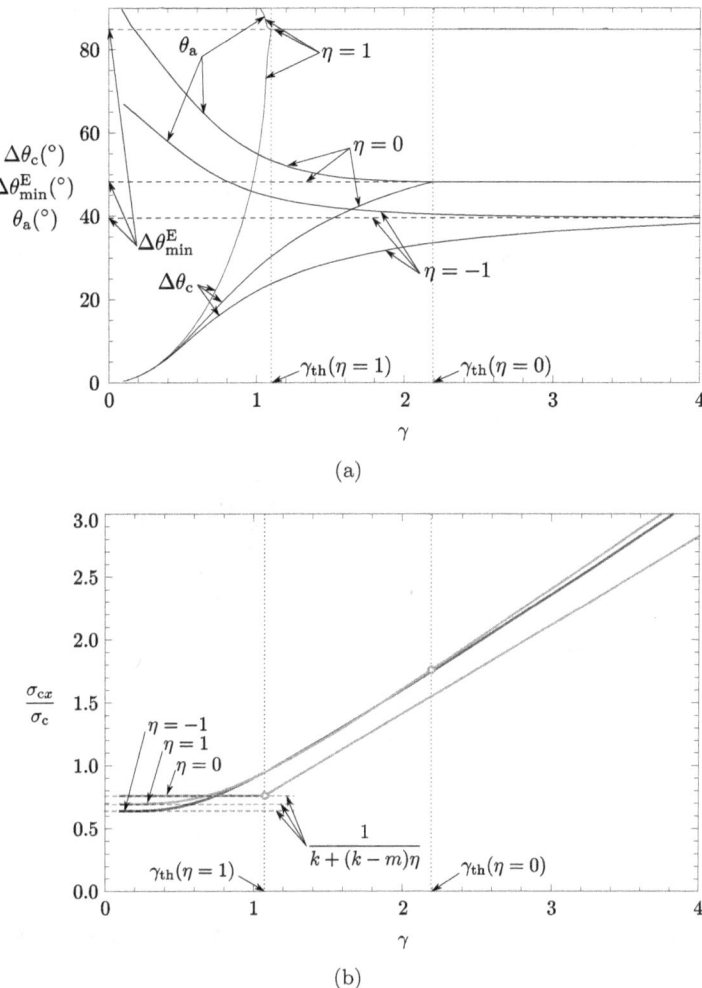

Figure 4.41: (a) Semiangles $\Delta\theta_c$, $\Delta\theta_{\min}^E$ and θ_a, and (b) Critical remote tension σ_{cx}^∞ as functions of the brittleness number γ, taking $\lambda = 0.3$, $\theta_l = 0.1°$ and glass/epoxy.

range of failure behavior governed by scenario B. On the contrary, the presence of a remote secondary compression $\sigma_y^\infty < 0$ leads to scenario A, for small and moderate values of γ.

Figure 4.44 studies the influence of α and β values on the biaxial safe region for $\gamma \to 0^+$, for selected theoretical (but possible) bimaterials and also for usual composites. From Dundurs' $\alpha - \beta$ parallelogram it is seen that the most common bimaterials have very similar properties in the debond onset problem. A more extensive list of $\alpha - \beta$ values for real bimaterials can be found in Suga et al. (1988) and Schmauder and Meyer (1992).

The safe region in the limit case $\gamma \to 0^+$ is defined by the intersection of

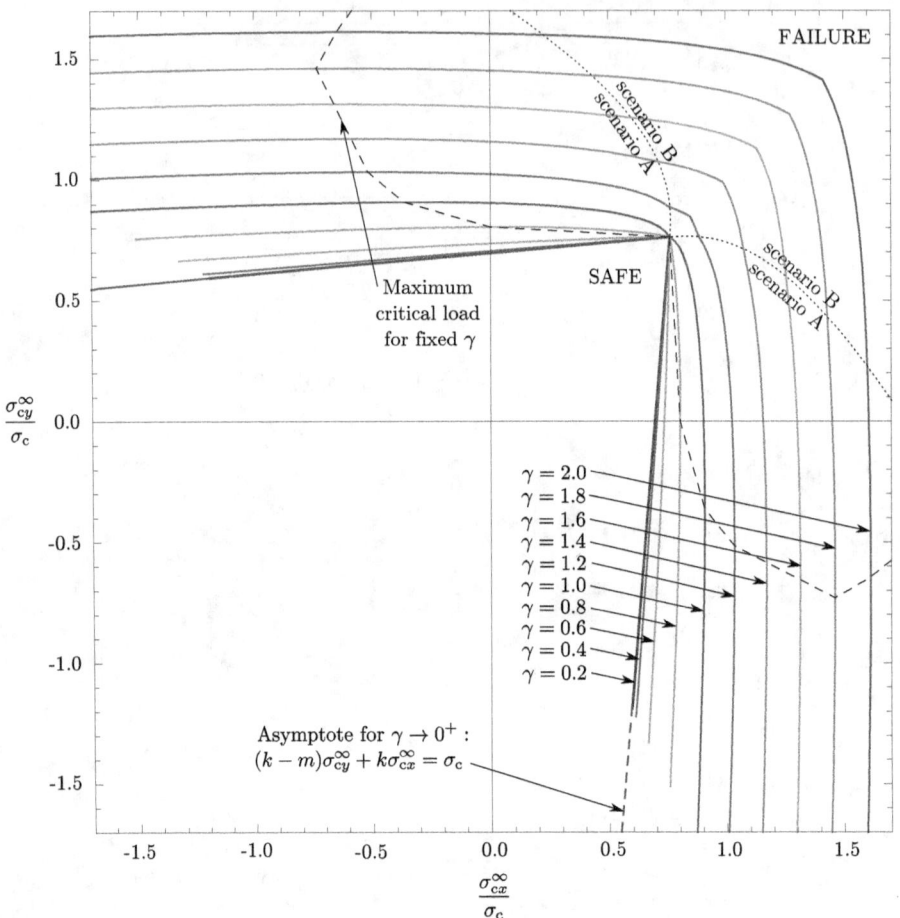

Figure 4.42: Critical biaxial loads originating a crack for different values of γ, taking $\lambda = 0.3$, $\theta_l = 0.1°$ and glass/epoxy.

the semiplanes including the origin of coordinates and limited by the straight line defined by (4.96) and the symmetric one with respect to the bisector of the coordinate axes. From (4.96), the position of the corner point of the safe region ($\eta = 1$) is given by

$$\frac{\sigma_{cx}^{\infty}}{\sigma_c} = \frac{\sigma_{cy}^{\infty}}{\sigma_c} = \frac{1}{2k - m}. \tag{4.97}$$

The slope of the linear relation in (4.96) characterizes the influence of the secondary load σ_y^{∞} on the critical load σ_{cx}^{∞}. This slope can be expressed as:

$$\lim_{\gamma \to 0^+} \left(\frac{\partial \sigma_{cx}^{\infty}}{\partial \sigma_{cy}^{\infty}} \right) = \eta_0 \tag{4.98}$$

where η_0 is defined in (4.116) and its range in (4.117), see also Figure 4.49.

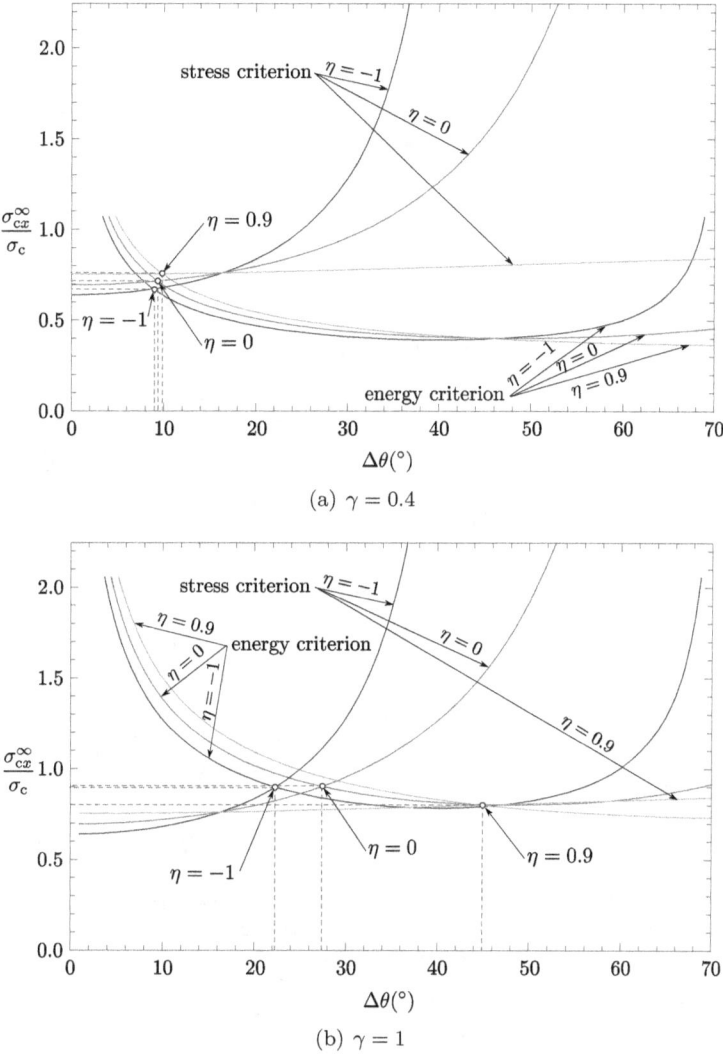

(a) $\gamma = 0.4$

(b) $\gamma = 1$

Figure 4.43: Stress and energy criteria curves, taking $\lambda = 0.3$, $\theta_l = 0.1°$ and glass/epoxy for two different values of γ. (a) $\gamma = 0.4$, (b) $\gamma = 1$

Hence, this slope is only dependent on the elastic bimaterial properties and does not depend on the interface properties. In view of the range of possible values for the slope (4.117), the safe region is always convex, cf. Figure 4.44(b).

For $k/m < 1$ the slope (4.98) is positive and an increase in the secondary load σ_y^∞ increases the critical load σ_{cx}^∞ necessary to originate a debond, see Figure 4.44. This is the case of the two bimaterials defined in Table 4.3, i.e. glass/epoxy and carbon/epoxy. However, an opposite effect is predicted for $k/m > 1$, see Figure 4.44. In fact, this dependence matches up with the effect of the secondary load

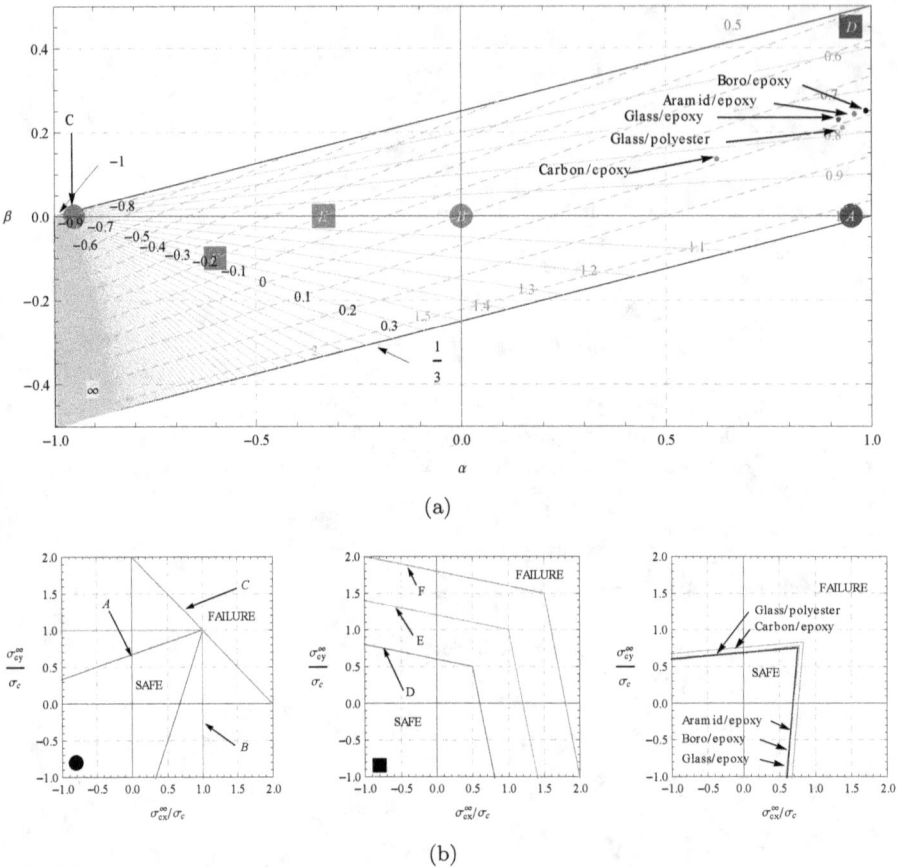

(a)

(b)

Figure 4.44: (a) $\alpha - \beta$ diagram for bimaterials in plane strain with isovalue curves for $\gamma \to 0^+$ corresponding to: solid lines with values of $\sigma_{cx}^\infty/\sigma_c = \sigma_{cy}^\infty/\sigma_c$, and dashed lines with values of $\partial\sigma_{cx}^\infty/\partial\sigma_{cy}^\infty$. (b) Some biaxial failure envelopes for $\gamma \to 0^+$ for selected points in the $\alpha - \beta$ diagram and usual composites. A: $k/m = 0.75$, B: $k/m = 1$, C: $k/m \to +\infty$, D: $k/m = 1.25$, E: $k/m = 1.25$, F: $k/m = 1.25$, Carbon/epoxy: $k/m = 0.9200$, Glass/polyester: $k/m = 0.9201$, Glass/epoxy: $k/m = 0.9205$, Aramid/epoxy: $k/m = 0.9217$, Boro/epoxy: $k/m = 0.9204$

σ_y^∞ predicted by Goodier's solution for the interface point $\theta = 0°$, see (4.53), which shows that a compression or tension is expected at $\theta = 0°$ when a secondary load $\sigma_y^\infty > 0$ is applied for $k/m < 1$ or $k/m > 1$, respectively.

The present results show that the influence of the secondary load σ_y^∞ on the critical load σ_{cx}^∞ is at most moderate in usual composites. Taking into account that in the case $\gamma \to 0^+$, analyzed in Figure 4.44, the values of σ_{cx}^∞ are the most sensitive to the values of σ_y^∞, in general for $k/m < 1$ the influence of σ_y^∞ on the value of σ_{cx}^∞ exists but it is small or at most moderate. For bimaterials as glass/epoxy and carbon/epoxy with $\frac{k}{m} < 1$, these results agree with París et al. (2003). According to these authors a secondary compression σ_y^∞ makes easier the debond onset for these bimaterials and reduces the critical load σ_{cx}^{∞} as shown in Figure 4.44(b). However, for bimaterials with $\frac{k}{m} > 1$ the effect is opposite, a secondary compression $\sigma_y^{\infty} < 0$ increases the critical load σ_{cx}^{∞}.

4.3.3.4 Post-crack-onset evolution

After the onset of a new crack, an unstable growth of the crack is possible depending on the relation between $G(\theta_d)$ and $G_c(\theta_d)$ for $\theta_d \geq \theta_c$ and according to the criterion of the classical (infinitesimal) interface fracture mechanics (see, e.g. París et al. (2007); Mantič et al. (2006)). Thus, the condition for the further crack growth will be

$$G(\theta_d, \eta) \geq G_c(\psi(\theta_d, \eta)), \qquad \theta_d \geq \theta_c. \qquad (4.99)$$

The crack will stop growing at an arrest angle $\theta_a \geq \Delta\theta_c$ verifying $G(\theta_a, \eta) = G_c(\psi((\theta_a, \eta))$ if for angles $\theta_d \gtrsim \theta_a$ criterion (4.99) is not fulfilled. The stability of the post-onset growth of the crack is different for two scenarios A and B separated by γ_{th} two post-onset scenarios being possible:

- For $\gamma < \gamma_{th}$, $G(\theta_c, \eta) > G_c(\psi(\Delta\theta_c, \eta))$ and the crack is expected to grow in an unstable manner up to an arrest angle $\theta_a > \Delta\theta_{min}^E$, which can be shown similarly as in Mantič (2009).

- For $\gamma \geq \gamma_{th}$, $\Delta\theta_c = \Delta\theta_{min}^E$, $G(\Delta\theta_c, \eta) = G_c(\psi(\Delta\theta_c, \eta))$ and the derivative $dG/d\theta_d|_{\theta_d=\Delta\theta_{min}^E} \leq dG_c/d\theta_d|_{\theta_d=\Delta\theta_{min}^E}$, see (4.83) and the related discussion in Section 4.3.3.2. Therefore, assuming strict inequality in (4.83) (which, in fact, has been verified in all present calculations), no unstable crack growth is usually expected after the crack onset and $\theta_a = \Delta\theta_{min}^E$.

Figures 4.41(a) and 4.45 were computed by implementing the above ideas. The values of G and G_c and their derivatives are compared for $\theta_d \geq \Delta\theta_c$ in order to find the arrest semiangle θ_a. According to these figures, a long unstable crack growth after the crack onset is predicted for small values of γ (brittle configurations), whereas short (or zero) unstable crack growth is predicted for large vales of γ (tough configurations).

4.3.3.5 Applicability of the open model of interface cracks

The applicability of the theoretical model developed is limited by its assumptions, perhaps the most restrictive being the usage of the open model of interface cracks. Toya's (1974) solution assumes negligible overlapping of traction-free crack faces. The angle of the overlapping zone at the crack tip can be estimated by the formula deduced by Hills and Barber (1993) and generalized by Graciani et al. (2007), which rewritten for the present case is defined as the largest value of

$$\theta_I(\theta_d, \eta) = \theta_l \cdot \exp\left[((2n - 1/2)\pi - \psi(\theta_d, \eta, \theta_l)\mathrm{sign}\varepsilon + \arctan(2|\varepsilon|))/|\varepsilon|\right],$$
$$(4.100)$$

lower than the semidebond angle θ_d, with n being an integer.

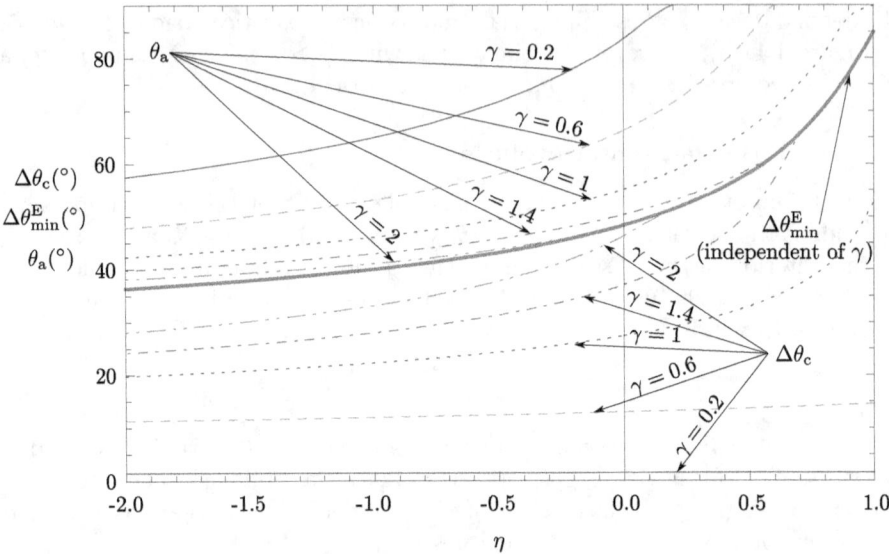

Figure 4.45: Semiangles $\Delta\theta_c$, $\Delta\theta_{min}^E$ and θ_a as functions of the biaxiality parameter η for different values of γ, taking $\theta_l = 0.1°$, $\lambda = 0.3$ and glass/epoxy.

Figure 4.46 shows the evolutions of $\hat{G}$ for glass/epoxy and different values of the load biaxiality parameter η. Additionally, for $\gamma = 1.5$, corresponding to relatively tough configurations, the values of $\Delta\theta_c$, $\Delta\theta_{min}^E$ and θ_a computed by the present model are indicated. Finally, angles $\theta_{I,1\%}$ for which the overlapping zone represent 1% of the crack length, i.e. $\theta_I/2\theta_d = 0.01$, providing a reasonable limit of validity of the open model, are also presented in Figure 4.46. This figure shows that all the values of $\Delta\theta_c$ are lower than the reference limit $\theta_{I,1\%}$, therefore the open model is acceptable for the evaluation of $\Delta\theta_c$ and the critical load σ_{cx}^∞. However, it might not be fully acceptable when computing θ_a for large negative values of η. Note that, the above discussed limit on the semiangle θ_d is mainly due to somewhat inaccurate evaluation of $\hat{G}$ because of a large overlapping zone at the crack tip, see also Figure 4.36 and the related discussion. A correct procedure for the evaluation of $\hat{G}$ is such cases would require employing the contact model of interface cracks as in París et al. (2007) and Correa (2008).

4.3.4   Size effect of the inclusion radius a on the crack onset and its variations with the load biaxiality

Analogously to the uniaxial case, see e.g. Section 4.2, a size effect in the present debond onset problem can be understood as a dependence of the critical remote load $(\sigma_{cx}^\infty, \sigma_{cy}^\infty)$ and critical semiangle $\Delta\theta_c$ on the only geometric parameter in the present problem, the inclusion radius a. The objective of this section is to study the variation of this size effect predicted by the uniaxial model in the biaxial case.

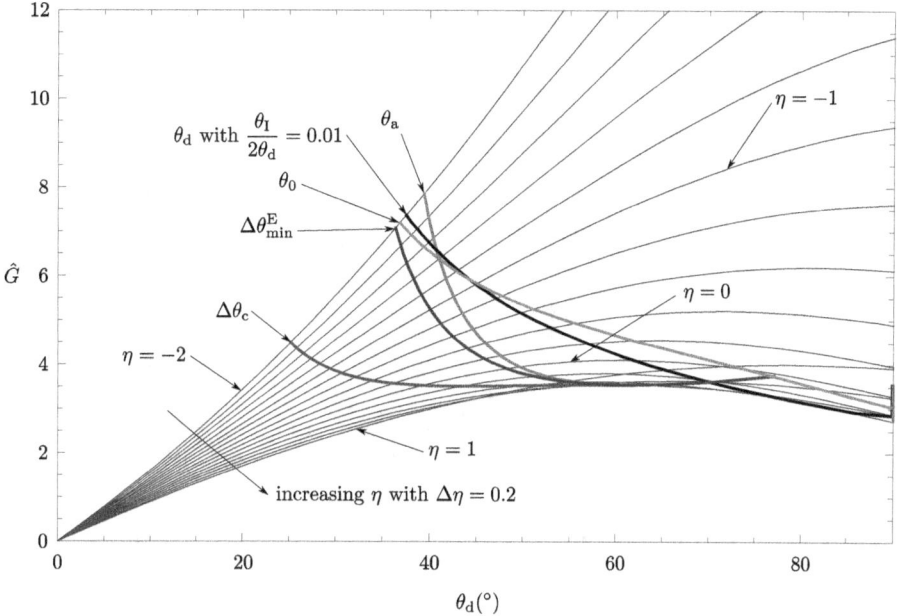

Figure 4.46: Plots of evolution of values of dimensionless ERR limited by the semiangles θ_d with $\frac{\theta_I}{2\theta_d} = 0.01$ which estimate the validity of the model, and representation of values of semiangles $\Delta\theta_c$, $\Delta\theta_{\min}^E$, θ_0 and θ_a, taking $\lambda = 0.3$, $\theta_l = 0.1°$, $\gamma = 1.5$ for glass/epoxy.

Let a bimaterial characteristic length a_0 be defined in terms of the interface properties σ_c and G_{1c} and the harmonic mean of effective Young moduli E^*,

$$a_0 = \frac{G_{1c}E^*}{\sigma_c^2}. \tag{4.101}$$

Then, the ratio a/a_0 and γ are related by

$$\gamma = \sqrt{\frac{a_0}{a}}. \tag{4.102}$$

A threshold value of a can be defined in terms of the threshold value γ_{th} (4.93),

$$a_{th} = \frac{a_0}{\gamma_{th}^2}. \tag{4.103}$$

Scenario A taking place for $a \geq a_{th}$ and B for $a < a_{th}$. In fact, all the above analyses of the results obtained by the coupled stress and energy criterion taking γ as a governing parameter could be rewritten in terms of the ratio a/a_0, see Martin et al. (2008) for a similar approach. For sufficiently large values of a, which correspond to small values of γ, the critical semiangle θ_c and the critical remote load σ_{cx}^∞ can be approximated by the following expressions, see Section 4.3.3,

$$\Delta\theta_c \cong \frac{2\cosh^2(\pi\varepsilon)}{\pi(1 + 4\varepsilon^2)} \frac{a_0}{a} \tag{4.104a}$$

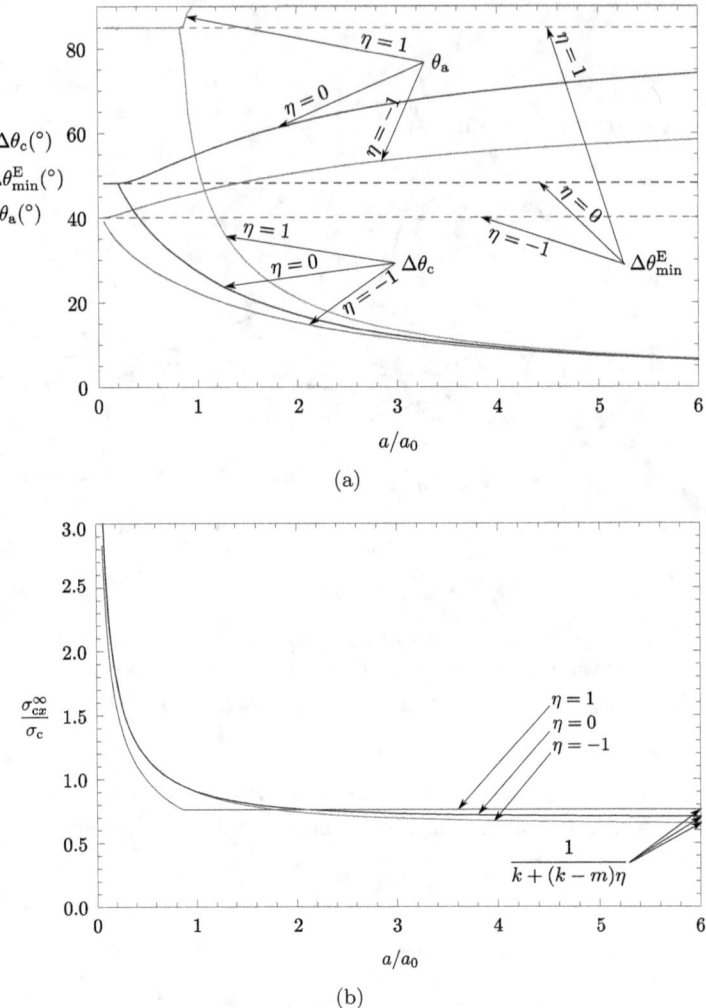

(a)

(b)

Figure 4.47: (a) Critical semiangle θ_c and arrest semiangle θ_a and (b) critical remote tension σ_{cx}^∞ as a function of the inclusion radius a, taking $\lambda = 0.3$, $\theta_l = 0.1°$ and glass/epoxy.

and

$$\frac{\sigma_{cx}^\infty}{\sigma_c} \cong \frac{1}{k + (k - m)\eta},$$

(4.104b)

whereas for $a \leq a_{\text{th}}$,

$$\Delta\theta_c = \Delta\theta_{\min}^E$$

(4.105a)

$$\frac{\sigma_{cx}^\infty}{\sigma_c} = \sqrt{g(\Delta\theta_{\min}^E, \eta)}\sqrt{\frac{a_0}{a}}.$$

(4.105b)

As follows from (4.104) and (4.105), the critical crack-semilength $a\Delta\theta_c$ is constant and independent of a and η for large a, whereas for small a it is linearly proportional to a.

The above described asymptotic behavior of θ_c and σ_{cx}^∞ can be easily identified in Figure 4.47 where the variations of $\Delta\theta_c$, θ_a and σ_{cx}^∞ as functions of a are plotted. In particular, in Figure 4.47(a) and (b) it is seen that $\theta_c = \theta_a = \theta_{min}^E$ and $\sigma_{cx}^\infty \approx 1/\sqrt{a}$, respectively, for $a \leq a_{th}$. Thus, σ_{cx}^∞ increases drastically for small inclusions whereas for large inclusions it tends to a constant value given by the stress criterion applied at $\theta = 0°$. As can be observed in Figure 4.47(b), the size effect on σ_{cx}^∞ is similar for different values of η, being quite independent of the combination of remote transverse loads.

4.3.5 Experimental procedure for the measurement of the brittleness number γ, interface tensile strength σ_c and fracture toughness G_{1c}

Fracture properties of the fiber-matrix interfaces are very important for the macroscopic behavior of fiber reinforced composites. However the experimental measurement of these is very difficult to be carried out. An indirect experimental procedure is proposed here for obtaining first the value of γ, and subsequently the values of σ_c and G_{1c}, for a bimaterial. Elastic properties of the bimaterial (E^*, α, β) are assumed to be known, as they can be measured by carrying out standard material tests for each material separately.

To determine the interface properties, the sole measure of the critical stress for the case of remote uniaxial tension $(\eta = 0)$ is not sufficient. The reason is that the critical stress depends not only on γ but also on σ_c, see the normalization used in (4.90). Nevertheless, if the critical stress is also measured for a biaxial load $(\eta \neq 0)$, then, the ratio of these critical stresses depends only on the value of γ due to the influence of γ value on the solution θ_c of (4.89). This is the key idea behind the experimental procedure proposed. This procedure employs plots of the ratio of critical stresses $\frac{\sigma_{cx}^\infty(\eta\neq0)}{\sigma_{cx}^\infty(\eta=0)}$ as a function of γ or η. As an example, Figure 4.48 shows values of $\frac{\sigma_{cx}^\infty(\eta\neq0)}{\sigma_{cx}^\infty(\eta=0)}$ for glass/epoxy, taking $\theta_l = 0.1°$ and $\lambda = 0.3$.

The steps of the experimental procedure are briefly explained in the following:

1. Determine the critical stress σ_{cx}^∞ in the uniaxial tension test $(\eta = 0)$. This is a relatively easy test, thus a good accuracy is expected.

2. Determine another critical stress σ_{cx}^∞ in a biaxial test for $\eta \neq 0$. Combining the plots in Figure 4.48(a) and a rough a priori estimation of the γ value, choose the most suitable value of η to test by looking for an invertible segment of the pertinent function plotted in Figure 4.48(a) and for its maximum slope. Capabilities of the testing machine may represent an additional constraint.

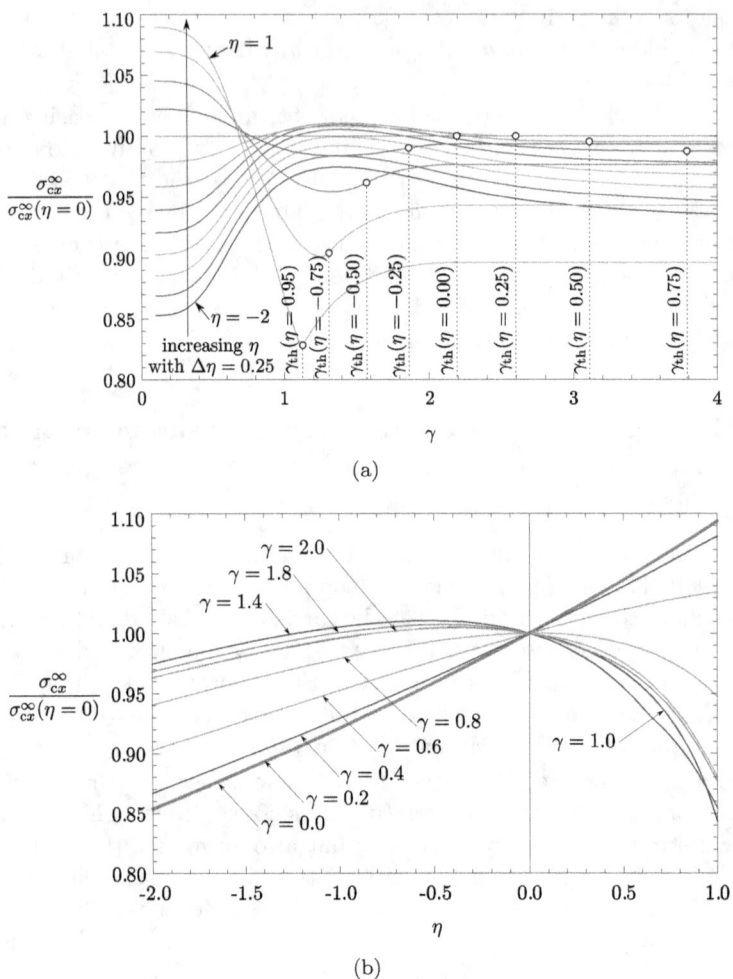

Figure 4.48: Graphs for determination of γ by applying the experimental procedure proposed. Predictions for $\frac{\sigma_{cx}^{\infty}(\eta)}{\sigma_{cx}^{\infty}(\eta=0)}$ as a function of (a) γ and (b) η for glass/epoxy, $\theta_l = 0.1°$ and $\lambda = 0.3$.

3. Evaluate the ratio of the measured critical stress $\frac{\sigma_{cx}^{\infty}(\eta\neq0)}{\sigma_{cx}^{\infty}(\eta=0)}$. Then, estimate a value of $\tilde{\gamma}$ from the measured ratio of the critical stresses for the chosen value of η from Figure 4.48(a).

4. Estimate the corresponding value of the critical semiangle $\Delta\tilde{\theta}_c$ by solving the following nonlinear equation, employing $\tilde{\gamma}$, see (4.89):

$$\tilde{\gamma}\sqrt{g(\Delta\tilde{\theta}_c, \eta)} = s(\tilde{\theta}_c, \eta) \qquad (4.106)$$

5. Estimate the interface tensile strength σ_c from (4.90),

$$\tilde{\sigma}_c = \frac{\tilde{\sigma}_{cx}^{\infty}(\eta)}{\tilde{\gamma}\sqrt{g(\Delta\tilde{\theta}_c, \eta)}},$$

(4.107)

where $\tilde{\sigma}_{cx}^{\infty}$ is one of the two values measured, either for $\eta = 0$ or $\eta \neq 0$.

6. Estimate the interface fracture toughness G_{1c} from (4.88)

$$\tilde{G}_{1c} = \frac{(\tilde{\sigma}_c\tilde{\gamma})^2 a}{E^*}.$$

(4.108)

A few comments follow with reference to a possible stumbling block in the 2nd step of the above procedure.

As can be observed from Figure 4.48(a), the ratio $\frac{\sigma_{cx}^{\infty}(\eta\neq0)}{\sigma_{cx}^{\infty}(\eta=0)}$ for a given η is not an injective (one-to-one) function of γ for the whole range of γ considered. Nevertheless, this ratio may become an injective function of γ when restricted to a suitable interval of γ, e.g. to small values of γ roughly in the range $0 < \gamma \lesssim 1$. In general, an a priori estimate of γ will be very useful in choosing a suitable $\eta \neq 0$ for the biaxial test and a pertinent interval of γ where the above ratio is an injective function.

Recall that for $\gamma > \gamma_{th}$ the critical semiangle is constant ($\Delta\theta_c = \Delta\theta_{min}^E$) and the energy criterion determines the critical remote tension, which is directly proportional to γ, see (4.92). Then, for a given value of η, the value of $\sqrt{g(\Delta\theta_{min}^E(\eta), \eta)}$ is fixed. Thus, the ratio shown in Figure 4.48(a) for $\gamma \geq \gamma_{th}(\eta)$ and $\gamma \geq \gamma_{th}(\eta = 0)$ is

$$\frac{\sigma_{cx}^{\infty}(\eta)}{\sigma_{cx}^{\infty}(\eta = 0)} = \frac{\sqrt{g(\Delta\theta_{min}(\eta), \eta)}}{\sqrt{g(\Delta\theta_{min}(\eta = 0), \eta = 0)}}.$$

(4.109)

Hence, this ratio is a constant independent of γ, as can be observed in Figure 4.48(a). Obviously for these rather large values of γ, the proposed experimental procedure (including the biaxial test for only one value of η) is not directly applicable.

Nevertheless, repeating the biaxial tests for several adequately chosen values of η and applying least square fitting to functions plotted in Figure 4.48(b) could provide a good estimation of γ, and subsequently of σ_c by (4.107) and G_{1c} by (4.108) as well.

4.4   Concluding Remarks

Initially, the coupled stress and energy criterion of the FFM has been applied to the classical problem of debonding at the fiber-matrix interface in order to solve the controversy about the symmetry of the post-failure configuration found in the literature. Two configurations, an asymmetric and a symmetric one, with the number of debonds $n = 1$ and 2, respectively, have been studied by comparing the critical remote tension originating debond onset leading to each configuration.

According to the present analysis, the asymmetric post-failure configuration ($n = 1$) is the preferential one as it requires a lower critical remote tension than the symmetric one ($n = 2$). This agrees with some experimental evidence found in the literature and also with some numerical works using cohesive interface laws. However, it disagrees with other numerical works typically based on smoother cohesive interface laws.

The source of difference between both post-failure configuration is the energy criterion. Whereas the stress criterion is equivalent for both post-failure configurations, the energy criterion shows that the mean energy released rate associated to the onset of a new interface crack of finite length is smaller in the symmetric configuration due to the shielding effect between both debonds. The percentage difference in the critical remote tension varies with brittleness number γ from null to a maximum, denoted as $\max \Delta\sigma_c^\infty$, with values between 2.53% and 15.09% for glass/epoxy, depending on the sensitivity of the fracture toughness on the fracture mode mixity. This variation with γ can be interpreted as a size effect of the fiber radius, a large difference between both post-failure configurations may be expected for small fibers and possibly negligible difference for sufficiently large fibers. In addition to a glass/epoxy composite, several other bimaterials have been studied confirming that, although there are some quantitative differences, which have been explained, from the qualitative point of view the results are similar for all the bimaterials studied.

The relevance of post-failure configuration is seen from the analysis of the arrest semiangle θ_a of the unstable growth after the debond onset, since θ_a may be very different between these configurations, although the critical semiangle $\Delta\theta_c$ is very similar. Obviously the value of θ_a determines the subsequent stages of the whole failure mechanism of matrix cracking.

Stress-strain curves predicted by the coupled criterion of the FFM have also been plotted for both post-failure configurations in order to graphically interpret the results obtained and highlight differences in several stages of the failure mechanism.

The effect of the fiber size on the fiber-matrix interface debonding under transverse load predicted by the the coupled criterion is analyzed by comparing with the results of a Cohesive Zone Model (CZM) for the fiber-matrix interface. Differences and similitudes in the foundations of both models have been highlighted taking into account that the CZM is a damage model based on a continuous change of the damage at the interface, whereas the coupled criterion predicts an abrupt onset of a debond of finite length.

The FFM model developed in Mantič (2009) to study the crack onset at the fiber-matrix interface has been generalized and its parameter values adapted to represent a similar problem model as by the CZM used. In this sense, a stress criterion based on the classic Mohr-Coulomb criterion has been proposed in order to take into account a coupling of tangential and normal cohesive tractions similarly as in the CZM. Moreover, the adequate parameters have been selected to define an interface fracture energy according to that used in the CZM. These quite easy modifications show the versatility of the FFM to adapt its formulation to different fracture behaviors.

The comparison of the size effect predicted by both approaches shows a good agreement in a range of fiber radii. In particular the asymptotic behavior of the critical stress predicted by both of them is quite similar for large fibers. Both approaches agree also that the most critical point at the interface is situated about $\theta \sim 20° - 30°$ for large fibers. This agreement is also shown to be acceptable for medium size fibers. However, for small fibers the predicted behavior by both approaches is not similar. Both show an increment of the critical stress for smaller fibers but whereas CZM predicts that $\sigma_{1c} \sim 1/a$, FFM gives a weaker size effect: $\sigma_{1c} \sim 1/\sqrt{a}$ for $a \to 0$.

A simplified model has been developed in order to interpret these differences for small fibers. For small fibers, the normal and tangential gaps at the cohesive elements are demonstrated to be roughly proportional to the product of the remote stress and the fiber radius. In contrast, the critical normal gap is a material property. As a consequence critical stress varies with the inverse of the fiber radius. On the contrary in the FFM energetic condition, which governs the problem for small fibers, the origin of the size effect is the incomplete self-similarity between the dimensions of the region where the energy is dissipated $\sim a$ (a line in 2D elasticity) and where the elastic potential energy is released $\sim \sigma_1^2 a^2$ (a surface in 2D elasticity) leading to the variation $\sigma_{1c} \sim 1/\sqrt{a}$ for $a \to 0$.

In addition, a new theoretical model has been proposed as an extension of the model by Mantič (2009) to take into account a biaxial transverse load. This model is able to predict the critical biaxial load leading to the onset of a debond symmetrically situated with respect to the dominating remote tension along with the debond size.

It is expected that the present work will contribute to the knowledge of the governing parameters and to overall understanding of the failure mechanism in the fiber composites under tension dominated transverse loads. A special attention has been given to the influence of a secondary compression/tension on the value of the critical (primary) tension. Although the present work is focused on a stiff inclusion embedded in a compliant matrix (glass/epoxy composite has been used as a representative example), most results are generally valid for any combination of elastic bimaterial parameters.

For the sake of simplicity a remote biaxial stress state $(\sigma_x^\infty, \sigma_y^\infty)$ with $\sigma_{xy}^\infty = 0$ is assumed. Nevertheless, the present model and results may be easily adapted to a general remote in-plane stress state with $\sigma_{xy}^\infty \neq 0$ by working in its principal coordinate system and assuming that at least one principal stress is tension.

The predictions of the present model are governed by the dimensionless brittleness number γ introduced for interface cracks in Mantič (2009). For this problem, it is interesting to notice that an alternative brittleness number given in terms of the critical Stress Intensity Factor (SIF) in fracture Mode I K_{1c}, instead of the critical ERR G_{1c}, can be proposed. Taking into account the relation between the complex Stress Intensity Factor K and Energy Release Rate G in interface fracture mechanics (see Malyshev and Salganik (1965))

$G = |K|^2/(E^* \cdot \cosh^2(\pi\varepsilon))$, this alternative brittleness number is expressed as

$$\gamma_K = \gamma \cdot \cosh(\pi\varepsilon) = \frac{K_{1c}}{\sigma_c} \frac{1}{\sqrt{a}}. \qquad (4.110)$$

Recall that the expression of γ_K in terms of K_{1c} reminds the classical definition of the brittleness number s in homogeneous materials by Carpinteri (1981). Note that, $\varepsilon = 0$ for a crack in a homogeneous material, thus $\gamma_K = \gamma$ in this case.

The way how the remote secondary load σ_y^∞ influences the critical value of remote tension σ_{cx}^∞ depends on the value of the ratio $k/m = \frac{1}{2}\frac{2+\alpha-\beta}{1+\alpha-2\beta}$, defined in terms of the Dundurs elastic bimaterial parameters α and β. In particular, for $\gamma \ll 1$, a remote secondary compression σ_y^∞ decreases or increases σ_{cx}^∞ if $k/m < 1$ or $k/m > 1$, respectively. This result, for $k/m < 1$, is coherent with the hypothesis proposed and experimentally verified by París et al. (2003) for a particular carbon/epoxy composite.

For moderate or larger values of γ, $\gamma \gtrsim 1$, this model predicts an almost negligible influence of the secondary compression σ_y^∞ on the critical tension σ_{cx}^∞ for the glass/epoxy bimaterial studied, having a slightly flat maximum for $\eta \lesssim 0$ (see Figure 4.42). This observation is related to the fact that the critical semidebond angles θ_c predicted for these values of γ are sufficiently large to make the influence of the secondary compression σ_y^∞ more complex. However, the latter conclusions should be accepted with a caution in view of the range of model applicability, which appears to be very suitable for brittle configurations but to a lesser extent for tough ones.

In addition to the inclusion-matrix debond onset mechanism studied in the present work, other failure mechanisms can occur in the inclusion-matrix system under remote transverse loads. This is, for example, the case of the dominating compressive load studied by the coupled stress and energy criterion in Quesada et al. (2009), where parallel cracks in the inclusion and the matrix are predicted. Another example of inclusion-matrix debond configurations not allowed by the present assumption of the debond symmetrically situated with respect to the principal directions of the remote load were studied in Correa et al. (2008). Experimental tests of specimens subjected to remote transverse compressions show debonds originating at interface positions with large shear stresses. Thus, in order to complete the picture of failure envelopes shown in Figure 4.42, such configurations should be studied in a similar way as done in the present work. Other model was proposed by Carraro and Quaresimin (2014) evaluating the influence of a out-of-plane shear stress on the main secondary transverse load. A full theoretical model could be generated if all of these partial models are combined.

Appendices to Chapter 4

4.A   Dimensional analysis of the energy release rate of a debond at the fiber-matrix interface

The value of the ERR of a debond at a loaded specimen is a functional depending on the stress tensor σ_{ij} and the displacement vector u_i in the whole specimen,

$$G = f(\sigma_{ij}(x, y), u_i(x, y)). \tag{4.111}$$

The solution for stresses and displacements depends on the problem geometry, which includes the debond angle, the elastic properties of the material and boundary conditions. Assuming that the length of the contact zone is independent of the remote tension $\sigma^\infty > 0$, since this is a frictionless receding contact problem, σ_{ij} can be rewritten as the product of σ^∞ and a function $\hat{\sigma}_{ij}$ depending on the point (x, y) and the bimaterial elastic properties (E_1, ν_1, E_2, ν_2)

$$\sigma_{ij} = \sigma^\infty \hat{\sigma}_{ij}(x, y, E_1, \nu_1, E_2, \nu_2, a), \tag{4.112}$$

and analogously

$$u_i = \sigma^\infty \hat{u}_i(x, y, E_1, \nu_1, E_2, \nu_2, a). \tag{4.113}$$

It is known from the LEFM theory that the ERR value is a homogeneous and linear function of a product of displacements and stresses,

$$G = (\sigma^\infty)^2 h(\hat{\sigma}_{ij}, \hat{u}_i) = (\sigma^\infty)^2 h(\theta_d, a, n, E_1, \nu_1, E_2, \nu_2), \tag{4.114}$$

where the terms x, y, i and j have been omitted since the ERR value cannot depend on them. Then, in order to obtain the dimensionless parameters which govern the value of G, two independent dimensional parameters to normalize this expression should be selected since this is a static mechanical problem, see Barenblatt (1996) for details. Taking the fiber radius a and E^* defined in (4.9) and taking into account that the elastic parameters can be reduced to two under the conditions discussed by Dundurs (1967), the value of G can be expressed as

$$G = \frac{(\sigma^\infty)^2 a}{E^*} \hat{G}(\theta_d, n, \alpha, \beta). \tag{4.115}$$

135

4.B   Domain of existence of the angle θ_0 where the normal traction vanishes

A threshold parameter η_0 can be defined as

$$\eta_0 \left(\alpha, \beta \right) = \left(\frac{k}{m} \right)^{-1} - 1. \tag{4.116}$$

It will be useful to know the range of this parameter given in terms of the values of k/m,

$$\eta_0 \in \underbrace{(-1,0]}_{\frac{k}{m} \geq 1} \cup \underbrace{(0,1/3]}_{3/4 \leq \frac{k}{m} < 1} . \tag{4.117}$$

Then, the expression on the right-hand side of (4.60) makes sense for $\eta < 1$ only if

$$\begin{cases} \eta \leq \eta_0 & \text{if } \frac{3}{4} \leq \frac{k}{m} \leq 1, \\ \frac{1}{\eta_0} \leq \eta \leq \eta_0 & \text{if } \frac{k}{m} > 1. \end{cases} \tag{4.118}$$

As can be seen in Figure 4.49, only tensions (compressions) take place along the whole interface for $\eta > \eta_0$ (for $\eta < 1/\eta_0$ and $\frac{k}{m} > 1$).

It will be convenient to extend the definition of θ_0 defining $\tilde{\theta}_0$ as follows

$$\tilde{\theta}_0 \left(\eta; \alpha, \beta \right) = \begin{cases} \theta_0, & \exists \theta_0, \text{ see } (4.118), \\ 90°, & \eta_0 \leq \eta \leq 1, \\ \nexists, & \eta < 1/\eta_0 \text{ and } k/m > 1. \end{cases} \tag{4.119}$$

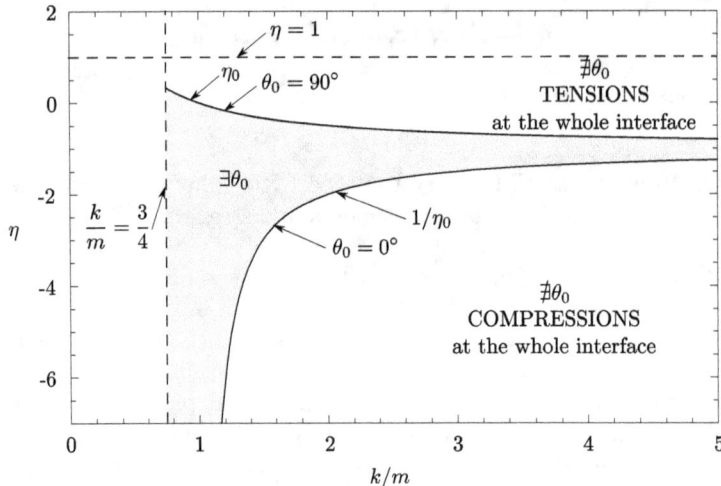

Figure 4.49: Domain of definition of an angle $\theta_0 \in [0°, 90°]$ along the interface where the normal tractions vanish as a function of the elastic bimaterial parameter k/m and the biaxiality parameter η for $\sigma_x^\infty > 0$.

4.C Functions χ and p of the expression of Toya for stresses along the interface with a partial debond

$$\chi(\theta, \theta_\mathrm{d}, \beta) = (e^{i\theta} - e^{i\theta_\mathrm{d}})^{-(1/2)-i\varepsilon}(e^{i\theta} - e^{-i\theta_\mathrm{d}})^{-1/2+i\varepsilon}, \tag{4.120}$$

$$p(\theta, \theta_\mathrm{d}, \eta, \beta) = q(\theta_\mathrm{d}, \eta, \beta)(e^{i\theta} - (\cos\theta_\mathrm{d} - 2\varepsilon\sin\theta_\mathrm{d})) - \frac{1+\alpha}{1-\alpha}(1-\eta)e^{2\varepsilon(\pi-\theta_\mathrm{d})-i\theta} \times$$

$$\times (\cos\theta_\mathrm{d} + 2\varepsilon\sin\theta_\mathrm{d} - e^{-i\theta}), \tag{4.121}$$

with

$$q(\theta_\mathrm{d}, \eta, \beta) = \left[(1+\eta)(1 - (\cos\theta_\mathrm{d} - 2\varepsilon\sin\theta_\mathrm{d})e^{2\varepsilon(\theta_\mathrm{d}-\pi)}) + \right.$$

$$\left. + \frac{1}{2}(1+\alpha)(1+4\varepsilon^2)(1-\eta)\sin^2\theta_\mathrm{d}\right] / \tag{4.122}$$

$$/ \left[3+\alpha - (1-\alpha)(\cos\theta_\mathrm{d} - 2\varepsilon\sin\theta_\mathrm{d})e^{2\varepsilon(\theta_\mathrm{d}-\pi)}\right] - \frac{1+\eta}{1-\alpha}.$$

where $i = \sqrt{-1}$ is the imaginary unit, and

$$\varepsilon = (1/2\pi)\ln(1-\beta)/(1+\beta) \tag{4.123}$$

is the so-called oscillation index (see Table 4.3).

4.D Expression of the dimensionless ERR $\hat{G}$

$$\hat{G}(\theta_\mathrm{d}, \eta, \alpha, \beta) = \frac{\pi}{4}e^{-2\varepsilon(\pi-\theta_\mathrm{d})}(\alpha-1)^2(1+4\varepsilon^2) \times$$

$$\times \|c(\theta_\mathrm{d}, \eta, \alpha, \beta) + d(\theta_\mathrm{d}, \eta, \alpha, \beta)\|^2 \cdot \sin\theta_\mathrm{d}, \tag{4.124a}$$

with

$$c(\theta_\mathrm{d}, \eta, \alpha, \beta) = \left[-0.5(1+\alpha)(1+4\varepsilon^2)(\eta-1)\sin(\theta_\mathrm{d})^2 + \right.$$

$$\left. + (1+\eta)\left(1 + e^{-2\varepsilon(\pi-\theta_\mathrm{d})}(2\varepsilon\sin(\theta_\mathrm{d}) - \cos(\theta_\mathrm{d}))\right)\right] / \tag{4.124b}$$

$$/ \left[3+\alpha + e^{-2\varepsilon(\pi-\theta_\mathrm{d})}(\alpha-1)(\cos(\theta_\mathrm{d}) - 2\varepsilon\sin(\theta_\mathrm{d}))\right],$$

and

$$d(\theta_\mathrm{d}, \eta, \alpha, \beta) = \frac{1 - e^{2\varepsilon(\pi-\theta_\mathrm{d})-i\theta_\mathrm{d}}(1+\alpha)(\eta-1) + \eta}{\alpha-1}. \tag{4.124c}$$

4.E Asymptotic analysis of $\hat{G}$ and G for a vanishing critical semiangle $\Delta\theta_\mathrm{c}$

For a vanishing crack semiangle, the derivatives of $\hat{G}$ and G with respect to the crack semiangle and semilength respectively are,

$$\left.\frac{\mathrm{d}\hat{G}}{\mathrm{d}\theta_\mathrm{d}}\right|_{\theta_\mathrm{d}=0} = (k + (k-m)\eta)^2 \frac{\pi(1+4\varepsilon^2)}{\cosh^2(\pi\varepsilon)}, \tag{4.125}$$

$$\left. \frac{\mathrm{d}G}{\mathrm{d}(a\theta_\mathrm{d})} \right|_{\theta_\mathrm{d}=0} = [(k + (k - m)\eta)\, \sigma_x^\infty]^2 \, \frac{\pi \left(1 + 4\varepsilon^2\right)}{E^* \cosh^2\left(\pi\varepsilon\right)}. \tag{4.126}$$

This expressions agree with the analogous derivative of ERR seen in Rice (1988) for a crack at a straight interface. It is instructive to notice that the factor $(k + (k - m)\eta)$ agrees with the concentration factor of normal tractions $\sigma(\theta)$ obtained from Goodier's solution at $\theta = 0°$, see (4.57a) and (4.59).

On the other hand, the function g in (4.80) can be approximated for small values of $\Delta\theta$. As $\psi(\theta_\mathrm{d})$ is small for small θ_d, thus $\tan^2(1 - \lambda)\psi$ is negligible with respect to the unity and $G_\mathrm{c}(\psi) \gtrsim G_\mathrm{1c}$. Then, using (4.125), for small $\Delta\theta$,

$$g(\Delta\theta, \eta; \alpha, \beta; \lambda, \theta_l) \gtrsim \tilde{g}(\Delta\theta; \eta; \alpha, \beta) = \frac{2}{\hat{G}'(0°; \eta; \alpha, \beta) \cdot \Delta\theta} =$$
$$= \frac{\cosh^2\left(\pi\varepsilon\right)}{\pi \left(1 + 4\varepsilon^2\right) (k + (k - m)\eta)^2} \frac{2}{\Delta\theta}, \tag{4.127}$$

where the asymptotic approximation function $\tilde{g}$ is smaller than the exact function g, see Figure 4.39.

Then, combining the approximation of g in (4.127) and (4.89)–(4.71), θ_c can be approximated by

$$\theta_\mathrm{c} \cong \frac{2(k + (k - m)\eta)^2}{\hat{G}'(0)}\gamma^2 = \frac{2\cosh^2(\pi\varepsilon)}{\pi(1 + 4\varepsilon^2)}\gamma^2 \quad \text{for} \quad \gamma \to 0^+. \tag{4.128}$$

Note that, the asymptotic solution is independent of the load biaxiality parameter η. Therefore, for small values of γ, the effect of the remote secondary load σ_y^∞ on the semiangle of the crack originated is negligible.

4.F   Partitioning of $\hat{G}$ into two components $\hat{G}_\mathrm{I}$ and $\hat{G}_\mathrm{II}$

ERR based measure of fracture mode mixity, angle ψ_G, is defined in terms of G_I and G_II as follows

$$\tan^2\psi_G = \frac{G_\mathrm{II}}{G_\mathrm{I}} = \frac{\hat{G}_\mathrm{II}}{\hat{G}_\mathrm{I}}. \tag{4.129}$$

According to Mantič and París (2004), the ERR and the SIF based measures of mode mixity, phase angles ψ and ψ_G, respectively, can be related by the following expression:

$$\cos(2\psi_G) = F(\varepsilon) \cos\left[2(\psi + \psi_0(\delta\theta/\theta_l, \varepsilon))\right], \tag{4.130}$$

where

$$F(\varepsilon) = \sqrt{\frac{\sinh(2\pi\varepsilon)}{2\pi\varepsilon(1 + 4\varepsilon^2)}}, \tag{4.131}$$

and

$$2\psi_0(\delta\theta/\theta_l, \varepsilon) = 2\varepsilon \ln(\delta\theta/2\theta_l) + \varphi(\varepsilon) - \arctan(2\varepsilon) \tag{4.132}$$

with

$$\varphi(\varepsilon) = \arg\left[\frac{\Gamma(1/2 + i\varepsilon)}{\Gamma(1 + i\varepsilon)}\right], \tag{4.133}$$

$\Gamma(\cdot)$ being the gamma function.

Combining (4.129), (4.130) and some trigonometric identities, the value of $\hat{G}_I(\theta_d, \delta\theta)$ and $\hat{G}_{II}(\theta_d, \delta\theta)$ can be expressed in terms of $\psi(\theta_l)$ as,

$$\hat{G}_{I,II}(\theta_d, \delta\theta) = \frac{1}{2}\hat{G}(\theta_d)(1 \pm F(\varepsilon)\cos(2(\psi(\theta_d, \theta_l) + \psi_0(\delta\theta/\theta_l, \varepsilon)))). \tag{4.134}$$

4.G   Expressions for computational implementation

Combining the upper limit given by θ_0 in (4.74), those for the existence of θ_0 in (4.118) with the stress condition in (4.71), and the condition of minimal load (4.72), a maximum semiangle of a debond for a given load according to the stress criterion is defined as

$$\theta_c^\sigma\left(\frac{\sigma_x^\infty}{\sigma_c}, \eta; \alpha, \beta\right) = \begin{cases} \nexists, & (\eta \leq 1/\eta_0 \text{ and } k/m > 1) \\ & \text{or } \frac{\sigma_x^\infty}{\sigma_c} < \frac{1}{k+(k-m)\eta}, \\ 180°, & \eta > \eta_0 \text{ and } \frac{\sigma_x^\infty}{\sigma_c} \geq \frac{1}{k(1+\eta)-m}, \\ \arcsin\sqrt{\frac{k+(k-m)\eta - \frac{\sigma_c}{\sigma_x^\infty}}{(1-\eta)m}}, & \text{in other cases.} \end{cases} \tag{4.135}$$

In the same sense, the dimensionless function $s(\theta)$ can be generalized for $\theta \in [0°, 180°]$ as

$$s(\theta, \eta; \alpha, \beta) = \begin{cases} \frac{1}{k+(k-m)\eta-(1-\eta)m\sin^2\theta}, & \exists\tilde{\theta}_0 \text{ and } \theta < \tilde{\theta}_0, \\ +\infty, & \exists\tilde{\theta}_0 \text{ and } \theta \geq \tilde{\theta}_0, \\ \frac{1}{k(1+\eta)-m}, & \eta > \eta_0 \text{ and } \theta \geq 90°, \\ +\infty, & \eta \leq 1/\eta_0 \text{ and } k/m > 1. \end{cases} \tag{4.136}$$

Then, the present stress criterion can be written in a general form as follows: an interface debond onset of an angle $2\Delta\theta$ (symmetrical with respect to the x-axis) is possible if

$$\frac{\sigma_x^\infty}{\sigma_c} \geq s(\Delta\theta, \eta), \tag{4.137}$$

θ_c^σ representing the upper limit for the semiangles $\Delta\theta$ verifying (4.137) for a given remote load.

4.H   Proof of inequality $\frac{dG}{d\theta_d} \leq \frac{dG_c}{d\theta_d}$ for $\theta_d = \Delta\theta_c^E = \Delta\theta_{min}^E$

For the sake of simplicity, the following notation will be used

$$G(\theta_d) = G(\theta_d; \sigma_x^\infty, \eta; a; E^*, \alpha, \beta), \tag{4.138a}$$

$$G_c(\theta_d) = G_c(\psi(\theta_d, \eta), G_{1c}, \lambda). \tag{4.138b}$$

Recall that $\Delta\theta_c^E$ is defined as the minimum positive angle for which the equality in (4.78), or equivalently in (4.79), hold. Then, if $\Delta\theta_c^E = \Delta\theta_{min}^E$,

$$\int_0^{\Delta\theta} G(\theta_d)d\theta_d < \int_0^{\Delta\theta} G_c(\theta_d)d\theta_d \quad \text{for} \quad 0 < \Delta\theta < \Delta\theta_{min}^E, \tag{4.139}$$

and

$$\int_0^{\Delta\theta_{min}^E} G(\theta_d)d\theta_d = \int_0^{\Delta\theta_{min}^E} G_c(\theta_d)d\theta_d. \tag{4.140}$$

By subtracting (4.139) from (4.140),

$$\int_{\Delta\theta}^{\Delta\theta_{min}^E} G(\theta_d)d\theta_d > \int_{\Delta\theta}^{\Delta\theta_{min}^E} G_c(\theta_d)d\theta_d \quad \text{for} \quad 0 < \Delta\theta < \Delta\theta_{min}^E. \tag{4.141}$$

Let $G(\theta_d)$ and $G_c(\theta_d)$ be approximated by Taylor polynomials centered at $\theta_d = \Delta\theta_{min}^E$,

$$G(\theta_d) \approx G(\Delta\theta_{min}^E) + \frac{dG}{d\theta_d}\bigg|_{\theta_d=\Delta\theta_{min}^E} (\theta_d - \Delta\theta_{min}^E) + O\left((\theta_d - \Delta\theta_{min}^E)^2\right),$$

$$\tag{4.142a}$$

$$G_c(\theta_d) \approx G_c(\Delta\theta_{min}^E) + \frac{dG_c}{d\theta_d}\bigg|_{\theta_d=\Delta\theta_{min}^E} (\theta_d - \Delta\theta_{min}^E) + O\left((\theta_d - \Delta\theta_{min}^E)^2\right).$$

$$\tag{4.142b}$$

Then, introducing (4.142) in (4.141) and integrating the Taylor polynomials, the following inequality is obtained, denoting $h = \Delta\theta_{min}^E - \Delta\theta$:

$$G(\Delta\theta_{min}^E)h - \frac{dG}{d\theta_d}\bigg|_{\theta_d=\Delta\theta_{min}^E} \frac{h^2}{2} + O\left(h^3\right) > G_c(\Delta\theta_{min}^E)h - \frac{dG_c}{d\theta_d}\bigg|_{\theta_d=\Delta\theta_{min}^E} \frac{h^2}{2} + O\left(h^3\right).$$

$$\tag{4.143}$$

for $0 < h < \Delta\theta_{min}^E$. Hence, taking into account that $G(\Delta\theta_{min}^E) = G_c(\Delta\theta_{min}^E)$ from (4.82) and considering (4.143) for vanishing $h > 0$,

$$\frac{dG}{d\theta_d}\bigg|_{\theta_d=\Delta\theta_{min}^E} \leq \frac{dG_c}{d\theta_d}\bigg|_{\theta_d=\Delta\theta_{min}^E}, \tag{4.144}$$

otherwise (4.143) could not hold for a sufficiently small $h > 0$.

CHAPTER 5

Failure initiation at spherical-particle reinforced composites

Particle-reinforced composites are extensively used in a variety of industrial applications. Their main advantages over the fiber-reinforced composites are cost-effectiveness and manufacturing flexibility since they can be machined with the majority of traditional manufacturing processes designed for metals. In particular, spherical particles are usually employed as a reinforcement to enhance tensile strength, stiffness, tear and abrasion resistances of polymeric matrices, see e.g. Fu et al. (2008); Shen et al. (2002), or to increase fracture toughness of ceramic matrices, see Dlouhy and Boccaccini (1996).

The manner in which the reinforcement modifies the failure properties of the unreinforced matrix is closely related to the reinforcement geometry, the contrast in elastic properties between reinforcement and matrix, and the mechanical behavior of the interface between them. The dependence between the macroscopic properties of the composite and these parameters have been observed in several experiments, see Fu et al. (2008) for a review. In particular, experiments show that the particle size affects significantly the tensile strength. In the specific case of reinforcements at a micro scale, composites with smaller particles present a higher tensile strength, see Gent and Park (1984); Yoshinobu et al. (1992); Cho et al. (2006). This is due to the influence of the microstructure on the first stages of the failure mechanisms. In fact, microscope observations by Cho et al. (2006) show that failure initiates as debonds at the particle-matrix interface and subsequently grow along the interface up to an angle for which the debonds kink out of the interface to coalesce with others.

The importance of these composites and the observed influence of the reinforcement and interface properties on the strength have encouraged a wide variety

141

of analysis at the reinforcement scale which is usually micro or nano. The problem of a spherical particle embedded in a matrix has been intensively studied since the seminal works by Goodier (1933) and Eshelby (1957) who introduced solutions and formalisms for this geometry assuming isotropic elastic materials. The solution of the basic problem of a spherical particle embedded in an infinite matrix, assuming a perfect particle-matrix interface, subjected to a uniaxial tension is fundamental in understanding micromechanics of particle-reinforced composites. When the spherical particle is stiffer than the matrix, this solution shows a stress concentration at the poles of the sphere axis parallel to the tension direction. As a consequence, these are *a priori* preferred points for a failure initiation in the form of a debond at (or a void near) the interface. A broad variety of works have been presented dealing with the interface debond onset and growth by different approaches, e.g. stress analysis assuming an elastoplastic matrix, see e.g. Gent (1980); Tszeng (2000), cohesive zone models for the interface, see e.g. Needleman (1987); Tan et al. (2007); Lee and Pyo (2007, 2008); Othmani et al. (2011) or using an energy criterion assuming a material-dependent critical debond angle by Tszeng (1993).

This chapter aims to develop a model to predict the debond onset at the particle-matrix interface as a function of the particle, matrix and interface properties employing the coupled criterion of the finite fracture mechanics (FFM) described in Chapter 3. Some works applying the coupled criterion dealing with similar problems to the present one have been presented for long fiber inclusions, see Chapter 4 and Carraro and Quaresimin (2014), for long fiber inclusions, and in Zappalorto et al. (2011), for nano-spherical inclusions subjected to hydrostatic-tension loading.

The chapter is organized as follows: first, the theoretical model is developed in Section 5.1 in accordance with the formulation proposed in Chapter 3, i.e. initially the conditions imposed by the stress and energy criteria are separately obtained in Sections 5.1.1 and 5.1.2 respectively, being combined in Section 5.1.3. The model is particularized for glass/vinylester employed by Cho et al. (2006) and a preliminary comparison with their experiments is carried out in Section 5.2.

5.1 A theoretical model based on the finite fracture mechanics

Consider a piece of a composite reinforced by spherical particles subjected to a remote uniaxial tension $\sigma^\infty > 0$ as shown in Figure 5.1(a). Under the assumption of dilute packing (with a low reinforcement volumetric fraction), the present work focuses on the debonding process at the interface of a single particle by neglecting the influence of the neighboring ones. Thus, the geometry under study is simplified to that shown in Figure 5.1(b): a spherical particle of radius a perfectly bonded to an infinite surrounding matrix. In the context of the present model, the spherical particles are assumed to be stiffer than the surrounding matrix. Both particles and matrix are assumed to be isotropic and linear elastic. For the sake of illustration of the model, a glass/vinylester com-

posite used in experiments in Cho et al. (2006) is taken as an example here, see
material properties in Table 5.1. As will be justified in Section 5.1.1, preferred
points for debonding are situated at the poles of the sphere axis parallel to the
load direction. In view of this, for a certain critical value of the remote load σ_c^∞,
two debonding modes will be studied in the following: the onset of either a single
debond at one of the poles or two symmetric debonds at both poles, see Figure
5.1(c). As both geometries and loads are axisymmetric, the problem is solved as
an axisymmetric one, see Graciani et al. (2005), measuring all the angles from
the symmetry axis.

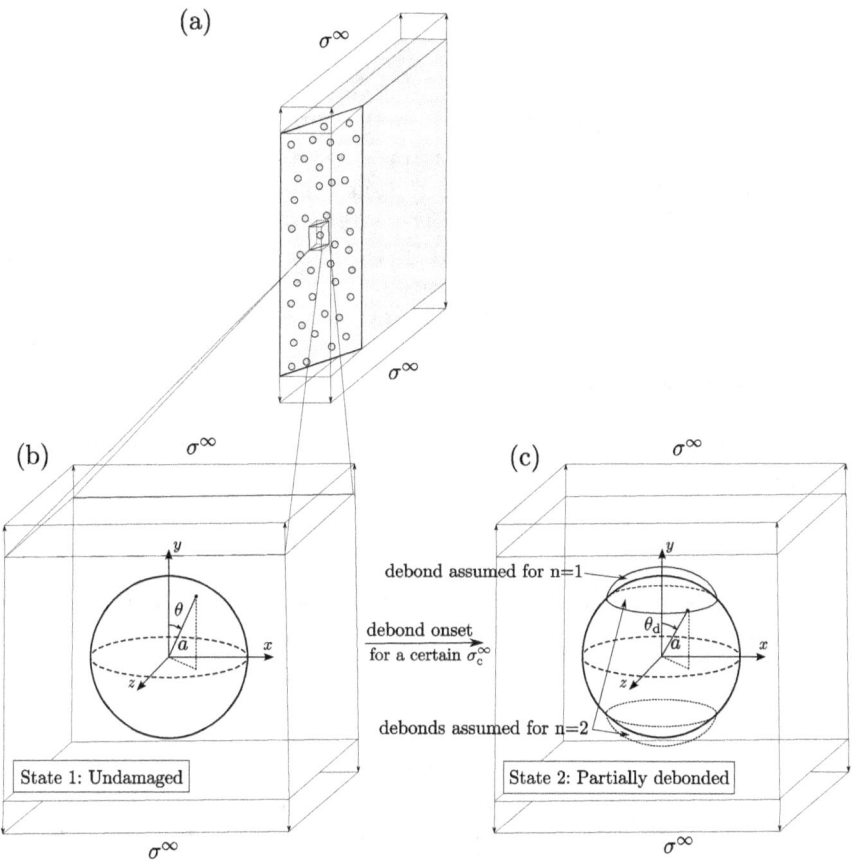

Figure 5.1: Schematic of the models employed to study the process of debond initiation at
microparticles-reinforced composites under the assumptions of the finite fracture mechanics.
(a) Particle-reinforced composite specimen subjected to tension, Spherical particle (b) perfectly
bonded to the surrounding matrix or (c) partially debonded at the interface.

Following the FFM approach Leguillon (2002); Hashin (1996); Cornetti et al.
(2006), next subsections will be devoted to obtain the independent conditions
imposed by the stress and energy criteria for the debond onset. Finally, and

Material	Young's modulus E_i (GPa)	Poisson's ratio ν_i
Glass ($i = 1$)	70	0.25
Vinylester ($i = 2$)	3.5	0.35

Table 5.1: Elastic properties of materials used in the numerical study.

according to Leguillon's hypothesis Leguillon (2002), the critical value for the remote load leading to the onset σ_c^∞ is obtained as the minimum value of σ^∞ fulfilling both criteria.

5.1.1 Stress criterion

In general, the stress criterion (sometimes referred to as strength criterion) imposes that stresses prior to the debond onset along the path assumed for it have to exceed a certain critical value. Several forms of such a stress criterion have been proposed to apply this condition. First tensile criteria considered only normal stresses pointwise Leguillon (2002); Mantič (2009) or in average sense Cornetti et al. (2006, 2012); Camanho et al. (2012); Weissgraeber and Becker (2013), see Cornetti et al. (2012) for a comparison. Subsequently and in order to take into account the influence of the shear stresses, several criteria combining normal and shear stresses were proposed Carraro and Quaresimin (2014); Mantič (2009). In view of some experimental evidences Ogihara and Koyanagi (2010); Koyanagi et al. (2012) showing a high influence of τ in similar problems, a combined stress criteria will be applied, see Section 3.3 for a review. First, an equivalent stress is defined as

$$\sigma_{eq}(\sigma, \tau) = \sqrt[p]{\left\langle \text{sgn}(\sigma) |\sigma|^p + \left(\frac{|\tau|}{\mu}\right)^p \right\rangle_+}, \tag{5.1}$$

where σ and τ are the normal and shear component of the interface stresses, respectively, and

$$\mu = \frac{\tau_c}{\sigma_c}, \tag{5.2}$$

with σ_c and τ_c being the interface tensile and shear strength, $p > 0$, and $\langle \cdot \rangle_+$ denotes the positive part of a real number. A similar proposal of an equivalent stress with $p = 2$ can be found in Lemaitre and Desmorat (2005) (Section 7.7.2). For the sake of illustration, $p = 2$ and $\mu = 1$ are chosen in the present work.

According to this criterion, a debond can appear at those points of the interface where

$$\sigma_{eq}(\sigma, \tau) \geq \sigma_c. \tag{5.3}$$

In view of this condition, it is necessary to know $\sigma(\theta)$ at the interface before the debond onset. For the sake of comparison, the stresses at the interface are obtained by two methods: i) employing the analytical solution derived by Goodier (1933) for an infinite matrix and ii) by computational modeling based on the BEM code used in the next section to solve the problem with a debond, for which no analytical solution is available. This BEM code was developed by

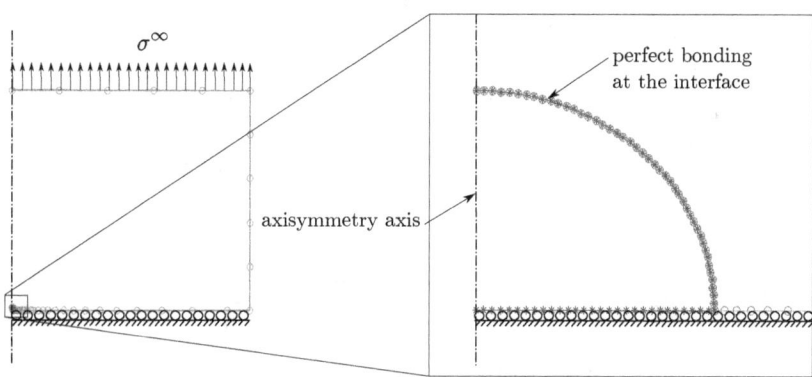

Figure 5.2: A BEM axisymmetric model used to compute the interface stresses at the perfectly bonded inclusion-matrix interface.

Graciani et al. (2005) and is used here since it enables the study, with a high accuracy, of axisymmetric problems of interface cracks with a possible contact zone at the crack tip, features necessary in next section.

The BEM model with linear elements used here is shown in Figure 5.2. Note that a half of the geometry is modeled due to the symmetry. The matrix is represented by a square with side 133 times larger than the inclusion radius a. The boundary element mesh is uniform at the interface and the outer matrix boundary. At the lower horizontal edge, where the symmetry conditions are applied, the mesh is not uniform, the element length decreasing when approaching the interface following a geometric series with ratio 1.2. The extreme length of this edge matches the element length at the interface at one extreme and the element length at the outer matrix boundary at the other extreme. A uniform remote tension σ^∞ is applied at the upper horizontal edge. Finally, the whole interface is defined as perfectly adhered.

The analytic solution of this problem, assuming an infinite matrix subjected to the remote tension σ^∞ and defining the angle θ measured from the axisymmetry axis, can be written as,

$$\sigma(\theta) = \sigma^\infty \hat{\sigma}(\theta) = \sigma^\infty \left(k + m \cos 2\theta\right), \tag{5.4a}$$

$$\tau(\theta) = \sigma^\infty \hat{\tau}(\theta) = \sigma^\infty m \sin 2\theta \tag{5.4b}$$

was obtained in Goodier (1933), from where the following explicit expressions of the dimensionless parameters $k(E_1, \nu_1, E_2, \nu_2)$ and $m(E_1, \nu_1, E_2, \nu_2)$ can be derived,

$$k = \frac{(1 - \nu_2)(3k_1\left(\nu_1, \nu_2\right) + 2\frac{E_2}{E_1}k_2\left(\nu_1, \nu_2\right))}{2k_3\left(\frac{E_2}{E_1}, \nu_1, \nu_2\right)k_4\left(\frac{E_2}{E_1}, \nu_1, \nu_2\right)}. \tag{5.5a}$$

where k_1, k_2, k_3 and k_4 are:

$$k_1\left(\nu_1, \nu_2\right) = \nu_1(\nu_2 + 1)(2\nu_2 - 1)(5\nu_2 - 7) \tag{5.5b}$$

$$k_2\left(\nu_1,\nu_2\right)=\left(2\nu_1-1\right)\left(\nu_1\left(\nu_2(15\nu_2-7)-7\right)+5\nu_2(2\nu_2-1)\right) \tag{5.5c}$$

$$k_3\left(\frac{E_2}{E_1},\nu_1,\nu_2\right)=\left(\nu_1(\nu_2+1)^2-2\frac{E_2}{E_1}\left(2\nu_1^2+\nu_1-1\right)\nu_2\right) \tag{5.5d}$$

$$k_4\left(\frac{E_2}{E_1},\nu_1,\nu_2\right)=\left(\left(20\nu_2^2-26\nu_2+8\right)+\frac{E_2}{E_1}(2\nu_1-1)(5\nu_2-7)\right) \tag{5.5e}$$

and the function m is,

$$m=\frac{1}{2}\frac{15(\nu_2-1)(2\nu_2-1)}{(20\nu_2^2-26\nu_2+8)+\frac{E_2}{E_1}(2\nu_1-1)(5\nu_2-7)}. \tag{5.6}$$

For the glass/vinylester bimaterial defined in Table 5.1, the values of these functions are: $k=0.80$ and $m=1.03$.

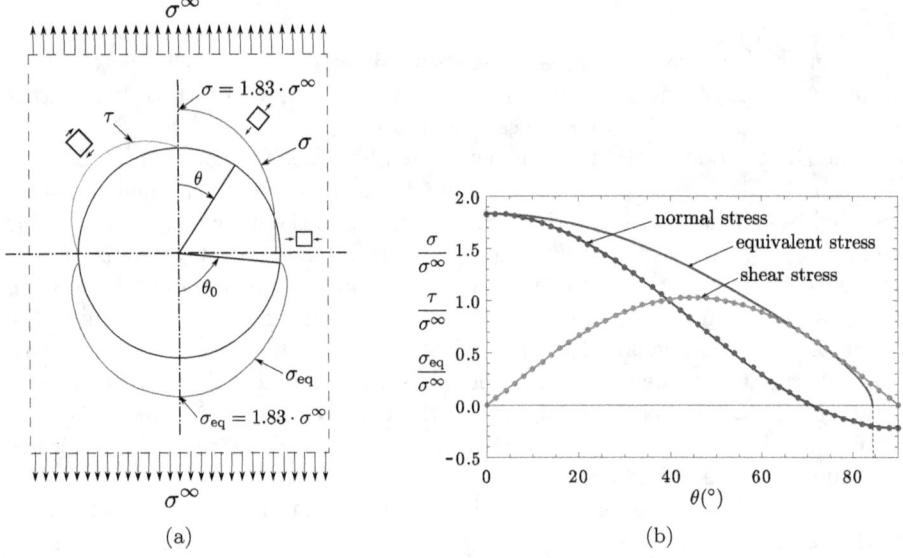

(a)   (b)

Figure 5.3: Normal, shear and equivalent stresses at the interface previous to the crack onset. (a) Polar representation of the analytical solution in schematic of the problem and (b) Interface stresses as a function of the polar angle θ obtained by employing the analytical solution (solid line) and the BEM model shown in Figure 5.2 (points)

Figure 5.3(a) shows polar plots of $\hat{\sigma}(\theta)$, $\hat{\tau}(\theta)$ and the corresponding $\hat{\sigma}_{\text{eq}}(\theta)=\sigma_{\text{eq}}(\hat{\sigma}(\theta),\hat{\tau}(\theta))$ given by the analytical solution. Figure 5.3(b) shows the interface stresses as a function of θ for both the analytical solution (solid line) and the results extracted from the BEM model (points). Observe that the results are very similar, thus, in the following the analytical solution is employed. Taking as reference the distributions of $\hat{\sigma}$ or $\hat{\sigma}_{\text{eq}}$, the most loaded points at the interface are situated at the poles. This justifies the adopted hypothesis about the problem geometry after the debond onset shown in Figure 5.1(c). Figure 5.4 also shows

that σ_{eq} is decreasing when increasing θ until it vanishes at $\theta = \theta_0$. As a consequence, no debond onset can include points with $\theta_0 \leq \theta \leq 90°$. From decreasing
monotonicity of $\hat{\sigma}_{eq}$ it follows that if the criterion (5.3) is satisfied for a certain
debond-angle $\Delta\theta < \theta_0$ at onset, it is also satisfied for all the points $0 \leq \theta \leq \Delta\theta$.
Therefore, the stress condition in the present case simplifies to

$$\frac{\sigma^\infty}{\sigma_c} \geq s(\Delta\theta) = \hat{\sigma}_{eq}^{-1}\left(\hat{\sigma}(\Delta\theta), \hat{\tau}(\Delta\theta)\right). \tag{5.7}$$

Thus, for a debond angle $\Delta\theta$ at onset, $s(\Delta\theta)$ gives a minimum value of the
normalized remote load required. Figure 5.4 shows the stress criterion curve
defined by (5.7) separating the "safe" and "unsafe" zones. According to this
criterion a larger angle of debond at onset requires a higher remote load.

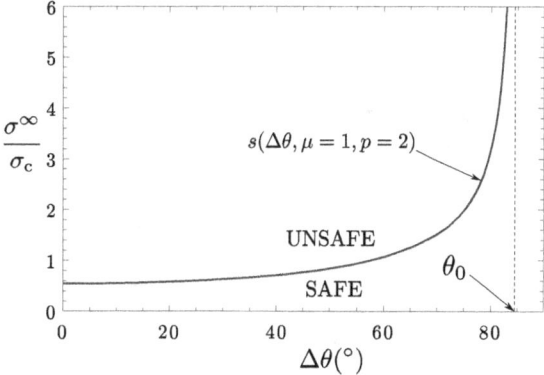

Figure 5.4: Graphical representation of the stress criterion for glass/vinylester, $p = 2$ and
$\mu = 1$. The schema on the left shows the normal, shear and equivalent stresses at the interface.

5.1.2 Energy criterion

Although (5.7) is satisfied for a certain $\Delta\theta$, this may not directly originate a
debond onset since it has to be energetically allowed as well. This is why the
FFM additionally requires that an incremental energy criterion, based on the
energy balance between the states before and after the onset of a debond of
angle $\Delta\theta$, is satisfied

$$\Delta\Pi(\Delta\theta) + \Delta\Gamma(\Delta\theta) + \Delta E_k(\Delta\theta) = 0, \tag{5.8}$$

where $\Delta\Pi$ and ΔE_k are the change in the elastic potential and kinetic energy
at the debond onset, respectively, and $\Delta\Gamma$ is the energy dissipated during the
debond onset. $\Delta E_k \geq 0$ since the initial state is assumed to be quasistatic. $\Delta\Pi$
can be related, using the linear elastic fracture mechanics (LEFM) theory, to the
energy release rate (ERR) by $G = -d\Pi/dA$ where dA is an increment of the

debonded area. Hence,

$$-\Delta\Pi(\Delta\theta) = n \int_0^{\Delta\theta} G(\theta_d, n) 2\pi a^2 \sin(\theta_d) d\theta_d, \tag{5.9}$$

where θ_d is an intermediate variable used only to obtain $\Delta\Pi$ and $n = 1, 2$ is the number of debonds assumed.

A solution for G is required when employing (5.9) to evaluate $\Delta\Pi$. Although some solutions for G in this problem are available in Altenbach et al. (1995); Martin (2001); Martynenko and Lebedyeva (2006), they are based on the open model of interface cracks with a zone of non-physical overlapping of the crack faces near the crack tip, and this zone can be quite relevant for large debond angles. This may lead to errors in computation of G for such angles. To avoid this, the contact model of interface cracks Comninou (1977), see also Mantič et al. (2006) for a review, is used here in the evaluation of G, the corresponding frictionless contact problem being solved by a numerical procedure. In order to reduce the amount of numerical models to compute, a dimensional analysis is carried out for G. It can be proven, see Section 4.D for a proof in the context of a similar problem, that a dimensionless ERR $\hat{G}$ can be defined by

$$G\left(\sigma^\infty, \theta_d, n, a, E_1, E_2, \nu_1, \nu_2\right) = \frac{(\sigma^\infty)^2 a}{E^*} \hat{G}\left(\theta_d, n, E_1/E_2, \nu_1, \nu_2\right), \tag{5.10}$$

where E^* is the harmonic mean of effective elastic moduli,

$$E^* = 2\left(\frac{1-\nu_1^2}{E_1} + \frac{1-\nu_2^2}{E_2}\right)^{-1}. \tag{5.11}$$

The definition in (5.10) enables to generalize a single numerical result for G to any other value of a, σ^∞ and E^*, provided that E_1/E_2, ν_1 and ν_2 are fixed.

Values of $\hat{G}$, plotted in Figure 5.6, are computed by a highly accurate axisymmetric boundary-element-method (BEM) code Graciani et al. (2005), which includes a contact algorithm to take into account a possible contact between the interface crack faces. An axisymmetric model is generated for each combination of values of $\theta_d = 1°, 2°, \cdots, 88°, 89°$ and $n = 1, 2$. In order to compute accurately the singular elastic solution near the interface crack tip, the mesh is refined there. The polar angles of boundary element goes from $0.0001°$ close to the crack tip to $2°$ in a zone far from the tip increasing in geometric progression with a ratio 1.2 between the length of adjacent elements. An example of the numerical model is shown in the schematic on the right in Figure 5.7.

Numerical solutions of the above described nonlinear models are post-processed to calculate $\hat{G} = \hat{G}_I + \hat{G}_{II}$, with $\hat{G}_m$ being the ERR associated to mode m=I and II. The axisymmetric Virtual Crack Closure Technique (VCCT), with a sufficiently small value of $\delta\theta = 0.5°$, is used

$$\hat{G}_m = \frac{E^*}{(\sigma_0^\infty)^2 a} \frac{1}{2\Delta A} \int_0^{\delta\theta_d} \sigma_{r\alpha}(\theta_d+\theta)\Delta u_\alpha(\theta_d-\delta\theta_d+\theta) 2\pi a^2 \sin(\theta_d+\theta) d\theta, \tag{5.12}$$

where $\alpha = r$ and θ for m=I and II, respectively, σ_0^∞ is the reference remote load applied in the BEM model, σ_{rr} and $\sigma_{r\theta}$ are the normal and shear interface stresses, and Δu_r and Δu_θ are the normal and tangential displacement jumps, respectively. These interface stresses and displacement jumps are extracted from the BEM results at the nodes near the crack tip. $\Delta A = \int_0^{\delta\theta_d} 2\pi a^2 \sin(\theta_d + \theta)d\theta$ is the area corresponding to the angle increment $\delta\theta_d$. The integrals in (5.12) are computed by a Chebyshev-Gauss quadrature.

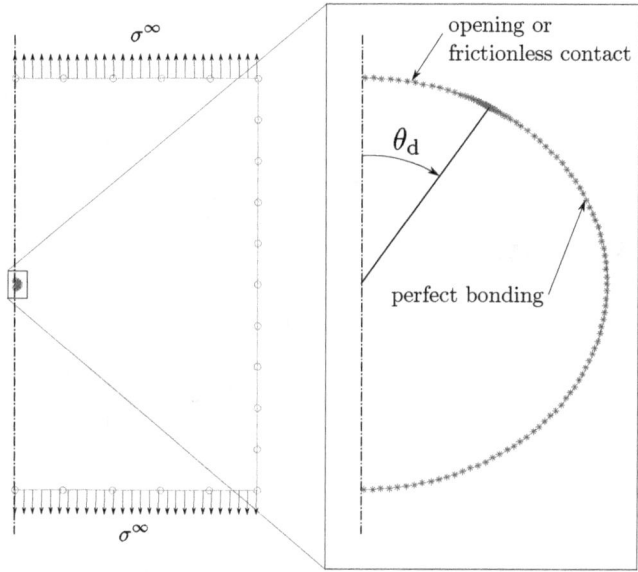

Figure 5.5: Example of a BEM axisymmetric model used to compute the ERR of a crack at the inclusion-matrix interface. The matrix nodes at the interface, which are coincident with the inclusion ones, have been removed from the representation to allow a clear visualization of the refined mesh in the vicinity of the crack tip.

Figure 5.6 shows $\hat{G}$ as a function of θ_d calculated for glass/vinylester for both hypotheses of debond geometry after the debond onset $(n = 1, 2)$. Observe that for both hypotheses the value of $\hat{G}$ increases with θ_d up to reach a maximum and then decreases. In particular, for $n = 2$, $\hat{G}$ vanishes for $\theta_d \to 90°$ since this case corresponds to two debonds approaching each other in mode 2. It is interesting to remark that $\hat{G}$ is larger for the asymmetric case $(n = 1)$ than for the symmetric case $(n = 2)$ for any value of θ_d. The difference between both increases for larger values of θ_d, this is due to the shielding effect between the two debonds for the symmetric case, see Section 4.1 for a detailed discussion in the context of a similar problem.

In Figure 5.6 the numerical results for $\Delta\Pi(\Delta\theta)$ are shown as points and their spline interpolations by lines. Introducing these interpolations in (5.9) and taking into account (5.10), the term $\Delta\Pi(\Delta\theta)$ in (5.8) can be determined.

The fracture mode mixity ψ, given by the angle between the interface stresses and the interface outer normal, computed at a distance of 0.1° ahead of the crack

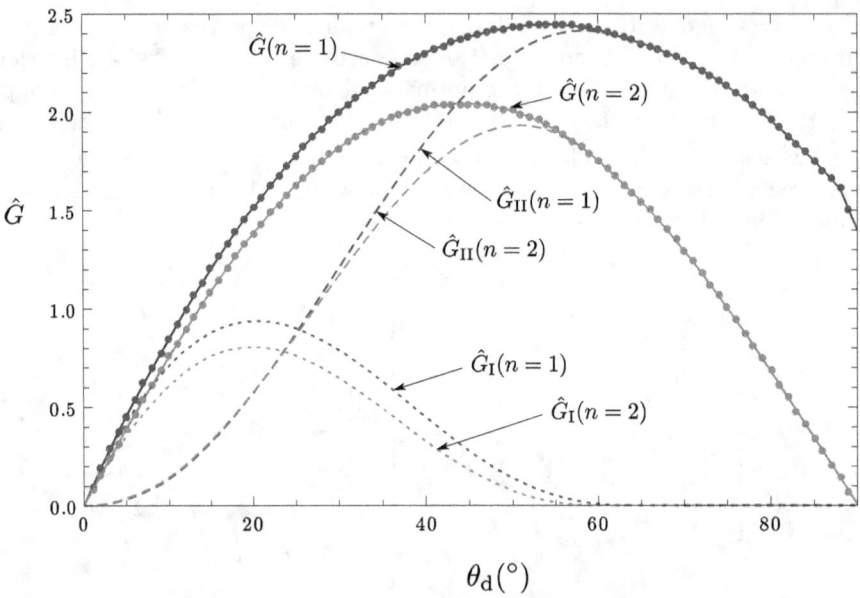

Figure 5.6: Points: dimensionless energy release rate $\hat{G}$ as a function of the debond angle θ_{d} for glass/vinylester and $n = 1, 2$ obtained from the BEM model showed in the schematic in Figure 5.7. Solid, dashed and pointed lines: Spline interpolation of the computational results used for the criterion application.

tip at angle θ_{d} is plotted in Figure 5.7. ψ is very similar for both values of n except for θ_{d} approaching $90°$. Additionally, the angle $\tilde{\psi}$ of the interface stress vector evaluated at θ before the debond onset given by (5.4a), $\tan \tilde{\psi} = \tau/\sigma$, is also plotted in Figure 5.7.

The dissipated energy $\Delta\Gamma(\Delta\theta)$ in (5.8) is estimated analogously to $\Delta\Pi(\Delta\theta)$ by

$$\Delta\Gamma(\Delta\theta) = n \int_0^{\Delta\theta} G_{\mathrm{c}}\left(\phi(\theta), G_{1\mathrm{c}}, G_{2\mathrm{c}}\right) 2\pi a^2 \sin(\theta)\mathrm{d}\theta, \qquad (5.13)$$

with the interface fracture toughness

$$G_{\mathrm{c}}\left(\phi, G_{1\mathrm{c}}, G_{2\mathrm{c}}\right) = G_{1\mathrm{c}}\hat{G}_{\mathrm{c}}\left(\phi, \frac{G_{2\mathrm{c}}}{G_{1\mathrm{c}}}\right), \qquad \phi = \psi \text{ or } \tilde{\psi}, \qquad (5.14)$$

expressed in terms of its values in mode I and II, $G_{1\mathrm{c}}$ ($\phi = 0°$) and $G_{2\mathrm{c}}$ ($\phi = 90°$), and a dimensionless function $\hat{G}_{\mathrm{c}}$ introducing the dependence of the fracture toughness on the mode mixity. According to a discussion in García and Leguillon (2012), $\tilde{\psi}$ is taken as the mode mixity measure in (5.13), see Section 3.4.2 for a detailed discussion on the influence of the fracture mode mixity. Thus, the the dissipated energy per unit fracture area is approximated as Hutchinson and Suo (1992),

$$\hat{G}_{\mathrm{c}}(\phi) = 1 + \tan^2\left[(1 - \lambda)\phi\right], \qquad (5.15)$$

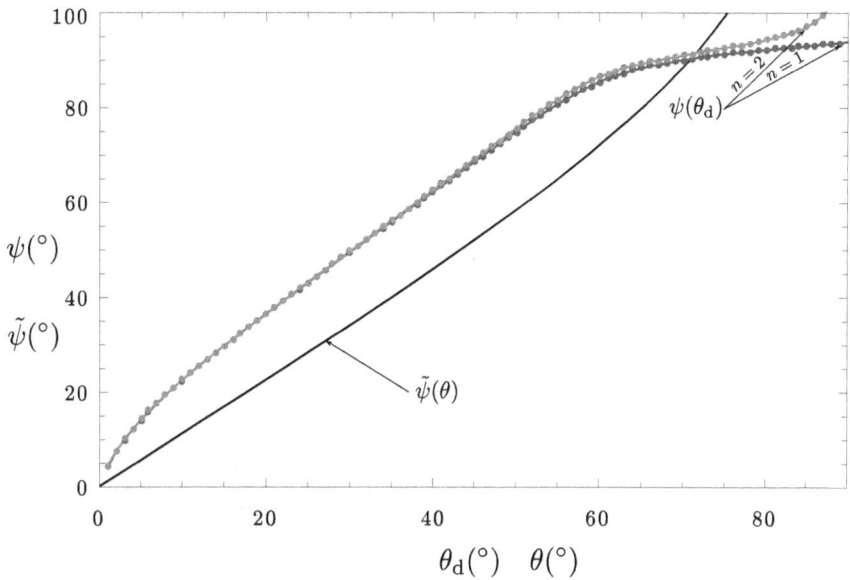

Figure 5.7: Fracture-mode mixity ψ based on stress mode-mixity at a distance $\theta_1 = 0.1°$ from the crack tip for glass/vinylester and $n = 1, 2$. The schematic on the right represents the BEM model used.

where λ ($0 \leq \lambda \leq 1$) is a mode-mixity sensitivity parameter defined by the ratio G_{2c}/G_{1c}.

For the sake of comparison of the present theoretical predictions of debond onset at a spherical particle with some experiments, the results of tests carried out in Cho et al. (2006) will be used in the following. First, the measurements of interface fracture toughness G_c for different debond angles θ_d in Cho et al. (2006) for stable growth are used to approximate $G_c(\psi)$. These measurements of G_c are plotted in Figure 5.8 against ψ by employing the function $\psi(\theta_d)$ extracted from Figure 5.7. Note that, since the experimental observations correspond to the values of a stable growth, the mixity employed here is ψ. Thus, G_{1c} and λ are estimated by fitting the interface fracture toughness law defined by (5.14) and (5.15) to these results by least squares. Figure 5.8 presents the comparison between the experimental results[1] obtained in Cho et al. (2006) and the fitted law with $G_{1c} = 7.29$ N/m and $\lambda = 0.11$.

Once expressions for $\Delta\Pi$ and $\Delta\Gamma$ in (5.8) have been obtained, the energy criterion can be expressed, analogous to that of the stress criterion (5.7), by introducing (5.9) and (5.15) into (5.8), using (5.10) and (5.14), taking into account

---

[1] Actually, results for G_c in Fig. 22 in Cho et al. (2006) are expressed there in "N/m$\times 10^{-3}$" but it is assumed here that this is a typographical error and should be "N/m" to be coherent with units in Figures 18 and 21 in Cho et al. (2006) from which these results are derived.

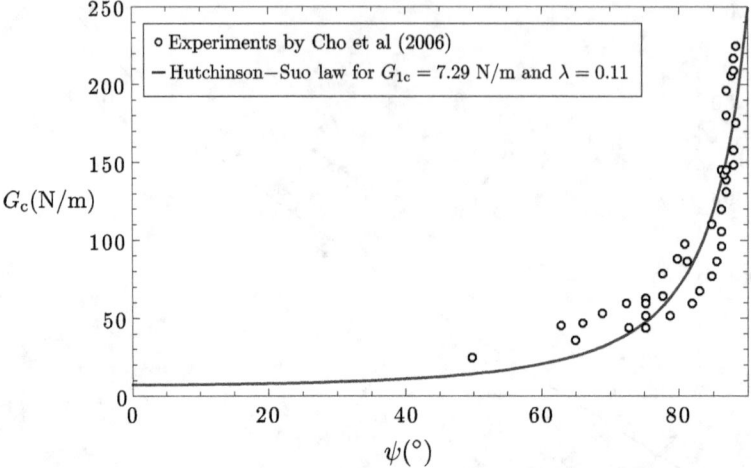

Figure 5.8: Circles: experimental results Cho et al. (2006) for the variation of the interface fracture toughness G_c with the fracture-mode-mixity angle ψ. Solid line: the Hutchinson-Suo law for $G_c(\psi)$ with G_{1c} and λ estimated by least-squares fitting.

that $\Delta E_k \geq 0$ and dividing by σ_c,

$$\frac{\sigma^\infty}{\sigma_c} \geq \gamma\sqrt{g(\Delta\theta, n, E_1/E_2, \nu_1, \nu_2, G_{2c}/G_{1c})}, \tag{5.16}$$

where

$$\gamma = \frac{1}{\sigma_c}\sqrt{\frac{G_{1c}E^*}{a}} \tag{5.17}$$

is the brittleness number defined by Mantič (2009), see Section 3.5.1 and

$$g(\Delta\theta, n, E_1/E_2, \nu_1, \nu_2, G_{1c}/G_{2c}) = \frac{\int_0^{\Delta\theta} \hat{G}_c(\theta)\sin(\theta)d\theta}{\int_0^{\Delta\theta} \hat{G}(\theta_d)\sin(\theta_d)d\theta_d} \tag{5.18}$$

is a dimensionless function characterizing the ratio of dissipated to released energy, see Mantič (2009).

In view of (5.16), the energy criterion requires a minimum remote load depending on the debond angle $\Delta\theta$ along with material and interface properties. Figure 5.9 shows the curves of the energy criterion divided by γ, i.e. $\sqrt{g(\Delta\theta)}$, for $n = 1$ and 2. Similarly to the stress criterion, these curves separate the "unsafe" and "safe" zones. Both curves have a vertical asymptote for $\Delta\theta \to 0^+$ in accordance with the impossibility of the energy criterion to predict onset of an infinitesimal crack at a stress concentration point. In general, these curves are decreasing up to reach a minimum at $\Delta\theta_{min}^E$. For larger values of $\Delta\theta$ the curve is increasing due to the strong influence of the steep part of function $\hat{G}_c(\tilde{\psi}(\theta))$.

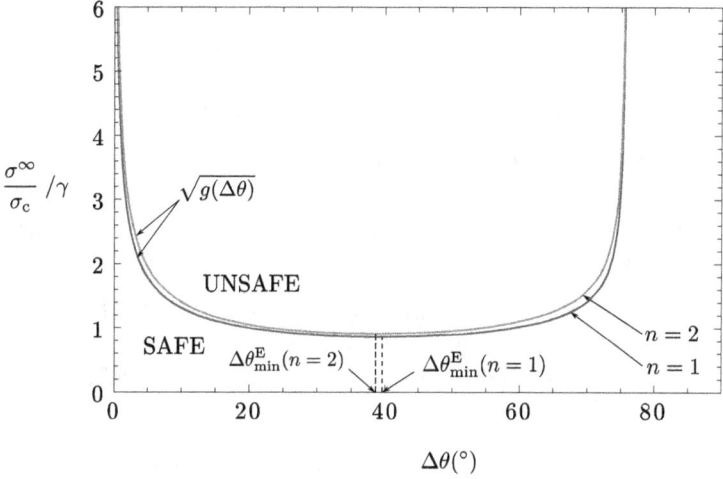

Figure 5.9: Graphical representation of the energy criterion for glass/vinylester, $G_{1c} = 7.29$ N/m, $\lambda = 0.11$, $n = 1$ and 2.

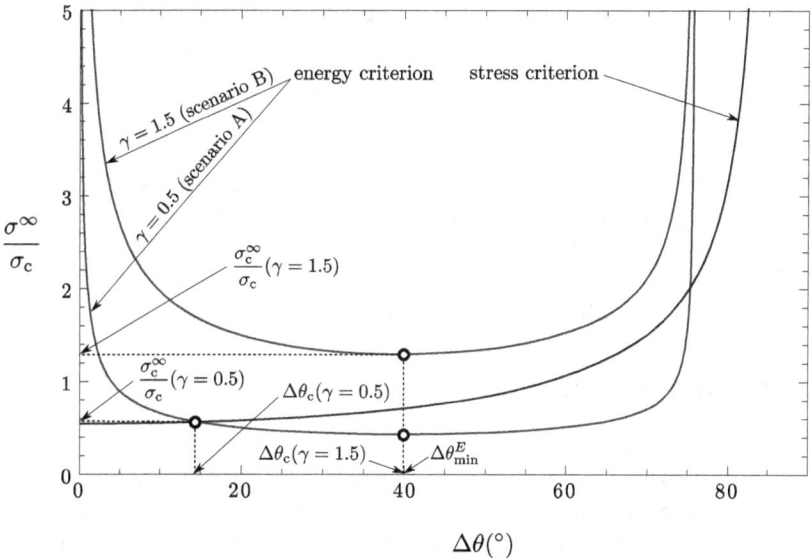

Figure 5.10: Examples on how the two criteria are combined for $\gamma = 0.5$ (scenario A) and $\gamma = 1.5$ (scenario B) taking glass/vinylester, $G_{1c} = 7.29$ N/m, $\lambda = 0.11$ and $n = 1$.

5.1.3   Coupled criterion

According to Leguillon's hypothesis Leguillon (2002), the critical remote load σ_c^∞ is given by the minimum value of remote load satisfying both criteria described above. The combination of these criteria gives the critical remote load σ_c^∞ and the critical angle $\Delta\theta_c$ of debond at onset. To clarify how both criteria are coupled, their expressions, (5.7) and (5.16), are plotted in Figure 5.10 for two values of γ. In view of this figure, two scenarios are possible depending on the value of γ:

- Scenario A: Curves of both criteria intersect at the critical angle of debond onset $\Delta\theta_c \leq \Delta\theta_{min}^E$. In this case the minimum load satisfying both criteria is given by $\Delta\theta_c$ whose value is computed by solving the nonlinear equation obtained by equaling the right-hand sides of (5.7) and (5.16).

- Scenario B: If the condition for scenario A does not hold (i.e. there is no common point of the criteria curves for angles less than $\Delta\sigma_{min}^E$), the minimum σ^∞ satisfying both criteria is found for the critical angle of debond onset $\Delta\theta_c = \Delta\theta_{min}^E$.

For both scenarios, once $\Delta\theta_c$ is obtained, the critical remote load is given by,

$$\frac{\sigma_c^\infty}{\sigma_c} = \gamma\sqrt{g(\Delta\theta_c)}. \tag{5.19}$$

The threshold value for γ for which the curves intersect at $\Delta\theta = \Delta\theta_{min}^E$, denoted γ_{th}, separates both scenarios, A for $\gamma \leq \gamma_{th}$ and B for $\gamma > \gamma_{th}$,

$$\gamma_{th} = \frac{s\left(\Delta\theta_{min}^E\right)}{\sqrt{g\left(\Delta\theta_{min}^E\right)}}. \tag{5.20}$$

As discussed in some previous works, e.g. Mantič (2009), increasing γ value represents brittle (scenario A) to ductile (scenario B) transition in the problem studied.

Stable or unstable debond growth after its onset can be studied by the classical LEFM using the above expressions of G and G_c.

5.2   Results and comparison with experiments

The theoretical model developed here can be used to study the dependence of σ_c^∞ and $\Delta\theta_c$ on the main parameters characterizing the inclusion, matrix and interface. For the sake of brevity, this section highlights only some of the most important results obtained by applying the model to the glass/vinylester tested by Cho et al. (2006).

The preferential debonding mode, i.e. either $n = 1$ or $n = 2$ in Figure 5.1, predicted by the procedure described in Section 5.1.3 is that requiring a lower value of σ_c^∞. Figure 5.11 shows σ_c^∞/σ_c predicted as a function of γ for glass/vinylester and $n = 1$ and 2. In both modes, σ_c^∞/σ_c increases with γ, the case $n = 2$ requiring always a larger σ_c^∞. The percentage difference in σ_c^∞ between both debonding modes, also shown in Figure 5.11, increases with γ, achieving a constant value about 5% for $\gamma \gtrsim 0.5$. This is because the energy criterion, the source of difference between the modes, cf. Figures 5.6 and 5.9, governs the solution for large values of γ. This result is confirmed by the majority of experimental observations presented in the literature, where the asymmetric debond mode, i.e. $n = 1$, is usually observed. See, e.g., the debonding process observed at inclusions on micro-scale by Cho et al. (2006), and similar observations reported by Gent and Park (1984) for an elastomeric matrix. Therefore, hereinafter, only predictions for $n = 1$ are presented.

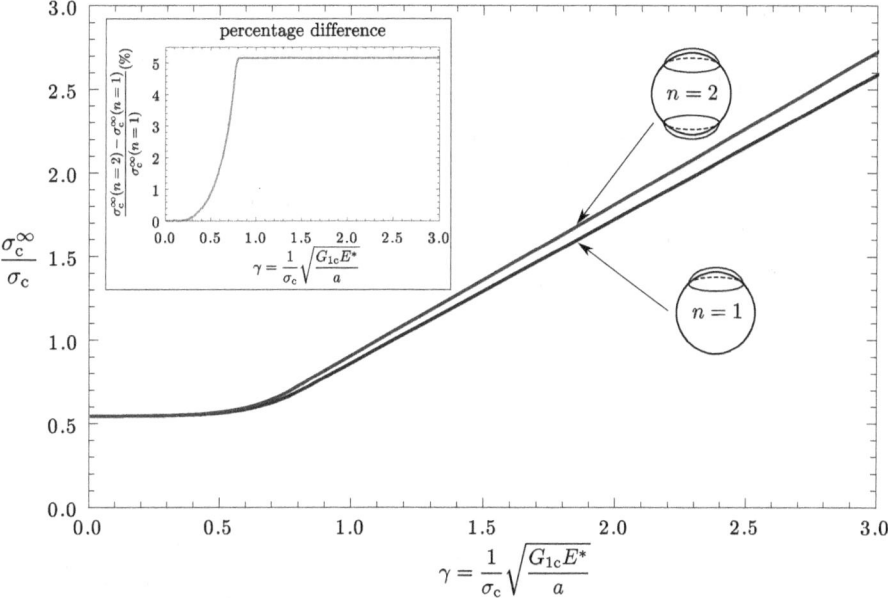

Figure 5.11: Predicted critical remote load σ_c^∞ for the debond onset as a function of γ for glass/vinylester, $G_{1c} = 7.29$ N/m, $\lambda = 0.11$ and for the two debonding modes, $n = 1$ and 2. Graph on the left represents the percentage difference between the predicted values of σ_c^∞ for $n = 1$ and 2.

Defining $a_0 = E^* G_{1c}/\sigma_c^2$ as a reference radius, the governing parameter γ defined in (5.16) can be expressed as $\gamma = \sqrt{a_0/a}$. Hence, for fixed material and interface properties a size effect of the particle radius a on σ_c^∞ and $\Delta\theta_c$ is predicted by the present model. The prediction for σ_c^∞ shown in Figure 5.11 is plotted again in Figure 5.12 as a function of a. For the sake of comparison with the experiments by Cho et al. (2006), the values predicted for the critical load are normalized by the ultimate strength of vinylester $\sigma_{\text{resin}} = 50$ MPa given by Cho et al. (2006). Since the interface tensile strength σ_c was not measured by Cho et al. (2006), the present prediction is computed for two extreme values of the ratio $\sigma_c/\sigma_{\text{resin}} = 0.5$, and 1. The part of the curve corresponding to $\sigma^\infty > \sigma_{\text{resin}}$ is plotted with a dashed line because it is not considered realistic since for this load a global failure of the resin or other failure mechanisms are expected to occur before the interface crack onset. The remote load for which the smallest debond is detected and reported by Cho et al. (2006) is associated here with the critical load. Unfortunately, only 4 single tests without replication were reported for 4 different values of the inclusion radius. As a consequence, the uncertainty cannot be estimated and the comparison presented cannot be considered concluding. Moreover, 2 of these 4 values are partially obtained from derived results presented by Cho et al. (2006) because the original measured values were not published by Cho et al. (2006). Analogously to γ_{th}, a threshold value for the particle radius a_{th} separating both scenarios (A and B) can be defined representing brittle-to-ductile transition as described above for γ_{th}, which reasonably agrees with results by cohesive zone models (CZM) by Needleman (1987); Ngo et al. (2010).

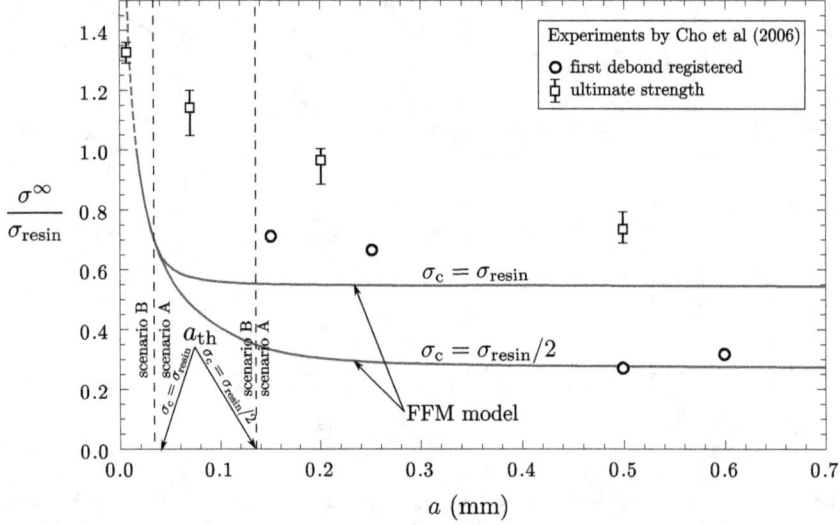

Figure 5.12: Critical remote load σ_c^∞ for the debond onset predicted as a function of the particle radius a for glass/vinylester, $G_{1c} = 7.29$ N/m, $\lambda = 0.11$ and $n = 1$, compared with the experimental results by Cho et al. (2006).

According to Figure 5.12, the debond onset at larger particles requires a lower remote load. This tendency agrees qualitatively with the behavior observed in the experiments by Cho et al. (2006). However, the quantitative agreement cannot be considered neither good in view of the moderate divergence found, nor concluding given the lack of knowledge about the dispersion of the results. The quantitative differences may be due to, among other reasons, a strain-rate dependence of results as reported for similar material-systems by Spanoudakis and Young (1984). Such a strain-rate dependence has not been taken into account in the present analysis, for the sake of simplicity. This size effect has also been reported and interpreted by other authors, e.g., by Sun et al. (2009).

The ultimate strength reported by Cho et al. (2006) is also plotted in Figure 5.12, for the sake of comparison with the critical load, although the prediction of

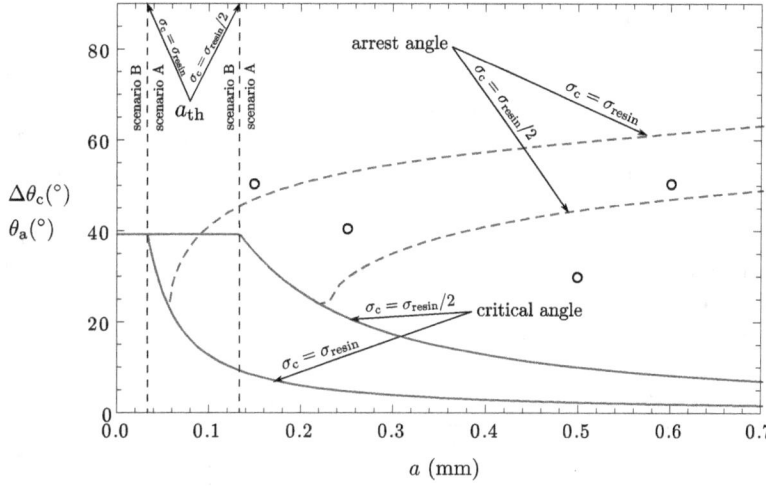

Figure 5.13: Critical and arrest debond angles, $\Delta\theta_c$ and θ_a, predicted as a function of the particle radius a for glass/vinylester, $G_{1c} = 7.29$ N/m, $\lambda = 0.11$ and $n = 1$, in comparison with the experimental results obtained by Cho et al. (2006).

the ultimate strength is out of the scope of the present work. It would require the analysis of the subsequent debond propagation, its arrest at the interface, kink out the interface and coalescence with other similar cracks, see París et al. (2007) for a discussion on the analogous problem for a cylindrical inclusion. A relevant difference between the ultimate strength of a composite and the critical remote load for the debond onset at one particle indicates that the intermediate steps add a significant strength reserve to the composite. This difference increases with a in accordance with the observations presented by Mower and Argon (1996).

The size effect of a on the debond and arrest angles, $\Delta\theta_c$ and θ_a, predicted by the present model is shown in Figure 5.13, taking $\sigma_c/\sigma_{resin} = 0.5$, and 1. The only value which can be compared with the experimental observation reported by Cho et al. (2006) is θ_a since the growth (if any) between $\Delta\theta_c$ and θ_a is unstable and as a result the value observed in an post-onset observation corresponds to θ_a. Thus, the first debond angle observed by Cho et al. (2006) for several particle radii is also presented in this figure. Analogously to previous results for the critical load, only 4 experimental values for θ_a without replication were reported by Cho et al. (2006). These results do not show a regular tendency. This fact can be a partial evidence of that the uncertainty, which is unknown due to the lack of replication of the experiments, is high as a consequence of the difficult measurement of θ_a on the one hand, and the strong influence of the variability of interface properties on the crack arrest after an unstable growth on the other hand. The comparison shows that the experimental values are relatively near the prediction but the comparison is not concluding.

5.3   Concluding remarks

A theoretical model, based on the coupled stress and energy criterion of the FFM, for the prediction of the critical remote tension originating a debond onset at the spherical particle embedded in an infinite matrix has been developed and analyzed. This model enables to study in a seminalytical manner the influence of the particle, matrix and interface properties. In particular, this work has focused on the influence of the particle radius showing that the strength increases with decreasing particle radius and is essentially constant for large particles. A preliminary comparison with experiments has been carried out by employing the results reported by Cho et al. (2006). The comparison shows that the model agrees with the tendency observed for the size effect on the critical load leading to the interface crack onset and the asymmetry observed in the debond mode. However, the quantitative comparison does not show a good agreement which is however not concluding due to the shortage of experimental results (only 4 points without replication could be extracted from the literature), and the lack of knowledge about the value of key parameters as the interface tensile strength. As a consequence of this analysis, new experiments should be carried out to obtain a full validation. The debond onset studied here is the first step of the failure mechanisms leading to the macroscopic failure of a particle reinforced composite. The result for the critical load and angle of debond obtained by the model can be used as a starting point for the analysis of the subsequent steps: unstable interface crack propagation, its kink towards the matrix and coalescence of several cracks initiated at different particles.

CHAPTER 6 ■

Transverse cracking in cross-ply laminates

Cracks appearing and growing in the transverse inner-ply of cross-ply laminates $[0_m/90_n]_s$ represent a classical problem, which has been studied for a long time. Whereas these cracks usually do not significantly reduce the global stiffness of the structure, they are the source of a more dangerous failure mechanism. Basic steps of this mechanism are well known, see reviews by Nairn (2000); Berthelot (2003). First, above a critical strain level, some cracks appear perpendicular to the load in the inner ply, see e.g. Varna et al. (2001). Then, some interface cracks appear when transverse cracks reach the interface (or before it) between the inner and outer plies. Finally, coalescence of interface cracks occurs leading to macroscopic delaminations.

The present chapter focuses on the study of the first transverse crack initiation. Actually, it is also applicable to the sufficiently distant transverse cracks which appear almost simultaneously with the first one. Crack initiation due to unusually large flaws is not taken into account in this chapter. This particular problem has been studied for decades. The experimental tests carried out by Garrett and Bailey (1977); Parvizi et al. (1978); Bailey and Parvizi (1981) showed a size effect of the inner-ply thickness on the critical applied strain originating the first transverse crack onset. These experimental results showed that the first failure of the transverse ply is not exclusively either stress-dependent or strain-dependent, which is in the basis of the majority of failure criteria. As a consequence, further theoretical models have been developed dealing with the problem of explaining the size effect found in experiments. Three models stand out among others:

- Incremental energy criterion model: It was initially proposed by Aveston et al. (1971) in the context of micromechanics of unidirectional fiber composites. Subsequently, it was applied to this problem by Garrett and Bailey

159

(1977). The model is based on assuming that a transverse crack appears abruptly spanning the entire inner-ply thickness. This occurs when the energy released at the crack onset is enough to cover the energy dissipated during the onset. Garrett and Bailey (1977) approximated the released energy by assuming oversimplifying hypotheses about the stresses after the transverse cracking, giving a quite poor approximation. Hashin (1996) applied this model, improving significantly the approximation, and obtained an analytical expression for the transverse crack density as a function of the external load. He assumed that stresses after the crack onset do not vary within each ply along the direction perpendicular to the ply interface. Employing this assumption, he implemented a variational approach to calculate the released energy. In both approaches, due to Garrett and Bailey (1977); Hashin (1996), the physical interpretation of the size effect provided by the model is the same: the released energy varies with the square of the inner-ply thickness, whereas the energy dissipated is linear with this thickness. This source of size effect will be discussed in depth in Section 6.1.5.1.

- Dvořák's model: It was proposed initially by Dvorak and Laws (1986, 1987), see also Dvorak (2013). The key hypothesis is that a certain "damaged zone" or "non-Griffith crack" grows stably when load is increased up to reaching a critical length. Then, the behavior of the crack becomes governed by the Griffith criterion causing its unstable growth. The critical length is assumed to be material-dependent. According to this assumption and depending on the inner-ply thickness, the "damaged zone" may reach the interface before having reached the critical length, its growth being arrested. Two possible scenarios lead to two very different behaviors. For inner plies sufficiently thick in comparison with the critical length, the crack growth is essentially driven by the unstable growth along the direction perpendicular to the interface. On the contrary, for thin inner plies, the "damaged zone" reaches the interface before having the critical length and subsequently it grows along the direction parallel to the fiber in the inner ply. After reaching the critical length along this direction, the crack grows unstably. As a consequence, for thin laminates, the unstable growth is governed by the behavior along the longitudinal direction, parallel to the interface. van der Meer and Dávila (2013) demonstrated that a different behavior for thick and thin laminates is also predicted when a circular transverse defect is introduced and the growth is simulated by prescribing a cohesive zone model at the surface containing the defect.

Therefore, whereas for thick plies, the critical strain predicted by this model is independent of the inner-ply thickness, a strong size effect is found for thin plies, which agrees with experiments. Following the work by Dvořák, this model has been widely analyzed and extended by other authors, see e.g. Maimí et al. (2011). Some of them assume the preexistence of a crack of a fixed material-dependent length instead of the original Dvořák's hypothesis of "non-Griffith crack" below a critical length. However, both hypotheses lead to equivalent analyses and results. Focusing on the design, a new matrix failure criterion taking into account this model was proposed by Camanho et al. (2006).

- Statistical model: It is based on the well-known Weibull's theory. The key assumption is the pre-existence of defects, e.g. microcracks, flaws or damaged zones with lengths distributed following a Weibull-like law in the whole inner-ply volume, see e.g. Li and Wisnom (1997); Wisnom (2000). The distribution is defined by two parameters which are assumed to be material-dependent. Since the defects density is statistically distributed, increasing the volume of material makes more likely to find a larger defect. As a consequence, critical strain for which the first defect becomes a crack decreases with volume. Thus, when the other laminate lengths are fixed, the critical strain is predicted to decrease with the inner-ply thickness as observed in experiments.

As described in Section 3.2, inspired by the experimental results due to Parvizi et al. (1978) among others, Leguillon (2002) proposed, in the framework of the Finite Fracture Mechanics (FFM) introduced by Hashin (1996), the coupled stress and energy criterion which is used in the context of this thesis, see Chapter 3. He explained qualitatively the Parvizi's observations by means of this criterion. This novel approach enables to understand the phenomenon observed by Parvizi et al. (1978) from the point of view of a general criterion which is applied to other problems of crack initiation. This is an advantage over the previous interpretation given except the statistical one that are based on hypothesis very attached to the problem.

This chapter aims to revisit the problem from the new point of view provided by the FFM and the coupled criterion. Following the generalized plane strain approach widely used to analyze this problem, see e.g. McCartney (1998), a novel 2D theoretical model is developed in Section 6.1 following the methodology introduced by Mantič (2009). This approach allows a "semianalytical" expression to be obtained in order to predict the critical strain leading to the first transverse crack onset. This expression is analytical except for a scalar value which is obtained by the computational analysis of the problem. A 3D FFM model is developed in Section 6.2 in order to evaluate the accuracy of the 2D approach to this problem. These two models are developed without considering the influence of the residual thermal stresses which can be relevant in some types of composites. Thus, a modification of the model is proposed in Section 6.3 to take into account this influence. Finally, the results from a new set of experiments carried out in the context of this thesis are presented in Section 6.4 and compared with the model.

6.1 2D generalized-plane-strain model of transverse crack initiation

This section describes the 2D model based on a generalized plane strain approach to this problem employing the coupled criterion. The stress criterion is presented as a direct and simple application of the laminate theory in Section 6.1.1. After a dimensional analysis of the energy criterion and some other considerations, a final expression of this criterion is presented in Section 6.1.2. Additionally, results related to the Energy Release Rate (ERR) of the transverse crack computed by the BEM, suitable for this kind of problem, are presented in this section. A combination of both criteria is introduced in Section 6.1.3. Critical applied strain leading to the onset of a transverse crack and the length of this crack at its onset predicted by the model are presented in Sections 6.1.3 and 6.1.4. A size effect predicted by the application of the present approach is described in Section 6.1.5, and compared with the experimental evidences from the bibliography in Section 6.1.6. Finally, a very simple indirect procedure for the measurement of the transverse fracture toughness of a ply is described in Section 6.1.7.

6.1.1 Stress criterion

The stress criterion is applied to the uncracked cross-ply laminate shown in Figure 6.1. The laminates under study here are symmetric and composed by an inner transverse-ply with a thickness $2t_{90}$ and two outer and longitudinal plies with an individual thickness t_0. The two other dimensions of the rectangular specimens under study are $W \times H$. Let (x, y, z) be a suitably defined cartesian coordinate system, the y-axis being coincident with the outer-ply fiber axis and with the load direction, the z-axis being coincident with the inner-ply fiber axis and the x-axis being perpendicular to the interface between the inner and outer ply.

All the plies are of the same orthotropic elastic material (Table 6.1) and are assumed to be perfectly bonded along their interface. Let E_{11} and E_{22} denote respectively the longitudinal and transverse Young's moduli ($E_{11} > E_{22} = E_{33}$), $\nu_{12} = \nu_{13}$ and ν_{23} the Poisson's ratios, and $G_{12} = G_{13}$ and G_{23} the transverse shear moduli, the other elastic parameters being defined by these properties. A uniform longitudinal strain ε_{yy} is applied along the y-axis. Geometry, loads and material properties enable to assume a generalized plane strain state in the xy-plane in the sense used by Blázquez et al. (2006).

Composite	E_{11}(GPa)	E_{22}(GPa)	ν_{12}	ν_{23}	G_{12}(GPa)	G_{23}(GPa)
carbon/epoxy	141.3	9.58	0.3	0.32	5	3.5
glass/epoxy	42	14	0.278	0.4	5.83	5

Table 6.1: Properties of the two materials used: the carbon/epoxy used in París et al. (010a,b) and the glass/epoxy used in Parvizi et al. (1978).

The tensile stress criterion applied here assumes the existence of a critical value σ_c as the tension required to originate fracture at a certain plane of the

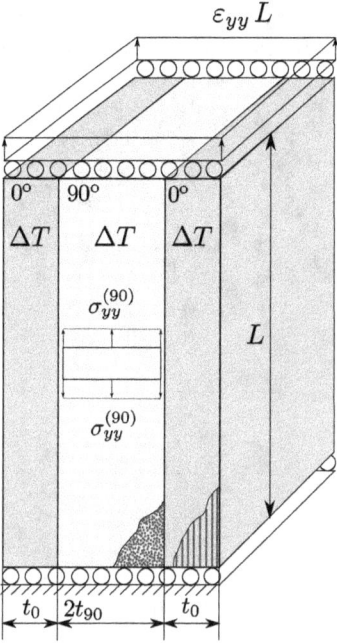

Figure 6.1: Schematic representation of the undamaged laminate.

lamina. This value depends on the ply orientation with respect to the load direction because of the material microstructure of the ply, given by approximately parallel long fibers embedded in matrix. In the present problem a crack perpendicular to the load direction is assumed to appear in the 90° lamina. Thus, the value of σ_c can be identified with the transverse strength under tension of the lamina Y_t. According to this criterion, the crack can appear at those points (x, y, z) where the value σ_{yy} of the tension normal to the plane of the crack verifies

$$\sigma_{yy}^{(90)}(x, y, z) \geq Y_t. \tag{6.1}$$

Note that the sign in the above condition is "exceeding or equaling" ($\geq$) instead of "reaching" ($=$) because the stress criterion is only a necessary condition. Therefore, the critical value for stresses could be exceeded without leading to a crack onset. An analytic solution for stresses is possible under the assumption of the laminate theory, see e.g. Jones (1999). Assuming that a uniform longitudinal strain ε_{yy} is applied, the value of $\sigma_{yy}^{(90)}$ is uniform along the thickness and width of the 90° lamina, $\sigma_{yy}^{(90)} = \tilde{E}_{22}\varepsilon_{yy}$, and the stress criterion leads to a very simple expression

$$\tilde{E}_{22}\varepsilon_{yy} \geq Y_t, \tag{6.2}$$

where $\tilde{E}_{22}$ is the apparent Young's modulus of the inner 90° ply in the y-direction. This value can easily be expressed as a function of the laminate properties by the laminate theory as demonstrated in 6.A,

$$\frac{E_{22}}{\tilde{E}_{22}} = \frac{1 - \nu_{12}\nu_{21}}{1 - \nu_{12}\nu_{21} \dfrac{1 + \frac{t_0}{t_{90}}}{1 + \frac{E_{22}}{E_{11}}\frac{t_0}{t_{90}}}}. \tag{6.3}$$

Figure 6.2 shows the value of this ratio as a function of the ratio of the outer to the inner ply thickness t_0/t_{90} for the materials in Table 6.1. Hereinafter, glass/epoxy and carbon/epoxy refer to the materials defined in this table. This plot demonstrates that both Young's moduli are very close as expected.

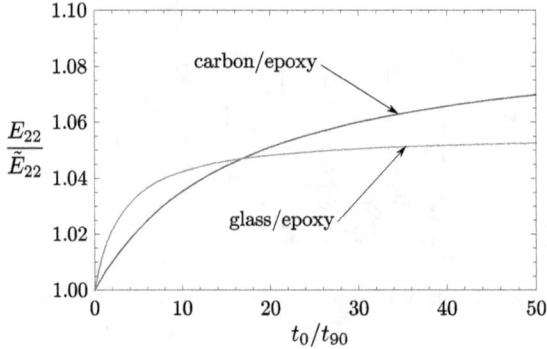

Figure 6.2: Ratio of the true (material) to the apparent transverse Young moduli in the inner ply as a function of the ratio of the outer to the inner ply thickness. These values have been computed for the composite laminates defined in Table 6.1.

Defining the maximum admissible longitudinal strain in a unidirectional (UD) lamina on the plane perpendicular to the fiber direction as

$$Y_{\varepsilon t} = \frac{Y_t}{E_{22}}, \tag{6.4}$$

the final expression of the stress criterion (6.2) takes the form:

$$\frac{\varepsilon_{yy}}{Y_{\varepsilon t}} \geq \frac{E_{22}}{\tilde{E}_{22}}, \tag{6.5}$$

where $\frac{E_{22}}{\tilde{E}_{22}} \gtrsim 1$ according to Figure 6.2. This inequality represents a necessary condition for rupture at any point in the inner 90° lamina. It can be noticed from the above expression that the stress criterion will be fulfilled for all the points of the 90° lamina for the same value of the applied strain ε_{yy} because the right-hand side does not depend on the point (x, y, z). Under present assumptions and according to this conclusion, stress criterion does not impose any restriction on the size of the originated crack because the stress criterion is simultaneously fulfilled either at all points or at none point of the 90° lamina.

6.1.2   Energy criterion

The energy criterion adopted here is based on the incremental Griffith criterion proposed by several authors as Garrett and Bailey (1977); Hashin (1996);
Leguillon (2002). The energy balance of an abrupt onset of a crack under the
assumptions presented above can be expressed as

$$\Delta\Pi + \Delta E_k + G_c 2\Delta a = 0, \tag{6.6}$$

where $\Delta\Pi$ and ΔE_k are, respectively, the change in elastic potential and kinetic
energy due to crack onset. $G_c 2\Delta a$ is the energy dissipated during the onset.
G_c and Δa are, respectively, the fracture toughness, also referred to as fracture
energy, and the semilength of the crack after the onset. In the present problem, a constant fracture toughness G_c during the crack onset and growth can
be assumed, because the transverse crack in the 90° ply grows in pure mode
1, therefore $G_c = G_{1c}$. Although the pure fracture mode 1 is assumed for a
crack growing in a plane without breaking fibers, different values for the fracture
toughness might be considered for the crack growth along the direction parallel
or perpendicular to the fibers. Nevertheless, in the present work both values are
assumed to be very close and are identified with the transverse fracture toughness
for transverse-crack propagation-direction perpendicular to the fibers. Additionally it is assumed that no damage at the interface takes place. The crack growth
after reaching the 90° and 0° ply interface is not studied here.

    As the initial state is quasistatic, then there is a production of kinetic energy
$\Delta E_k \geq 0$ and the above expression leads to

$$-\Delta\Pi \geq G_c 2\Delta a. \tag{6.7}$$

Introducing in this inequality the relation between the value of the energy release
rate (ERR) G and the derivative of the potential energy with respect to the crack
length $2a$, known from the linear elastic fracture mechanics (LEFM) theory,

$$G(a) = -\frac{d\Pi}{d(2a)}, \tag{6.8}$$

implying that

$$-\Delta\Pi = 2 \int_0^{\Delta a} G(a) da, \tag{6.9}$$

the energy balance in (6.6) finally leads to

$$\int_0^{\Delta a} G(a) da \geq G_c \Delta a. \tag{6.10}$$

This expression is a relation between the elastic energy available to be released
during the crack onset (left-hand side) and the energy necessary for the abrupt
onset (right-hand side). Values for G are required to fulfill the criterion (6.10).
Unfortunately any analytical solution of the problem with cracked geometry, see
Figure 6.3, is not available. Thus, G is obtained by computational methods.

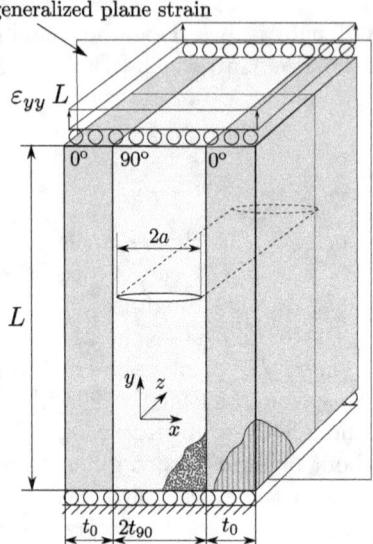

plane where the 2D analysis is carried out
by assuming generalized plane strain

Figure 6.3: Schematic representation of the cracked laminate.

Values of G have been calculated by Blázquez et al. (2009) for a particular
configuration of the problem by means of the Boundary Element Method (BEM).
These results can be exploited for many configurations of the present problem
by means of dimensional analysis. Thus, as demonstrated in 6.D a dimensionless
ERR $\hat{G}$ can be defined by means of

$$G = E_{22}\varepsilon_{yy}^2 t_{90}\hat{G}\left(\hat{a}, \frac{t_0}{t_{90}}, \text{E.P.}\right), \tag{6.11}$$

where $\hat{a} = a/t_{90}$ and "E.P." are the dimensionless elastic properties of the lamina. Thanks to this expression, the amount of models to be solved is reduced
drastically since values obtained can be exploited for any values of E_{22}, ε_{yy} and
t_{90}.

Values of G obtained by Blázquez et al. (2009) are normalized here according
to (6.11) and plotted in Figure 6.4. This figure represents the values of $\hat{G}(\hat{a})$ for
$t_0/t_{90} = 1$ and carbon/epoxy versus $\hat{a}$.

In view of Figure 6.4, three regions with different behavior of $\hat{G}(\hat{a})$ can be
differentiated:

1. In the first region (I) where the crack is short enough, $\hat{G}(\hat{a})$ is approximately linear. The reason is that when the crack tip is far enough from the
   interface, $\hat{G}$ can be approximated by its value corresponding to an infinite
   cracked plate under a remote uniaxial loading. Thus, at this extreme, $\hat{G}$

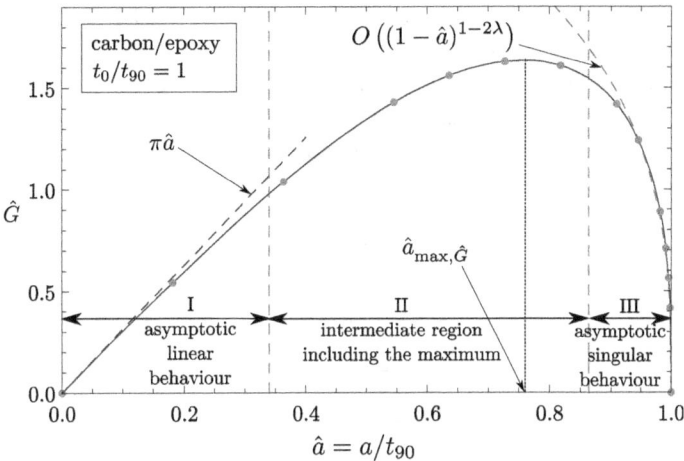

Figure 6.4: Dimensionless ERR $\hat{G}$ of the transverse crack in the inner ply for carbon/epoxy as a function of its dimensionless length $\hat{a} = a/t_{90}$.

has an asymptotic linear behavior,

$$\hat{G}(\hat{a} \to 0) = \frac{G(a \to 0)}{E_{22}\varepsilon_{yy}^2 t_{90}} \cong \pi\hat{a}. \tag{6.12}$$

2. The second region (II) can be considered around the crack length $\hat{a}_{\max,\hat{G}}$ where the function $\hat{G}(\hat{a})$ reaches its maximum value. In this region, the influence of the stiffer material in the $0°$ ply cannot be neglected.

3. In the third region (III), there is a strong transition between the maximum and the necessary zero value of $\hat{G}(\hat{a})$ for $\hat{a} = 1$ governed by the asymptotic singular behavior of $\hat{G}(\hat{a} \to 1)$, conditioned by the dominating influence of the outer ply.

   At this extreme, where the length of the transverse crack is close to the thickness of the inner-ply ($\hat{a} \to 1$), jump of the mechanical properties (becoming stiffer) across the interface implies that $\hat{G} \to 0$ for $\hat{a} \to 1^-$. In the present particular problem configuration, the asymptotic behavior was described by Blázquez et al. (2008) as

$$\hat{G} = O\left((1 - \hat{a})^{1-2\lambda}\right) \quad \text{for } \hat{a} \to 1^-, \tag{6.13}$$

where λ is the stress singularity order of the trimaterial corner defined by the transverse crack ending at the interface. The value of λ for carbon/epoxy $\lambda = 0.330111$ has been computed by the analytic procedure developed by Barroso et al. (2003).

The position of the maximum of the curve is affected by the ratio E_{22}/E_{11}. The maximum moves towards the interface, when the ratio E_{22}/E_{11} increases, diminishing the region III affected by the jump in stiffness across the interface.

Figure 6.5 shows the variations of $\hat{G}$ for glass/epoxy computed by the same BEM code. Unlike the carbon/epoxy laminate with $E_{22}/E_{11} = 0.0678$, the glass/epoxy laminate has a ratio $E_{22}/E_{11} = 0.33$. This difference in values of E_{22}/E_{11} generates the effect predicted above: the crack length corresponding to the maximum of ERR is situated closer to the outer-ply because the influence of this ply is lower.

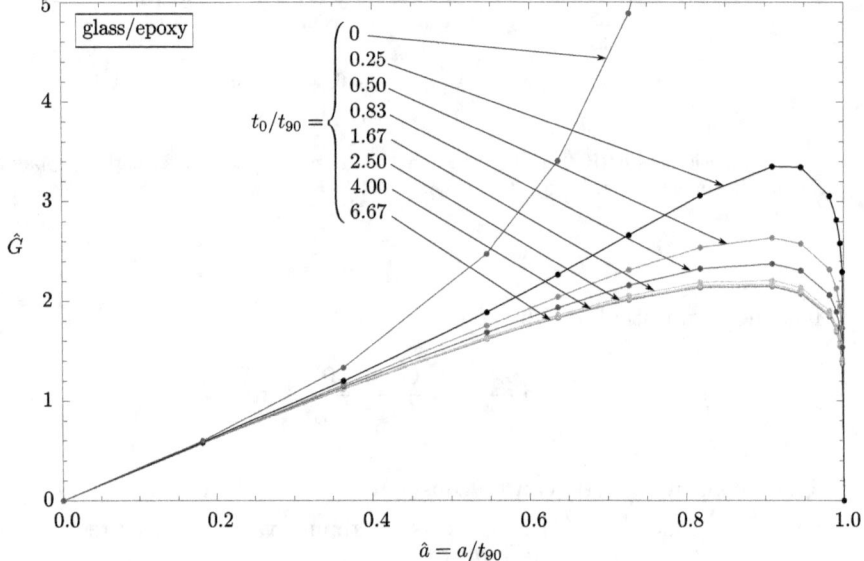

Figure 6.5: Dimensionless ERR $\hat{G}$ of the transverse crack in the inner ply as a function of its dimensionless length $\hat{a} = a/t_{90}$ for glass/epoxy and laminates used by Parvizi et al. (1978).

The influence of the ratio of the inner and outer ply thickness, t_0/t_{90}, can also be observed in Figure 6.5. Values of the parameter t_0/t_{90} shown in this figure correspond with the laminate configurations tested by Parvizi et al. (1978). On the one hand, this figure shows that the maximum value of $\hat{G}$ grows and slightly moves toward the interface when $t_0/t_{90} \to 0$. In the limit case, where the outer ply does not exist ($t_0/t_{90} = 0$), the value of $\hat{G}$ increases to infinity for $a/t_{90} \to 1$. On the other hand, Figure 6.5 shows that for $t_0/t_{90} \to \infty$, which corresponds to very thick outer plies, the values of $\hat{G}$ tend to a limit curve. The reason for this is that for large values of t_0/t_{90}, the influence of an addition of material to the outer ply (corresponding to an increase of the value of t_0/t_{90}) on the behavior of the crack in the inner ply is negligible because the tip of the crack is far from the new material added (due to the large value of t_0/t_{90}).

Introducing the expression of $\hat{G}$ (6.11) in the expression of the incremental energy criterion (6.10) gives

$$E_{22}\varepsilon_{yy}^2 t_{90} \int_0^{\Delta\hat{a}} \hat{G}(\hat{a})\mathrm{d}\hat{a} \geq G_c\Delta\hat{a}, \tag{6.14}$$

which after a rearrangement writes as

$$\frac{\varepsilon_{yy}^2 E_{22}t_{90}}{G_c} \geq g(\Delta\hat{a}), \tag{6.15}$$

where g is a dimensionless positive function defined as

$$g\left(\Delta\hat{a}; \frac{t_0}{t_{90}}, \mathrm{E.P.}\right) = \frac{\Delta\hat{a}}{\int_0^{\Delta\hat{a}} \hat{G}(\hat{a})\mathrm{d}\hat{a}}. \tag{6.16}$$

Some arguments of g will be omitted in the following for the sake of simplicity. The function g can be understood as the dimensionless ratio of the necessary to available energy for the onset of a crack of semilength $\Delta\hat{a}$. It means, function g measures the "resistance against crack onset", from an energetic point of view, for the crack semilength $\Delta\hat{a}$ at the onset. Figure 6.6 shows the function $g(\Delta\hat{a})$ for carbon/epoxy.

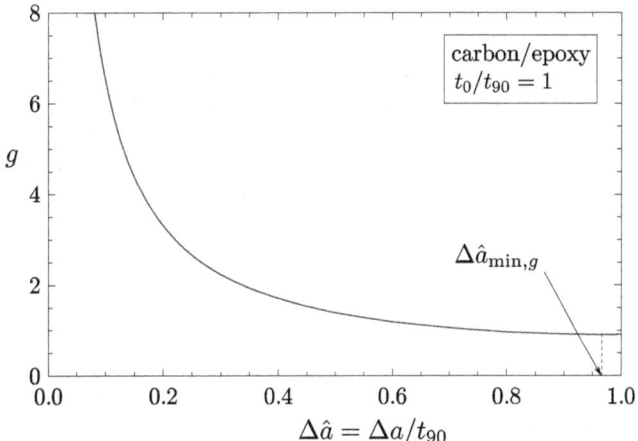

Figure 6.6: Example of the universal dimensionless function $g(\Delta\hat{a})$ for carbon/epoxy and $t_0/t_{90} = 1$.

Apparently, looking at Figure 6.6 by a naked eye, the function $g(\Delta\hat{a})$ is decreasing in the whole interval $(0,1)$. However, as will be seen, g has a quite shallow minimum at $\Delta\hat{a}_{\min,g} < 1$ typically close to 1. Thus, g is decreasing for a major part of the interval $(0,1)$, and generally speaking a lower applied strain is necessary for larger lengths of the crack at the onset. Figure 6.7 shows a detailed plot of g around its minimum very close to the interface between the plies. This result could erroneously be attributed to some errors in BEM results or in their interpolation. However, this property of the universal function g can be demonstrated independently of the computational results, see 6.E. Moreover, as demonstrated there, this minimum is situated closer to the interface than the maximum of $\hat{G}$ (6.134).

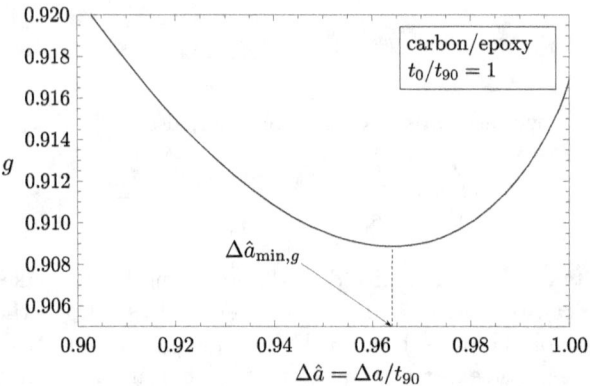

Figure 6.7: Universal dimensionless function $g(\Delta\hat{a})$ around the minimum point for carbon/epoxy and $t_0/t_{90} = 1$.

After a rearrangement of (6.15), the condition for the onset given by the energy criterion can be rewritten in an analogous manner to that defined by the stress criterion (6.5) as

$$\varepsilon_{yy} \geq \sqrt{\frac{G_c}{E_{22}t_{90}}\, g\,(\Delta\hat{a})}. \tag{6.17}$$

Note that, unlike the stress criterion, the energy criterion defines a condition for the strain required for the onset which depends on the semilength of the crack at the onset $\Delta\hat{a}$.

6.1.3   Coupled stress and energy criterion

In this section the coupled criterion by Leguillon (2002) will be applied to the present problem following the procedure used by Mantič (2009); Mantič and García (2012). The key idea of the coupled criterion, referred to as Leguillon's hypothesis see Section 3.2, assumes that a sufficient condition for the transverse crack onset is the simultaneous fulfillment of the two necessary conditions described in previous sections.

Both stress criterion in (6.5) and energy criterion in (6.17) define a required value for the applied strain ε_{yy} which can originate a crack. In order to have both conditions expressed for the same magnitude, (6.17) is divided by the critical strain transverse to the fibers $Y_{\varepsilon t}$, see (6.4). Thus, an expression of the energy criterion comparable to the expression of the stress criterion is obtained

$$\frac{\varepsilon_{yy}}{Y_{\varepsilon t}} \geq \frac{1}{Y_{\varepsilon t}} \sqrt{\frac{G_{\mathrm{c}}}{E_{22} t_{90}} g\left(\Delta\hat{a}, \frac{t_0}{t_{90}}, \mathrm{E.P.}\right)}. \tag{6.18}$$

Following Mantič (2009), a dimensionless brittleness number γ can be introduced as

$$\gamma = \frac{1}{Y_t} \sqrt{\frac{E_{22} G_{\mathrm{c}}}{t_{90}}}. \tag{6.19}$$

Note that γ is a structural parameter dependent on some material properties along with a geometric parameter. Then, the final expression of the energetic criterion adopts the form

$$\frac{\varepsilon_{yy}}{Y_{\varepsilon t}} \geq \gamma \sqrt{g\left(\Delta\hat{a}, \frac{t_0}{t_{90}}, \mathrm{E.P.}\right)}. \tag{6.20}$$

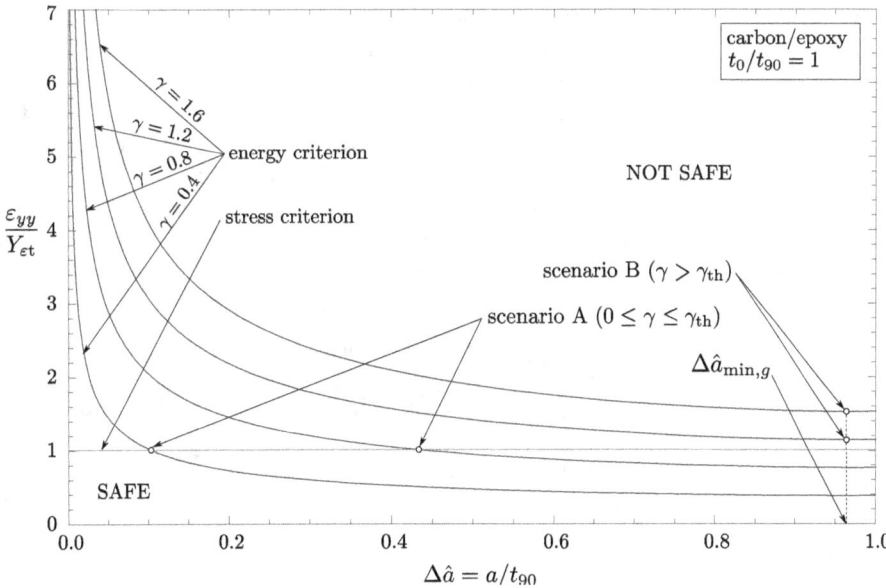

Figure 6.8: Coupled stress and energy criteria for several values of the structural parameter γ for carbon/epoxy laminates with $t_0/t_{90} = 1$.

Assuming Leguillon's hypothesis, the critical strain leading to the crack onset is given by the minimum value of ε_{yy} which fulfills both criteria. Figure 6.8 shows the curves of both criteria, (6.5) and (6.20), for several values of the dimensionless structural parameter γ, for carbon/epoxy and $t_0/t_{90} = 1$. This figure shows that there are two possible scenarios depending on the value of γ:

- Scenario A: This scenario corresponds to the values of γ for which the stress and energy criteria curves have some common point. In this case, the critical strain ε_{yy}^c and the length of the crack $\Delta\hat{a}_c$ are determined by this common point. In fact, as the stress criterion is a horizontal straight line, ε_{yy}^c does not depend on $\Delta\hat{a}_c$. Consequently, in scenario A, ε_{yy}^c will be defined by the constant value imposed by the stress criterion,

$$\frac{\varepsilon_{yy}^{c,A}}{Y_{\varepsilon t}} = \frac{E_{22}}{\tilde{E}_{22}}. \tag{6.21}$$

Thus, for the values of γ leading to scenario A, the critical strain depends only on the values of $E_{22}/\tilde{E}_{22}$ and $Y_{\varepsilon t}$. However, the extension of the crack length at the onset does depend on the value of γ and can be determined by finding the common point of the two criteria curves

$$\gamma\sqrt{g(\Delta\hat{a}_c^A)} = \frac{E_{22}}{\tilde{E}_{22}}, \tag{6.22}$$

where $\Delta\hat{a}_c^A$ is obtained by solving this non-linear equation.

- Scenario B: This scenario corresponds to the values of γ for which no common point exists between the stress and energy criteria curves. In this case, as can be seen in Figure 6.8, the stress criterion is fulfilled for a value of the applied strain lower than the minimum value defined by the energy criterion corresponding to its value at $\Delta\hat{a} = \Delta\hat{a}_{\min,g}$. Consequently, the coupled criterion will be fulfilled when the energy criterion is fulfilled. Therefore, scenario B is, in fact, governed only by the energy criterion. The lower value of the critical strain imposed by the energy criterion is determined by the minimum value calculated above

$$\frac{\varepsilon_{yy}^{c,B}}{Y_{\varepsilon t}} = \gamma\sqrt{g(\Delta\hat{a}_{\min,g})}, \tag{6.23}$$

where the value $\Delta\hat{a}_{\min,g}$ has been defined in Section 6.1.2. Since g does not depend on γ, therefore neither $\Delta\hat{a}_{\min,g}$ does. As a consequence $g(\Delta\hat{a}_{\min,g})$ is a constant which depends only on the dimensionless elastic properties and on the relation between layer thicknesses t_0/t_{90}. For given values of these parameters, the value of $g(\Delta\hat{a}_{\min,g})$ becomes a constant and consequently a linear relation exists between the critical strain $\varepsilon_{yy}^{c,B}$ and the value of γ, see (6.23). This is an important consequence because it reduces the calculus of this dependence to the minimization of the function g to obtain $\Delta\hat{a}_{\min,g}$. Having computed $g(\Delta\hat{a}_{\min,g})$, the value of the critical strain for any value of γ can be obtained by (6.23).

The value of the critical length of the debond $\Delta\hat{a}_c$ is determined directly by the minimum value for the energetic criterion. Therefore

$$\Delta\hat{a}_c^B = \Delta\hat{a}_{\min,g}, \tag{6.24}$$

a value which is independent of the structural parameter γ.

Both scenarios are separated by a threshold value of γ, denoted as γ_{th}, for which both criteria curves are tangent. Hence,

$$\gamma_{\text{th}} = \frac{\frac{E_{22}}{\bar{E}_{22}}}{\sqrt{g\left(\Delta\hat{a}_{\text{min},g}\right)}}. \tag{6.25}$$

Then, according to above definitions and as can also be seen in Figure 6.8, scenario A corresponds to $0 \leq \gamma \leq \gamma_{\text{th}}$ whereas scenario B corresponds to $\gamma > \gamma_{\text{th}}$. Therefore, in view of (6.19), scenario A, where the critical strain corresponds approximately to $Y_{\varepsilon t}$, is associated to low values of G_{c} and E_{22}, to high values of Y_{t} and to thick laminates.

Combining both scenarios, the critical strain leading to the first transverse crack onset is given by,

$$\frac{\varepsilon_{yy}^{\text{c}}}{Y_{\varepsilon t}} = \max\left(\frac{E_{22}}{\bar{E}_{22}}, \gamma\sqrt{g(\Delta\hat{a}_{\text{min},g})}\right). \tag{6.26}$$

Variation of $\varepsilon_{yy}^{\text{c}}$ as a function of the dimensionless structural parameter γ, for carbon/epoxy and $t_0/t_{90} = 1$, is shown in Figure 6.9. This plot as well as expression in (6.26) show the extreme simplicity of the critical strain prediction if the geometrical parameter t_0/t_{90} is fixed. From the computational model, only a scalar value $\sqrt{g(\Delta\hat{a}_{\text{min},g})}$, independent of the strength and fracture toughness properties, is necessary to obtain this curve.

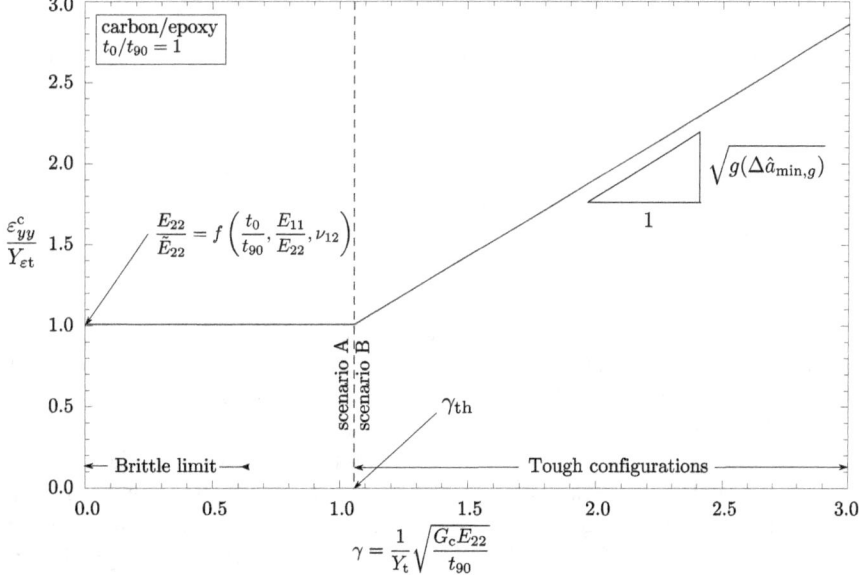

Figure 6.9: Critical applied strain $\varepsilon_{yy}^{\text{c}}$, normalized by the ultimate transverse strain $Y_{\varepsilon t}$ of the lamina, as a function of the structural parameter γ, for carbon/epoxy and $t_0/t_{90} = 1$.

These results predict that the transverse-ply can support strains quite higher than its nominal ultimate strain $Y_{\varepsilon t}$. The reason is that the stress criterion is not a sufficient condition but just a necessary one for the crack onset.

The progressive damage models of laminates (see e.g. Matzenmiller et al. (1995)) predict a failure of the $90°$ ply using a stress or equivalent strain criterion. However, according to the present results, under the hypothesis of the first failure associated to a transverse crack in the $90°$ ply, the first failure of the laminate is not expected to appear for nominal ultimate strain value in the direction perpendicular to the fibers, $Y_{\varepsilon t}$, if $\gamma > \gamma_{th}$. In these cases, predictions based on the traditional stress (or strain) criteria can become too conservative because the critical strain can become significantly larger than $Y_{\varepsilon t}$, as can be seen in Figure 6.9.

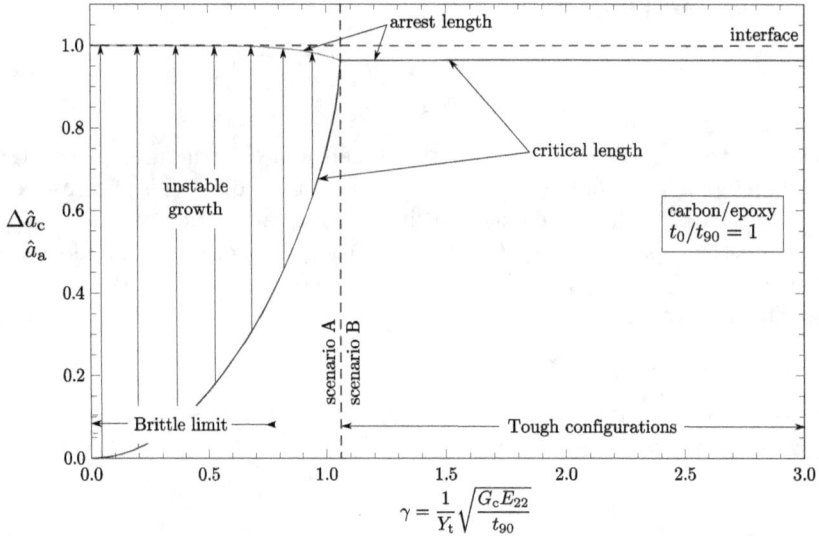

Figure 6.10: Critical length $\Delta\hat{a}_c$ and arrest length $\hat{a}_a$ as a function of the structural parameter γ for carbon/epoxy and $t_0/t_{90} = 1$.

The critical length of the crack at the onset $\Delta\hat{a}_c$ is shown in Figure 6.10 as a function of γ for carbon/epoxy. This figure shows that $\Delta\hat{a}_c$ is larger for higher values of γ and that it is constant in scenario B. In this scenario, $\Delta\hat{a}_c$ corresponds to about 96.4 % of the inner-ply thickness for carbon/epoxy. This value corresponds, as has been described above, to the position of the minimum value of the function g, $\Delta\hat{a}_{min,g}$. This position depends strongly on the relation between the longitudinal and transverse Young's moduli: E_{22}/E_{11} and the ply-thickness ratio t_0/t_{90}. A lower value of E_{22}/E_{11} corresponds to a larger distance between the crack tip at the crack onset and the interface.

The reason for this is that a stiffer outer ply makes more difficult approaching the interface by a crack. For instance, in the case of the glass/epoxy lamina (that corresponds normally to a lower difference between transverse and longitudinal properties than a carbon/epoxy lamina), the minimum of the function g corresponds to a 99.83 % of the inner-ply thickness for $t_0/t_{90} = 0.83$.

6.1.4   Post crack-onset evolution

The length of the crack at the onset, $\Delta\hat{a}_c$, predicted by the coupled criterion is very difficult to verify experimentally. The reason is that this original crack could grow unstably after the onset, for a fixed $\varepsilon_{yy} = \varepsilon_{yy}^c$. Therefore, a visible crack in a later post-failure exploration of the specimen, could be larger than predicted by $\Delta\hat{a}_c$, its final length being referred to as the dimensionless arrest length of the crack, $\hat{a}_a$.

Following the proof proposed by Mantič (2009), in this case it is possible to demonstrate that $G(\Delta\hat{a}_c) > G_c$ for $\gamma < \gamma_{th}$. Therefore, the originated crack will grow unstably after the onset for $\gamma < \gamma_{th}$ to the so-called arrest length where $G(\hat{a}_a) = G_c$ and $G'(\hat{a}_a) < 0$. The proof begins by evaluating the logarithmic derivative of the function $g(\Delta\hat{a})$ as

$$\frac{d\log g}{d\Delta\hat{a}} = \frac{1}{g(\Delta\hat{a})}\frac{dg}{d\Delta\hat{a}} = \frac{G_c}{G_c\Delta\hat{a}} - \frac{G(\Delta\hat{a})}{\int_0^{\Delta\hat{a}} G\,d\hat{a}}. \tag{6.27}$$

As discussed previously, in both scenarios, the energy criterion is fulfilled as an equality, then it holds

$$G_c\Delta\hat{a}_c = \int_0^{\Delta\hat{a}_c} G\,d\hat{a}. \tag{6.28}$$

From the monotony analysis of $g(\Delta\hat{a})$,

$$\frac{dg}{d\Delta\hat{a}} < 0 \quad \text{for} \quad \Delta\hat{a} < \Delta\hat{a}_{min,g},$$

$$\frac{dg}{d\Delta\hat{a}} = 0 \quad \text{for} \quad \Delta\hat{a} = \Delta\hat{a}_{min,g}. \tag{6.29}$$

Introducing conditions (6.28)–(6.29) and $g > 0$ in the expression of the logarithmic derivative (6.27), the next outcome is obtained

$$G(\Delta\hat{a}_c) > G_c \text{ for } \Delta\hat{a}_c < \Delta\hat{a}_{min,g},$$
$$G(\Delta\hat{a}_c) = G_c \text{ for } \Delta\hat{a}_c = \Delta\hat{a}_{min,g}. \tag{6.30}$$

The first result implies that the crack grows unstably after the onset for $\gamma < \gamma_{th}$. The second result implies that it is necessary to evaluate the derivative of G with respect to the crack length $2a$ to know the post-onset evolution for $\gamma \geq \gamma_{th}$.

The relation between the sign of the second derivative of function g and that of the first derivative of G for the critical points of g is given in the expression (6.133). Then, taking into account that $g''(\Delta\hat{a}_{min,g}) > 0$, and because g has a minimum value at $\Delta\hat{a}_{min,g}$, it leads to,

$$G'(\Delta\hat{a}_{min,g}) < 0. \tag{6.31}$$

Combining this expression with (6.30), it is obtained that no unstable growth is expected for $\gamma \geq \gamma_{\text{th}}$. Hence, for these values of γ,

$$\hat{a}_{\text{a}}^{\text{B}} = \Delta\hat{a}_{\text{c}}^{\text{B}} = \Delta\hat{a}_{\text{min},g}. \tag{6.32}$$

It is useful to observe that for $\gamma < \gamma_{\text{th}}$, after the initial crack onset, the value of the arrest length $\hat{a}_{\text{a}}$ of the crack can be calculated as

$$G(\hat{a}_{\text{a}}, \varepsilon_{yy}^{\text{c}}) = G_{\text{c}} \text{ and } G'(\hat{a}_{\text{a}}, \varepsilon_{yy}^{\text{c}}) < 0, \tag{6.33}$$

where $\varepsilon_{yy}^{\text{c}}$ is the critical strain generating the crack onset. This expression is a non-linear equation which can be solved by computational methods. The value of the arrest length of the crack for both scenarios has been computed and the results are also shown in Figure 6.10 versus γ for carbon/epoxy. This plot shows that the developed theoretical model predicts that the arrest length is always very close to the interface between the inner-ply and the outer-ply. This observation is demonstrated in general in 6.F. In particular for carbon/epoxy and any value of γ

$$\hat{a}_a \geq \Delta\hat{a}_{\text{min},g} \left(\frac{t_0}{t_{90}} = 1, \text{E.P.} \right) \approx 96.4 \text{ \%}, \forall\gamma. \tag{6.34}$$

In view of this result, the distance between the crack tip and the interface may be only a few times larger than the fiber radius, cf. París et al. (010b). Thus, the hypothesis of homogeneous material for the inner-ply might not be *a priori* coherent because of the influence of its inhomogeneity on the crack growth. Outcomes of this observation will not be studied here but it is important to take it into account when discussing the range of validity of the results presented.

6.1.5 Effect of the laminate geometry

Results presented above depend on the two geometric parameters of the laminate: t_{90} and t_0/t_{90}. It is interesting to study their effect on the results predicted. In particular, the variation of the critical strain originating a crack as a function of the geometry can have some interesting consequences for the design of laminates.

6.1.5.1 Influence of the inner-ply thickness for a fixed t_0/t_{90} - size effect

A size effect in the problem is associated to the dependency of results on γ as shown in Figure 6.9. According to the definition of γ in (6.19), the inner-ply thickness is the only material-independent parameter in γ. Following this idea, a reference length $t_{90,\text{r}}$ can be defined in terms of the material properties which appear in the definition of γ, i.e.

$$t_{90,\text{r}} = \frac{G_{\text{c}}E_{22}}{Y_{\text{t}}^2}. \tag{6.35}$$

Thus, γ can be rewritten as a function of t_{90} and $t_{90,\text{r}}$

$$\gamma = \sqrt{\frac{t_{90,\text{r}}}{t_{90}}}. \tag{6.36}$$

Hence, a threshold thickness given by the threshold value of γ can be defined as

$$t_{90,\text{th}} = \frac{t_{90,\text{r}}}{\gamma_{\text{th}}^2}. \tag{6.37}$$

This threshold thickness $t_{90,\text{th}}$ has a very important physical meaning. This value separates two ranges of the inner-ply thickness with a very different behavior. Keeping fixed $t_{90,\text{r}}$ (which corresponds to keep fixed a relation of E_{22}, Y_t and G_c), a critical strain ε_{yy}^c close to the ultimate UD transverse strain $Y_{\varepsilon t}$ of the lamina is predicted for laminates with $t_{90} > t_{90,\text{th}}$ (corresponding to $\gamma < \gamma_{\text{th}}$). However, a value of strain significantly higher than the nominal value $Y_{\varepsilon t}$ of the lamina is expected to be withstand by laminates with a lower inner-ply thickness than the threshold value, $t_{90} < t_{90,\text{th}}$. Therefore, a size effect associated to the inner-ply thickness is predicted by the present theoretical model. In fact, this size effect is strong, as can be seen when the results are presented as a function of γ in Figure 6.9 and as a function of the dimensionless inner-ply thickness $t_{90}/t_{90,\text{r}}$ in Figure 6.11. Representation of these results as a function of $t_{90}/t_{90,\text{r}}$ shows that the change of the apparent strength of the inner-ply is very abrupt. Figure 6.11 shows the importance, discussed above, of the threshold value of the inner-ply thickness $t_{90,\text{th}}$. Qualitatively similar curves were presented by Leguillon (2002); Ladevèze et al. (2006).

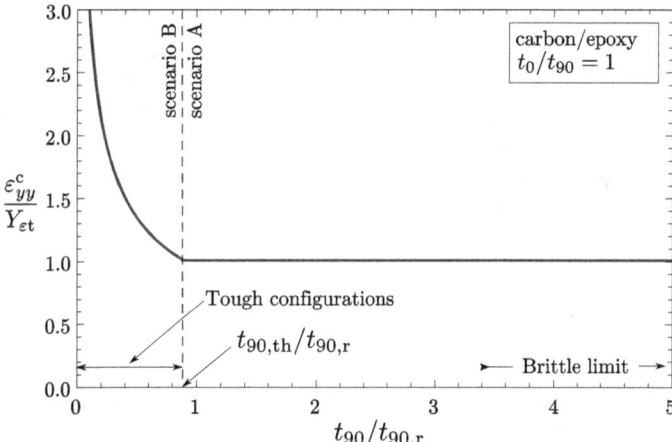

Figure 6.11: Critical strain ε_{yy}^c normalized by the unidirectional ultimate UD transverse strain $Y_{\varepsilon t}$ as a function of the inner-ply thickness t_{90} normalized by the reference thickness $t_{90,\text{r}}$ for carbon/epoxy and $t_0/t_{90} = 1$.

Taking into account the expression (6.36), it is possible to express the critical strain in scenario B as a function of the dimensionless inner-ply thickness

$$\frac{\varepsilon_{yy}^c}{Y_{\varepsilon t}} = \sqrt{\frac{t_{90,\text{r}}}{t_{90}}} \sqrt{g(\Delta\hat{a}_{\text{min},g})}. \tag{6.38}$$

Therefore, in view of this result, in scenario B, the critical strain is strongly increasing for $t_{90}/t_{90,r} \to 0^+$.

Physical interpretation of this abrupt change in the slope and curvature of the critical strain curve in Figure 6.11 is due to two different reasons. The first one is the change of criterion governing the failure as described above. The second one is the size effect inherent to the energy criterion. Introducing the dimensionless expression of $G(a)$ from (6.122) in the inequality of the energy criterion in (6.10) gives,

$$E_{22}\varepsilon_{yy}^2 t_{90}^2 W \int_0^{\Delta \hat{a}} \hat{G}(\hat{a}) \, \mathrm{d}\hat{a} \geq G_c t_{90} W \Delta \hat{a}. \tag{6.39}$$

The left-hand side represents the released energy when an onset of half-length Δa appears. In scenario B, where exclusively the energy criterion governs the crack onset, half-length of the crack at the onset is fixed $\Delta a_c = \Delta \hat{a}_{\min,g} t_{90} \sim t_{90}$. As a consequence, the integral term in (6.39) is fixed and the change in potential energy

$$-\Delta \Pi \sim E_{22}\varepsilon_{yy}^2 t_{90}^2 W, \tag{6.40}$$

where W is the width of the laminate. In view of this result, released energy is proportional to the product of stress $E_{22}\varepsilon_{yy}$ and strain ε_{yy} and to the volume $t_{90}^2 W$ of the material where the energy is released. In the case of the energy dissipated due to the onset,

$$\Delta \Gamma \sim G_c t_{90} W, \tag{6.41}$$

thus, it is proportional to the fracture toughness and the new crack surface created after the onset. In view of the above results and unlike the energy which is released in a volume, the energy necessary for the crack onset is dissipated at a surface. This is actually the origin of the size effect. When the laminate is thinner, then the volume where the available energy is stored near the crack ($\sim t_{90}^2 W$) decreases faster than the surface where the energy will be dissipated ($\sim t_{90} W$). Therefore, an increase of strain ε_{yy}^c is necessary to compensate this effect.

6.1.5.2   Influence of the geometric parameter t_0/t_{90}

The influence of t_0/t_{90} is pointed out by some evidences:

- Expression of the stress criterion depends directly on the value of t_0/t_{90} by the expression of $E_{22}/\tilde{E}_{22}$ (6.3). In the following, this dependence is studied by evaluating the value of $E_{22}/\tilde{E}_{22}$ at the extremal values of t_0/t_{90}.

  At the extreme $t_0/t_{90} \to 0^+$ which corresponds to a laminate without the longitudinal ply, it is fulfilled that

$$\lim_{t_0/t_{90}\to 0} \frac{E_{22}}{\tilde{E}_{22}} = 1. \tag{6.42}$$

  Therefore, the stress criterion (6.5) for the extreme case without outer ply ($t_0/t_{90} = 0$) leads to a foreseeable result:

$$\frac{\varepsilon_{yy}}{Y_{\varepsilon t}} \geq 1, \tag{6.43}$$

which corresponds to the typical prediction for a unidirectional transverse lamina.

At the other extreme, when $t_0/t_{90} \to \infty$, then the outer ply is much thicker than the inner-ply, the next result is obtained for the stress criterion, assuming $E_{22} < E_{11}$ and $\nu_{12}^2 \leq 1$

$$\lim_{t_0/t_{90} \to \infty} \frac{E_{22}}{\tilde{E}_{22}} = \frac{1 - \frac{E_{22}}{E_{11}}\nu_{12}^2}{1 - \nu_{12}^2} > 1. \tag{6.44}$$

Hence, the effect of a large thickness of the outer-ply compared to the inner-ply thickness is to increase the apparent strength in the stress criterion of the inner-ply. This increase depends strongly on the lamina properties ν_{12} and E_{22}/E_{11}. For example, the limit in the expression (6.44) is equal to 1.0558 for glass/epoxy and 1.092 for carbon/epoxy. In general, for typical values of ν_{12}, this limit value will be quite close to the unity. Therefore, it can be concluded from these results that the influence of the outer ply in the stress criterion is quite limited for the majority of usual composites.

• Energy criterion depends on the value of t_0/t_{90} through the dependence of $\hat{G}$ on it, see (6.11) and Figure 6.5. In general, a pair of thinner outer plies makes the energy to be released in an easy way when the crack grows. Thus, the universal dimensionless function g is expected to depend on this parameter as well. However, Figure 6.12 shows that the variation of g with t_0/t_{90} is not significant for the values of the geometric parameter t_0/t_{90} tested by Parvizi et al. (1978). Consequently, critical strain predicted does not change significantly either. Figure 6.13 confirms this result for the range of t_0/t_{90} tested by Parvizi et al. (1978).

As can be expected from the influences described above for the two independent criteria, t_0/t_{90} introduces a small variation of the critical strain predicted by the coupled criterion. Figures 6.13 and 6.14 show the critical strain originating a crack as a function of, respectively, $2t_{90}$ and γ for the values of t_0/t_{90} tested by Parvizi et al. (1978). In both scenarios, a higher t_0/t_{90} increases the critical strain originating a crack. The reason is that in scenario A, ε_{yy}^c is determinated directly by the stress criterion which is slightly more restrictive for higher values of t_0/t_{90} as demonstrated above. On the other hand, in scenario B, ε_{yy}^c is determinated by the energy criterion which is also a bit more restrictive for higher values of t_0/t_{90} because of the reasons presented above.

In summary, the effect of an increase of t_0/t_{90} is a slight increase of the critical strain that originates a crack in the inner-ply.

6.1.6 Comparison with experimental evidences from the literature

Two important and experimentally verifiable results have been obtained from the present theoretical model:

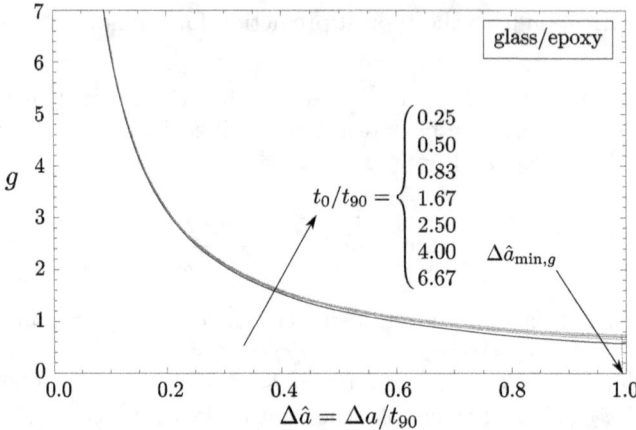

Figure 6.12: Universal dimensionless function $g(\Delta\hat{a})$ for glass/epoxy and the laminates used by Parvizi et al. (1978). The arrow shows the direction of increasing values of t_0/t_{90} in the list.

- A strong size effect of the inner-ply thickness on the critical strain ε_{yy}^c leading to the first crack onset.

- Arrest length of the crack originated in the inner-ply is very close to the inner-ply thickness.

Whereas the first prediction will be compared with experimental results obtained by Parvizi et al. (1978), the second result will be verified qualitatively by the observation of the cracks found in the specimens tested by París et al. (010b).

6.1.6.1 Experimental evidences for the size effect in the critical strain

Results for the critical strain that originates the initial failure in a cross-ply laminate are compared with the experimental results by Parvizi et al. (1978). They tested glass/epoxy cross-ply laminates (epoxy resin Shell Epikote 828 reinforced with E-glass fiber rovings Silenka 1200 tex). These laminates were made with a transverse-ply thickness ranging from 0.1 to 4 mm, keeping constant the longitudinal-ply thickness at 0.5 mm. Note that these laminates are not geometrically similar since t_0/t_{90} is not fixed. Therefore, the theoretical model developed here does not predict exactly a two-straight-lines law for ε_{yy}^c versus γ as shown in Figure 6.14.

To apply the model, elastic properties are taken from the information given by Parvizi et al. (1978). However, the authors did not give all the elastic properties necessary to characterize the laminate in accordance with our model. In particular, no data about the Poisson's coefficients are available in their article. As a consequence, the Poisson's coefficients are taken from Nurhaniza et al.

(2010) who use a very similar material. A collection of the elastic properties assumed for the comparison has been presented in Table 6.1.

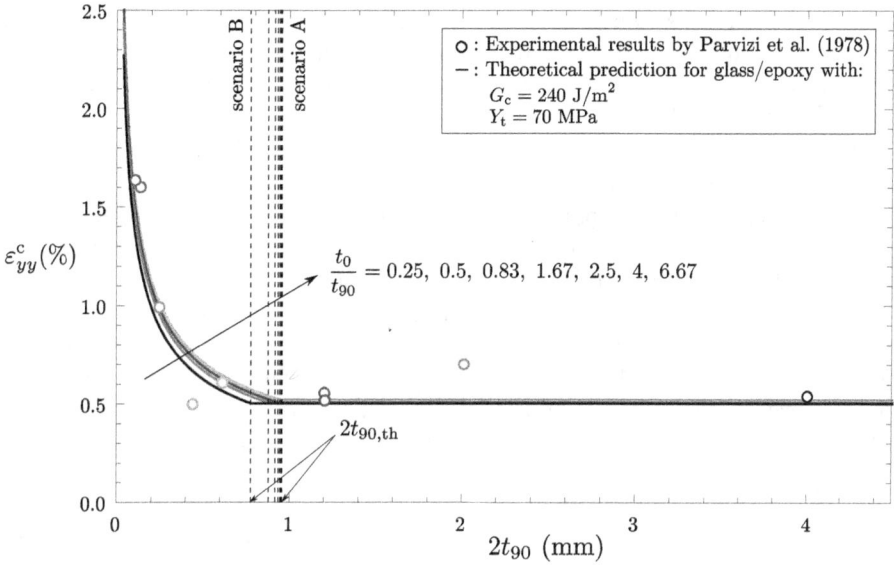

Figure 6.13: Experimental values of ε^c_{yy} for various ply thickness in glass fiber/epoxy cross-ply laminates used by Parvizi et al. (1978) and the present theoretical curves predicted for several values of t_0/t_{90}. The arrow shows the direction of increasing values of t_0/t_{90} in the list.

On the other hand, both the transverse fracture toughness G_c and the unidirectional ultimate UD transverse strain $Y_{\varepsilon t}$ are experimentally obtained by Parvizi et al. (1978) as,

$$G_c = 240 \pm 60 \ \text{J/m}^2, \tag{6.45a}$$

$$Y_{\varepsilon t} = 0.005 \pm 0.001, \tag{6.45b}$$

which corresponds to $Y_t = 70 \pm 14$ MPa.

First, a comparison based on the central values of the parameters G_c and $Y_{\varepsilon t}$ is carried out. Subsequently, a comparison taking the whole range of validity into account is done to observe the variability of results. Both comparisons have been calculated neglecting the residual stresses due to the different temperature of curing and service. Generally, in glass/epoxy laminates, residual stresses are much lower than in carbon/epoxy laminates due to the following reasons: lower contrast between the longitudinal and transverse properties and the similar sign of the dilatation coefficient for the longitudinal and transverse direction in glass/epoxy laminates. An estimation of the error introduced by neglecting residual stresses was carried out here using the methodology described by Nairn (2000) for the approximation of the variation of G with the curing temperature.

The maximum relative error expected was about 8% which is low in comparison with the dispersion in Parvizi et al. (1978) results.

Figure 6.13 shows the comparison between the experimental results and the theoretical model developed here for the critical strain ε_{yy}^c originating a crack as a function of the inner-ply thickness. Note that the colors of the points representing the experimental results correspond to the colors of the different curves for several values of t_0/t_{90}.

A satisfactory agreement between the experiments and theoretical predictions is remarkably evident from the results presented in Figure 6.13.

It is interesting to compare these experimental results with the theoretical predictions as a function of γ where the theoretical predictions have a very simple expression based on two-straight-lines law for geometrically similar laminates.

In order to plot the critical strain as a function of γ, its definition (6.19) is used but, as explained above, a high level of uncertainty exists concerning the values of the strength and fracture properties: Y_t and G_c, see (6.45). As a consequence, this uncertainty comes into the representation of the experimental results when they are expressed as a function of γ since it depends on Y_t and G_c.

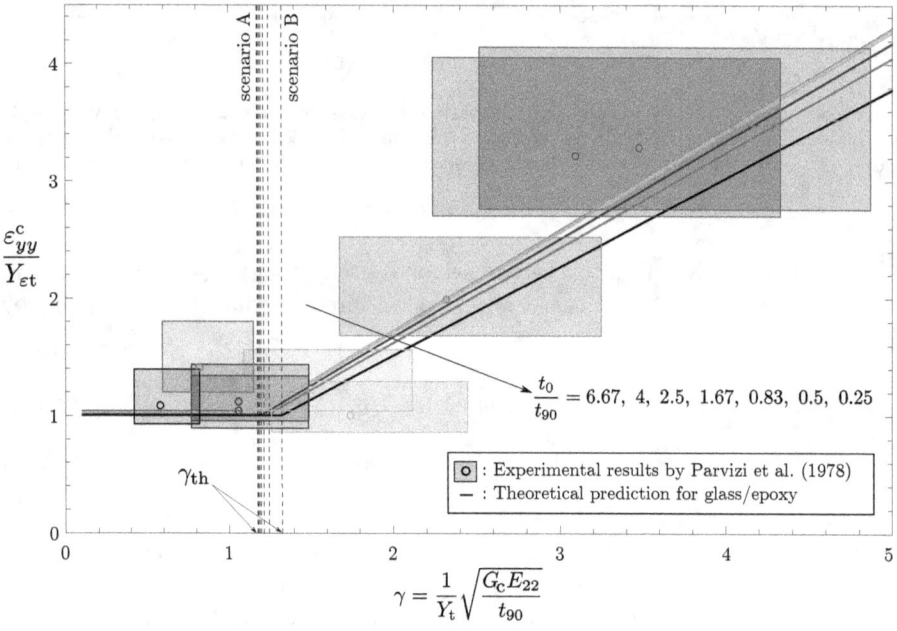

Figure 6.14: Experimental values of ε_{yy}^c normalized by the unidirectional ultimate UD transverse strain $Y_{\varepsilon t}$ as a function of inner-ply thickness in glass/epoxy cross-ply laminates from Parvizi et al. (1978) and the theoretical curves predicted for several values of t_0/t_{90}. The arrow shows the direction of decreasing values of t_0/t_{90} in the list.

Figure 6.14 shows experimental results with the rectangles of uncertainty associated to the operation of expressing the original experimental results in a dimensionless manner. The corners of the rectangles are calculated using the extreme values for Y_t and G_c. First, agreement between the experiments and theoretical model is satisfactory and supports that the simple expression of two straight lines approximates well the experiments. Second, the figure shows that these results, plotted as a function of the structural parameter γ, are more sensitive to the inaccuracy in G_c and Y_t for small values of the inner-ply thickness. In these cases, the rectangles of uncertainty are considerably larger.

The following two particular observations of agreement between the experimental results and theoretical prediction should be pointed out. First the threshold length of the inner-ply thickness for which the failure behavior changes abruptly is very well predicted. Second, the behavior according to scenario B is also well approximated as can be seen in Figures 6.13 and 6.14.

6.1.6.2 Experimental evidences for the arrest length of cracks

An arrest length of cracks very close to the inner-ply thickness has been predicted by the present theoretical model. In general, all the observed cracks have to satisfy

$$\hat{a}_a \geq \Delta\hat{a}_{\min,g}. \tag{6.46}$$

From the values of $\hat{G}$ computed numerically by Blázquez et al. (2009) and then obtaining the minimum of an approximation of the function g, it is possible to find the value of $\Delta\hat{a}_{\min,g}$, which depends only on the elastic properties of the plies and on the ply thickness ratio. The value obtained for carbon/epoxy and $t_0/t_{90} = 1$ is

$$\hat{a}_a \geq 0.9641. \tag{6.47}$$

For usual inner-ply thicknesses and composites, the distance between the arrested crack tip and the interface could be of the same order of magnitude as that of the micromechanical length of reference: the fiber diameter. As a consequence, a more accurate estimation of the crack length than its lower bound cannot be expected. The reason is that the hypothesis of the homogeneous material for the inner-ply is not plausible to predict phenomena associated to this small scale. It would be necessary to take into account the heterogeneous microstructure of the inner ply, which is out of the scope of this work.

Figure 6.15 shows a micrograph of a transverse crack in the inner-ply of a laminate tested by París et al. (010b). This microscopic observation shows that the crack is arrested after approximately reaching the lower bound predicted by the theoretical model developed. A lot of micrographs of cracks in cross-ply laminates can be found in the literature. The majority of those show a length of crack very close to the inner-ply thickness. This fact has been described, observed and justified by other approaches arriving to similar results, see e.g. París et al. (010a,b).

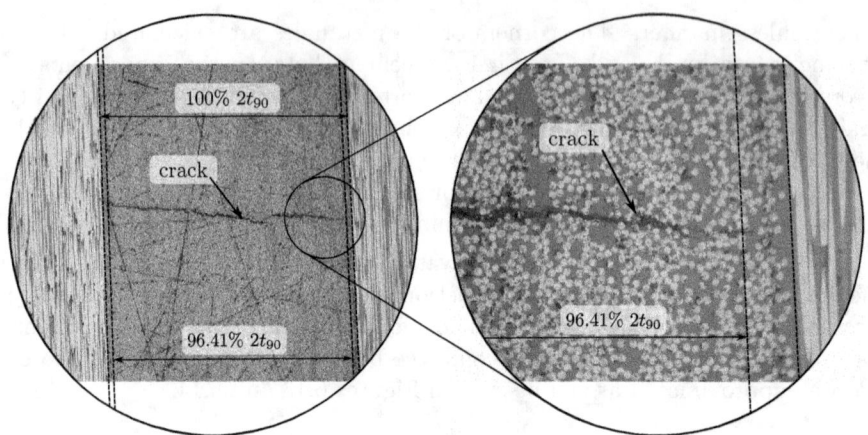

Figure 6.15: Example of a transverse crack arrested around the lower bound of the 96.4% on the inner-ply thickness as predicted. Microscope observations from the set of experiments carried out for Blázquez et al. (2009).

6.1.7   An experimental procedure for the measurement of the transverse fracture toughness

A new indirect experimental procedure to obtain G_c can be proposed assuming the validity of the theoretical model developed. First, recall the expression of the critical strain originating a crack as a function of γ in scenario B in (6.23). A test can be carried out for a value of $t_{90} < t_{90,\text{th}}$, obtaining a value of the critical strain when the first crack originates. Then, it is necessary to know the elastic properties of the lamina (which has been previously characterized), then $g(\Delta\hat{a}_{\min})$ is easily computed by the procedure presented above. If the value of Y_t is known, the approximated value of γ can be obtained by the following expression:

$$\gamma = \frac{E_{22}\varepsilon^{c}_{yy}}{Y_t\sqrt{g(\Delta\hat{a}_{\min,g})}},$$

(6.48)

and subsequently in view of definition (6.19), the value of G_c can be obtained by

$$G_c = \frac{\gamma^2 Y_t^2 t_{90}}{E_{22}}.$$

(6.49)

Note that this procedure requires an election *a priori* of an inner-ply thickness corresponding to scenario B, $t_{90} < t_{90,\text{th}}$ at testing. The problem is that the value of $t_{90,\text{th}}$ is not known without the value of G_c. The obvious solution can be to use an estimation of the expected result for G_c and a posterior verification that the result obtained for ε^{c}_{yy} is sufficiently different from the value of $Y_{\varepsilon t}$ which would correspond to scenario A.

Schematically, the procedure is:

Test	t_{90} (mm.)	ε^c_{yy} (%)	$g(\Delta\hat{a}_{min,g})$	G_c estimated (J/m^2)
1	0.0565	1.657	0.7415	292.89
2	0.0717	1.622	0.7415	356.15
3	0.1269	1.01	0.7404	244.77
4	0.3003	0.625	0.7368	222.89

Table 6.2: Results of the application of the proposed indirect procedure to obtain the transverse fracture toughness from the experimental results by Parvizi et al. (1978).

1. A test is carried out on a cross-ply laminate with a determinated value of t_0/t_{90}. A value of ε^c_{yy} is obtained for which the first crack originates. Information about the instant of the onset of the first crack can be obtained by attaching a contact microphone to the specimen and verified later on.

2. In order to verify the scenario associated to this inner-ply thickness, it is necessary that the critical strain obtained from the test is sufficiently higher than the unidirectional ultimate UD transverse strain $Y_{\varepsilon t} = Y_t/E_{22}$. If both values are very close, it is possible that the value of the inner-ply thickness corresponds to scenario A. Therefore it is necessary to reject this result and test a laminate with a thinner inner ply.

3. If the scenario has been verified successfully, an estimation of the fracture toughness value can be obtained as

$$G_c = \frac{E_{22}t_{90}\left(\varepsilon^c_{yy}\right)^2}{g(\Delta\hat{a}_{min,g})}, \tag{6.50}$$

result from substituting (6.48) in (6.49). The value of $g(\Delta\hat{a}_{min,g})$ can be computed from the elastic properties of the lamina and t_0/t_{90}. Note that the value of Y_t, whose dispersion is typically high, is not necessary to obtain G_c using this method.

As usual, this procedure can be repeated for several specimens or several configurations, obtaining a statistical estimation of the value.

It is interesting to obtain an estimation of G_c for the 4 points in scenario B obtained by Parvizi et al. (1978). Thus, a first evaluation of the validity of this procedure is carried out. Values of the present estimation for G_c are presented in Table 6.2. These results present a similar dispersion (279 ± 59 J/m^2) if compared with the dispersion of experimental results obtained by Parvizi et al. (1978): 240 ± 60 J/m^2.

6.2 3D model of transverse crack initiation

The 2D model developed previously allows estimating the critical strain leading to the onset of the first transverse cracks. This model provides a very simple semianalytical expression. However, this simplicity is at the expense of neglecting the importance of all possible phenomena associated to the axis perpendicular to the plane of study. The 2D modeling of this problem is very common and has been invoked in many relevant works dealing with it, see e.g. Wang (1984); Wang et al. (1985); Hashin (1996); McCartney (1998). However, other works, with a high impact, have highlighted the importance of the 3D effects on the understanding of the process of transverse cracking, see e.g. Dvorak and Laws (1986). In fact, fatigue experiments by Kobayashi and Takeda (2002) showed that the crack propagation as a "tunneling crack" could be important, see Hutchinson and Suo (1992) for additional information about the term "tunneling crack". Thus, this section develops and applies a 3D version of the FFM model employed previously in Section 6.1 with the aim of evaluating the accuracy of the 2D model and the effect of neglecting the third geometric dimension.

A 3D FFM model is developed here following the same steps used for the 2D model. For each step, the partial results are compared with those obtained by the 2D analysis in order to understand the cause of the observed differences in the final results. Since the main objective is to compare the results of this analysis with the 2D model, the material properties for carbon/epoxy, see Table 6.1, and the ratio $t_0/t_{90} = 1$, used as an example in the previous Section 6.1, will be utilized here.

First, the crack geometry and situations studied in the 3D analysis are presented in Section 6.2.1. Subsequently, in order to evaluate how the prediction of the coupled criterion is modified by the new 3D analysis, the stress and energy criteria are again evaluated separately in Sections 6.2.3 and 6.2.5, respectively. In both cases, the necessary elastic solutions have been obtained by means of a computational analysis. For the stress criterion, the stresses prior to the onset at the inner ply is obtained in Section 6.2.2 by means of a FEM analysis using Abaqus. The released energy is computed in Section 6.2.4 as a function of the crack geometry and the other problem parameters by combining Abaqus with FRANC3D NG, a pre- and post-processor for simulating crack growth, developed within the Cornell Fracture Group, see Wawrzynek et al. (2009). The combination of both criteria is studied in Section 6.2.6, and finally the results of the comparison with the 2D analysis are discussed in Section 6.2.7.

6.2.1   Crack geometry and locations studied

The crack geometry and location at its onset is a determining assumption in a typical FFM analysis. In the 2D problems studied in previous Chapters 4 and 5, the presence of an interface weaker than the bulk material has motivated the crack location and the existence of a stress-concentration point has determined the crack geometry. Under these assumptions, the geometry was fixed by a free scalar parameter, which defines the length of the crack at onset. In the case of the 2D analysis of the problem at hand, the symmetry and the constraint effect of the outer plies justifies that the crack onset contains the central point, therefore the crack geometry can be determined by a free scalar variable. However, the 3D analysis enlarges the set of possible geometries. In the context of this analysis, the crack is assumed to be contained in the plane normal to the load. In addition, the symmetry of the crack onset with respect to the symmetry plane parallel to the interface between plies is assumed. Two different locations, see Figure 6.16, are studied:

- Crack onset in the vicinity of the free edge: This location corresponds to a crack onset located at one of the extremes of the laminate width where the inner ply has a free edge.

- Crack onset inside the laminate: In this case, the crack onset is assumed to be located inside the laminate, sufficiently far from the free edge. In this case, an additional symmetry plane for the crack geometry is supposed: the xy-plane.

Under these assumptions, an infinite amount of crack geometries is still possible. Since this 3D analysis aims to study the influence of the crack length along the direction normal to the plane used for the 2D analysis, a geometry with two free scalar variables corresponding to the lengths along the $x-$ and $z-$direction is considered. This assumption simplifies the analysis significantly even though the geometry predicted by the analysis could not be exactly the optimum in terms of the critical load predicted.

Within the plane geometries with two free parameters, the ellipse seems the most natural geometry to be taken, due to its smoothness and simplicity. However, for high-aspect ratios, i.e. when both semilengths are very different, the curvature radius varies strongly. In these cases, a suitable introduction of the crack in a finite element analysis can be difficult because a high aspect ratio complicates the adequate meshing around the crack front. The reason is that elements have to be extremely small in the vicinity of the poles corresponding to the major semiaxis.

Since the objective here is the study of the influence of the crack length along one of the directions, cracks with high aspect ratios have to be studied to obtain the asymptotic results for the extreme cases. To avoid the problem with elliptical cracks described previously, long-shallow cracks are assumed instead of elliptical cracks. Long-shallow cracks are implemented in FRANC3D to be used instead of high-aspect ratio elliptical cracks, see Fracture Analysis Consultants, Inc (2011). These cracks are specified using the two main crack lengths Δa and Δb, and a corner radius r, see Figure 6.16. The corner radius is introduced only to avoid a sharp corner. In the subsequent computational analysis, since the influence of r is not in the scope of this work, r will be kept as small as possible within reasonable limits to avoid problems in the meshing process. This radius will be computed as a function of Δa and Δb. Thus, in what follows, the variations of different results studied due to the crack geometry will be presented only as a function of two free parameters Δa and Δb, assuming that the influence of r is sufficiently low.

Figure 6.16 summarizes the assumptions described previously. This figure presents the comparison between the cracks studied here and the crack geometry assumed in the 2D analysis as a consequence of the generalized plane strain assumption. Note that, the crack assumed in the 2D analysis can be seen as the limit case of those studied here for large Δb. As a consequence the value

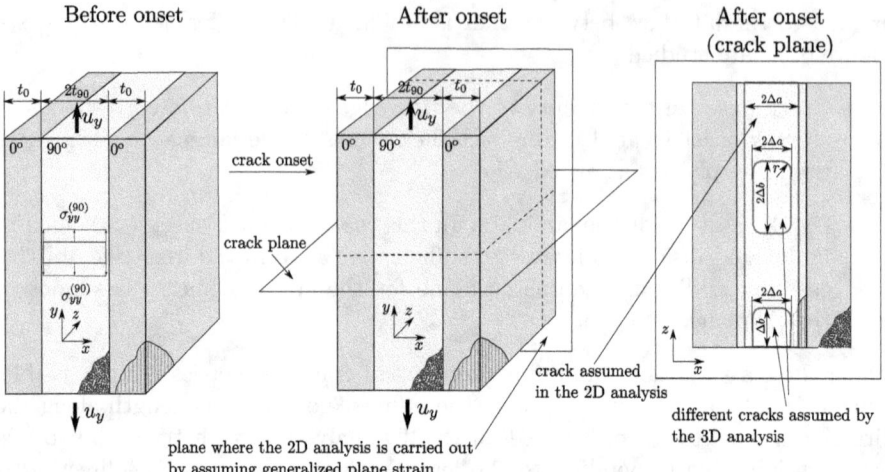

Figure 6.16: Schematic of the problem studied in the 3D analysis. The different crack shapes studied in 2D and 3D analyses are represented.

predicted for Δb at onset will be key for the evaluation of the simplified 2D analysis from the perspective of the 3D results.

6.2.2 Stresses in the inner ply before crack initiation according to a 3D analysis

In the 2D analysis, the stresses prior to the crack onset have been approximated by the laminate theory. In the present analysis, the result of uniform stresses within each ply is revisited by modeling the plies as homogeneous elastic anisotropic solids. In particular, it is interesting to study the influence of the free edge on the stress $\sigma_{yy}^{(90)}$ because it could affect significantly the stress criterion when the crack onset is situated at the free edge.

In order to approximate the stresses prior to the crack onset, a FEM analysis is carried out by Abaqus. Figure 6.17(a) shows the model used for this study. The FEM model is composed by three parts, each one corresponding to a ply. All these parts are modeled as solids of an orthotropic material which corresponds to the carbon/epoxy used by París et al. (010a,b), see Table 6.1. Local axes are defined for all the parts with the aim of defining the adequate material orientation. Following the assumption used for the 2D analysis, the interfaces are considered perfect. Thus, the parts are tied to each other along their common interfaces. In order to avoid the influence of other edges except that corresponding to $z = 0$, the length L and width W of the laminate are sufficiently large in comparison with the inner-ply thickness $2t_{90}$, see Table 6.3. The plane where stresses are evaluated is situated in the middle of the length L. A uniform displacement, $u_y = u_y^0$, along the y-direction is prescribed at the top edge ($y = L/2$). On the

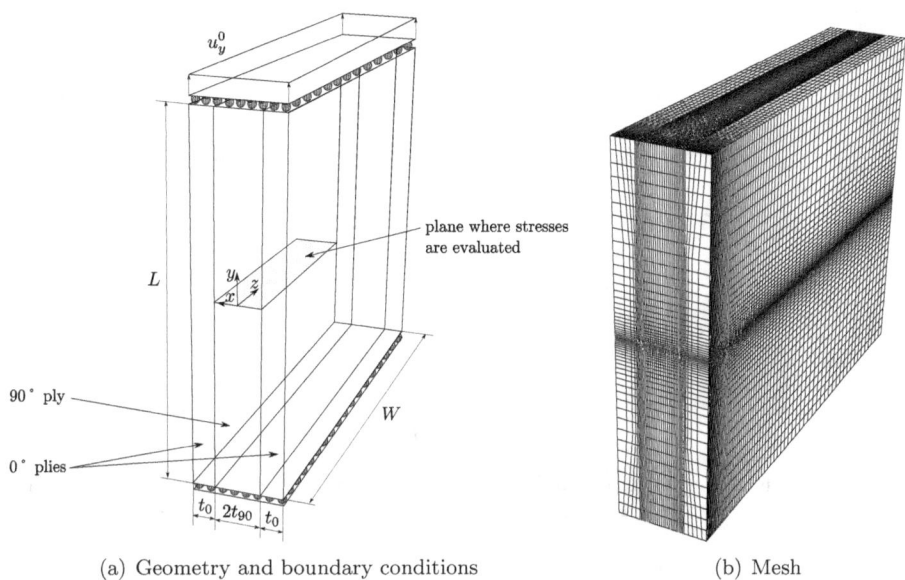

(a) Geometry and boundary conditions   (b) Mesh

Figure 6.17: Details of the FEM model used to obtain the value of the stresses in the inner ply.

other extreme, at the bottom edge this displacement is zero, see Figure 6.17. In addition, the conditions necessary to avoid rigid motion are prescribed using two arbitrary points of the bottom edge. A homogenized strain $\varepsilon_{yy}^0 = u_y^0/L$ can be defined in an analogous manner to the strain defined in the 2D analysis. The mesh used is composed by quadratic hexahedral elements and is refined in the vicinity of the plane where the stresses will be evaluated, see Figure 6.17(b).

t_0	L	W	u_y^0	ε_{yy}^0
$1t_{90}$	$16\ t_{90}$	$20\ t_{90}$	$0.02\ t_{90}$	$0.00125\ t_{90}$

Table 6.3: Values for the main geometric parameters of the 3D model used to obtain the stresses prior to the crack onset in a cross-ply laminate

Stress $\sigma_{yy}^{(90)}$ at the plane highlighted in Figure 6.17(a) is obtained from the FEM analysis. Due to the linearity of the problem, the stresses obtained can be normalized by ε_{yy}^0 in order to be employed for any value of ε_{yy}. In addition, $\sigma_{yy}^{(90)}/\varepsilon_{yy}$ is divided by the apparent Young modulus $\tilde{E}_{22}$ to obtain a dimensionless value to compare with the results from the 2D analysis which predicts $\sigma_{yy}^{(90)} = \tilde{E}_{22}\varepsilon_{yy}$, see Section 6.2.3. The dimensionless stress $\sigma_{yy}^{(90)}/\tilde{E}_{22}\varepsilon_{yy}^{(90)}$ is represented in Figure 6.18 as a function of $\hat{x} = x/t_{90}$, for several values of the distance to the edge $\hat{z} = z/t_{90}$. According to Figure 6.18, $\sigma_{yy}^{(90)}$ is very well approximated by the laminate theory for depths of the order of the inner-ply thickness. For smaller depths, the presence of the free edge is not negligible and $\sigma_{yy}^{(90)}$ is higher than the value predicted by the laminate theory, see also Figure

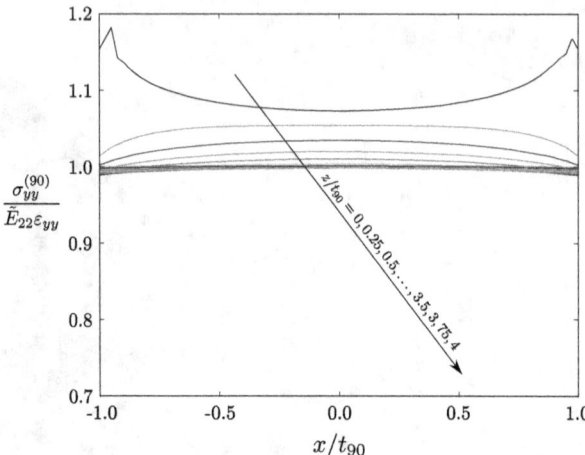

Figure 6.18: Dimensionless stresses σ_{yy} at a distance z from the free edge, for $t_0/t_{90} = 1$ and carbon/epoxy, obtained by a FEM analysis with Abaqus.

6.19 for the isovalue lines at the xz-plane. In general terms, $\sigma_{yy}^{(90)}$ rises when approaching the edge. Moreover, for the majority of values of $\hat{z}$ plotted, $\sigma_{yy}^{(90)}$ is higher for the central points of the inner ply (small $|\hat{x}|$). However, the curve for $\hat{z} = 0$ presents an opposite tendency: $\sigma_{yy}^{(90)}$ is higher when approaching the ply interfaces ($\hat{x} = \pm 1$). This is a consequence of the singularities found at the intersection between each interface and the free edge, see Zwiers et al. (1982). These singularities are in the origin of other very common failure mechanism in laminates: free-edge delamination. In fact Martin et al. (2010) studied this failure mechanism by using the coupled criterion employed here. Both models are complementary and can be combined to predict the preferential failure as a function of the laminate properties. The irregular values found for $\hat{z} = 0$ when approaching the interfaces are due to the fact that the mesh is not refined enough to approximate accurately the singularity. To capture well the singularity, it would be necessary a very fine mesh around the entire lines where the interfaces and the free edge intersect. Since the failure by edge delamination is out of the scope of this work, the approximation obtained is considered accurate enough for the present purpose.

It is interesting to remark that a qualitative difference is found with respect to the laminate theory: whereas the 2D analysis does not predict any condition on the crack length derived from the stress criterion, the 3D analysis of stresses predicts preferential points for the onset as a consequence of the non-uniformity of $\sigma_{yy}^{(90)}$ in the vicinity of the free edge. Note that this consequence is only found for the hypothesis of crack onset occurring at the free edge. In the case of a crack onset inside the laminate, Figure 6.18 shows that for large depths, $\sigma_{yy}^{(90)}$ is very uniform in the xy-plane, therefore the laminate theory in this case is accurate

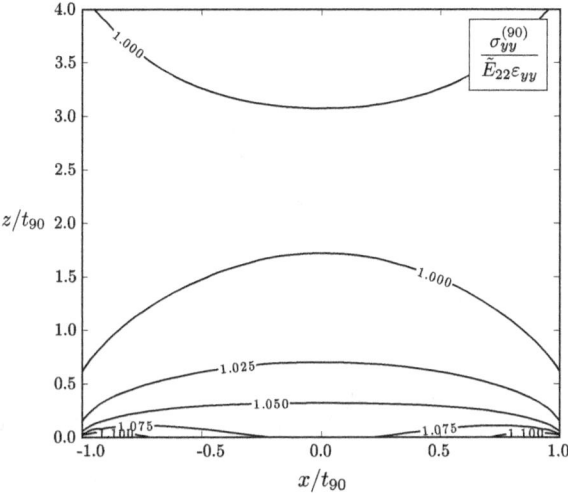

Figure 6.19: Isovalue lines for the dimensionless stress $\sigma_{yy}/E_{22}\varepsilon_{yy}$ at the plane $y = 0$ for t_0/t_{90} and carbon/epoxy obtained by a FEM analysis with Abaqus.

enough and the stress condition is almost independent of the the crack lengths Δa and Δb.

6.2.3 Stress criterion for the 3D model

The stress criterion used here requires that stresses at every point $x = (x, y, z)$ situated at the surface ΔS, which will occupy the crack after the onset, exceed a certain tensile strength. As justified in Section 6.1.1, this tensile strength is identified for this problem with the unidirectional transverse strength Y_t of the lamina. Under the hypothesis of crack geometry assumed in Section 6.2.1, the surface ΔS is a rectangle given by $\Delta a \times \Delta b$ with rounded corners of radius r. As discussed in Section 6.2.1, r is taken as a function of Δa and Δb, which implies that the surface is given directly by $\Delta S(\Delta a, \Delta b)$. Thus, the stress criterion writes in the 3D model as,

$$\sigma_{yy}^{(90)}(x) \geq Y_t, \quad \forall x \in \Delta S, \tag{6.51}$$

where $\sigma_{yy}^{(90)}$ is obtained from the previous FEM model. This condition can also be expressed as a function of the homogenized strain ε_{yy} in a similar manner to the 2D expression for the stress criterion,

$$\frac{\varepsilon_{yy}}{Y_{\varepsilon t}} \geq s\left(\Delta\hat{a}, \Delta\hat{b}\right) = \max_{x \in \Delta S(\Delta\hat{a}, \Delta\hat{b})} \left(\frac{E_{22}\varepsilon_{yy}^0}{\tilde{\sigma}_{yy}^{(90)}(x)}\right), \tag{6.52}$$

where $\tilde{\sigma}_{yy}^{(90)}$ is the value for $\sigma_{yy}^{(90)}$ extracted from the FEM analysis for a homogenized strain of ε_{yy}^0. These values are plotted in Figure 6.18 scaled with the ratio $E_{22}/\tilde{E}_{22}$.

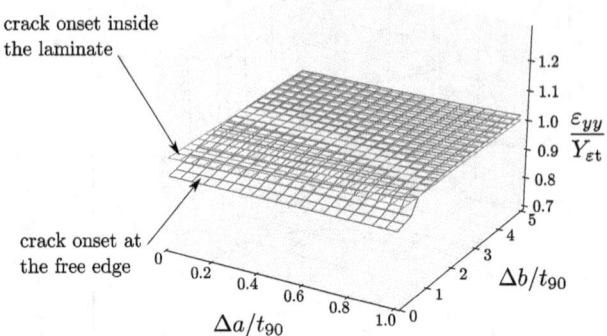

Figure 6.20: Stress criterion for the 3D analysis: value of ε_{yy} required for the crack onset as a function of $\Delta a/t_{90}$ and $\Delta b/t_{90}$ when the crack is situated either at the free edge or inside the laminate for carbon/epoxy and $t_0/t_{90} = 1$.

Note that the previous formulation of the stress criterion limits the evaluation to the crack geometries assumed in Section 6.2.1. However, the set of possible geometries is much wider. In fact, if only the stress criterion was taken into account, the set of preferential crack geometries would be given by surfaces contained by the different isovalue lines of $\sigma_{yy}^{(90)}$ on the xz-plane, see Figure 6.19. A rigorous analysis would require the evaluation of every crack geometry given by these lines. Moreover, if the energy criterion is also taken into account, other crack geometries could be optimum and a very complex analysis would be necessary. The simplification proposed here by the assumption of a certain crack geometry reduces significantly the complexity without affecting noticeably the results. The reason is that, except for very low values of z/t_{90}, the crack geometries assumed are qualitatively similar to those defined by the isovalue lines. Worst-case scenario, the inaccuracies due to the crack geometry assumption will be of the order of the variation of $\sigma_{yy}^{(90)}$, which is very uniform in the whole domain. Its highest variation is of the order of the typical variation of the strength properties as Y_t. Moreover, $\sigma_{yy}^{(90)}$ is almost constant for $\hat{z} \gtrsim 2$, therefore, above this depth the inaccuracies would be insignificant.

Introducing the values obtained by the FEM analysis in (6.51) gives the minimum value of ε_{yy} required by the stress criterion for the onset as a function of Δa and Δb. This condition is plotted in Figure 6.20 for the two locations of the crack at onset. When the crack is assumed to appear inside the laminate, the value of ε_{yy} required for the onset is independent of Δa and Δb and is given by the ratio of Young moduli $E_{22}/\tilde{E}_{22}$ as described in Section 6.1.1. When the crack is assumed to appear at the free edge, a slightly lower value of ε_{yy} is required for very low values of $\Delta b/t_{90}$, i.e. for shallow cracks. For carbon/epoxy and $t_0/t_{90} = 1$, the highest difference on ε_{yy} between shallow and deep cracks is below 10%. The comparison between the 2D and the present 3D stress criterion is equivalent because the value required for deep cracks corresponds exactly to that predicted by the 2D criterion. Thus, regarding to the stress criterion, the error of the 2D model is, in the worst case, about 10%.

6.2.4 Energy released at the crack onset

The energy released at the crack onset $\Delta\Pi$ is a key factor in the energy criterion. In the previous 2D model, it was computed by assuming that the crack spans the whole width W of the inner ply with a constant length Δa. Thus, the released energy was obtained as a function of Δa. This section aims to evaluate how the released energy varies with the crack depth Δb and with the location, see Figure 6.16.

The value of the released energy is estimated here by means of a FEM analysis for the set of values of Δa and Δb shown in Table 6.4. These values have been chosen with the aim of spanning a sufficiently wide domain and within the reasonable computational limits to keep the numerical models able to be managed. As is discussed below, in view of the results obtained, this set can be considered wide enough to capture the different tendencies.

The FEM analysis is carried out here in Abaqus with the aid of FRANC3D, see e.g. Wawrzynek et al. (2009). In the context of this work, Abaqus is used to generate the models without crack and FRANC3D to introduce the crack and remeshing the models, which are solved finally by Abaqus Standard. The entire process used in this work to obtain the released energy is outlined in Figure 6.21 along with the tools used in each step and the inputs and outputs.

$\Delta\Pi$ depends, *a priori*, on the set of parameters which define the elastic problems before and after the crack onset. Assuming the crack geometry described in Section 6.2.1, $\Delta\Pi$ depends on the external load, the laminate geometry, the crack geometry at the onset, its location and the elastic properties characterizing the laminae,

$$\Delta\Pi = \Delta\Pi(\underbrace{\varepsilon_{yy}}_{\text{load}}, \underbrace{\Delta a, \Delta b, \text{Location}}_{\text{crack geometry}}, \underbrace{t_0, t_{90}, W, L}_{\text{laminate geometry}}, \underbrace{E_{11}, E_{22}, \nu_{12}, \nu_{23}, G_{13}, G_{23}}_{\text{elastic properties}}),$$

$$(6.53)$$

where "Location" refers to the position of the crack, i.e. if this is at the free edge or inside the laminate. Analogously to the 2D model, see Section 6.D, a dimensional analysis is carried out here for $\Delta\Pi$ in order to reduce, without loss of generality, the number of parameters to be studied in the computational analysis. Thus, as can be demonstrated, in analogy to the demonstration for G, see Section 6.D, a dimensionless released energy $\Delta\hat{\Pi}$ can be given by the following expression

$$\Delta\Pi = E_{22}\varepsilon_{yy}^2 t_{90}^3 \Delta\hat{\Pi}\left(\Delta\hat{a}, \Delta\hat{b}, \text{Location}, \frac{t_0}{t_{90}}, \frac{E_{11}}{E_{22}}, \nu_{12}, \nu_{23}, \frac{G_{12}}{E_{22}}, \frac{G_{23}}{E_{22}}\right), \quad (6.54)$$

where $\Delta\hat{a} = \Delta a/t_{90}$ and $\Delta\hat{b} = \Delta b/t_{90}$. Thank to this analysis, if the material and t_0/t_{90} is fixed as in the 2D analysis, the parameters to be varied are reduced to those defining the crack geometry ($\Delta a/t_{90}, \Delta b/t_{90}$) and its location. The variation with the rest of parameters is given directly by (6.54).

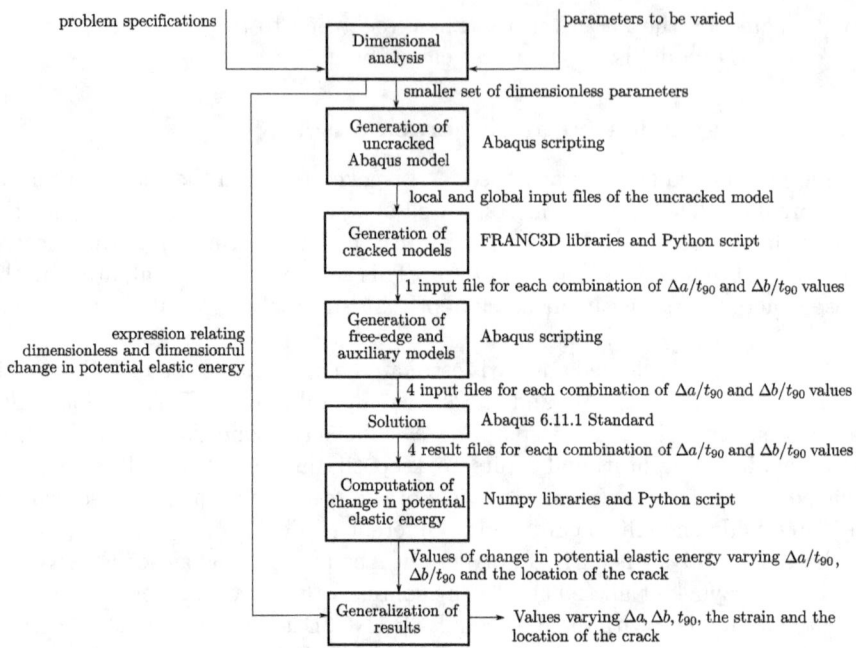

Figure 6.21: Outline of the steps followed to obtain the released energy as a function of the problem parameters.

Once the free parameters are defined, a model without crack is generated in Abaqus, see Figure 6.22(a). The model is similar to that generated in Section 6.2.2 to evaluate the stresses before the crack onset. Since the model is composed by several plies, which correspond to different parts in Abaqus, FRANC3D requires a local/global analysis. In this analysis, the local model will be the part of the model which will be remeshed, being afterwards coupled with the global model. The part to be cracked is the inner ply, therefore, the entire inner ply is identified with the local model. However, due to the large aspect ratio of the inner ply, it is unnecessary to remesh the whole ply when the crack is introduced. To avoid this, the inner ply is divided into four parts, see Figure 6.22(a). The central part of dimensions $2t_{90} \times W_c \times L_c$ is taken as the local model, whereas the other three parts are included in the global model. Every part is tied with the other parts along their common interfaces to simulate perfect interfaces at the interfaces between different plies, or being parts of the same ply at the interfaces between different parts of the inner ply. The material orientation at each part is defined with the aid of local axes. The boundary conditions are the same to those prescribed in Section 6.2.2, a uniform vertical displacement u_y^0 at the whole top edge and vertical displacements vanishing at the bottom edge. In addition, the conditions necessary to avoid rigid motion are prescribed using two arbitrary points of the bottom edge.

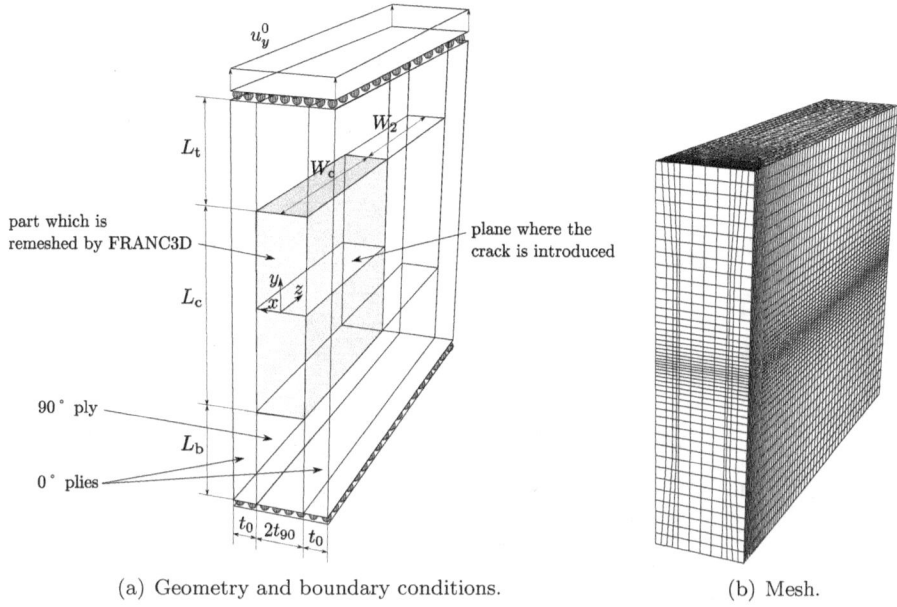

(a) Geometry and boundary conditions. (b) Mesh.

Figure 6.22: FEM model generated with Abaqus to be exported to FRANC3D.

The model described previously is meshed as represented in Figure 6.22(b). Regarding to the local model, this mesh is only relevant with respect to the interfaces between the local and global models. The cause is that the rest of the local model will be remeshed by FRANC3D. However, these surfaces should not be remeshed by FRANC3D to keep conforming the meshes at the interface in order to avoid inaccuracies. Since for large values of $\Delta\hat{a}$ the crack approaches the interface, the mesh is refined at these zones in both inner and outer plies, see Figure 6.22(b). Regarding to the outer ply and the global parts of the inner ply, the entire mesh is kept after the remeshing.

The local model generated by Abaqus is imported by FRANC3D to introduce different cracks and remeshing the models. With the aid of FRANC3D scripting, a model for each pair of values of $\Delta\hat{a}$ and $\Delta\hat{b}$ in Table 6.4 is automatically generated. A long-shallow crack is introduced in each model by meshing the crack front with quadratic wedge elements using quarter point nodes, see Fracture Analysis Consultants, Inc (2011) for additional details. The surface meshing is carried out by FRANC3D, whereas the volume meshing is accomplished by an Abaqus script called by FRANC3D. Observe in Table 6.4 that the pairs corresponding to values of $\Delta\hat{a} \geq 0.8$ and $\Delta\hat{b} \geq 2.5$ are not generated. The reason is that the combinations of values of $\Delta\hat{a}$ corresponding to cracks very close to the interface and very large values for $\Delta\hat{b}$ require an extremely fine mesh at the interface in the uncracked model, whose solution would have a very expensive computational cost. Consequently, this set is excluded from the present analysis since the other models are considered enough to capture the limit behavior. This hypothesis is

$\Delta\hat{b}$ \ $\Delta\hat{a}$	0.1	0.2	0.4	0.6	0.7	0.75	0.85	0.9	0.95
0.05	×	×	×	×	×	×	×	×	×
0.1	×	×	×	×	×	×	×	×	×
0.3	×	×	×	×	×	×	×	×	×
0.4	×	×	×	×	×	×	×	×	×
0.6	×	×	×	×	×	×	×	×	×
0.8	×	×	×	×	×	×	×	×	×
1.0	×	×	×	×	×	×	×	×	×
1.5	×	×	×	×	×	×	×	×	×
2.0	×	×	×	×	×	×	×	×	×
2.5	×	×	×	×	×				
3.0	×	×	×	×	×				
3.5	×	×	×	×	×				
4.0	×	×	×	×	×				
4.5	×	×	×	×	×				
5.0	×	×	×	×	×				

Table 6.4: Set of pairs of values of $\Delta\hat{a} = \Delta a/t_{90}$ and $\Delta\hat{b} = \Delta b/t_{90}$ for which a cracked model has been generated to obtain the released energy.

confirmed by the numerical results.

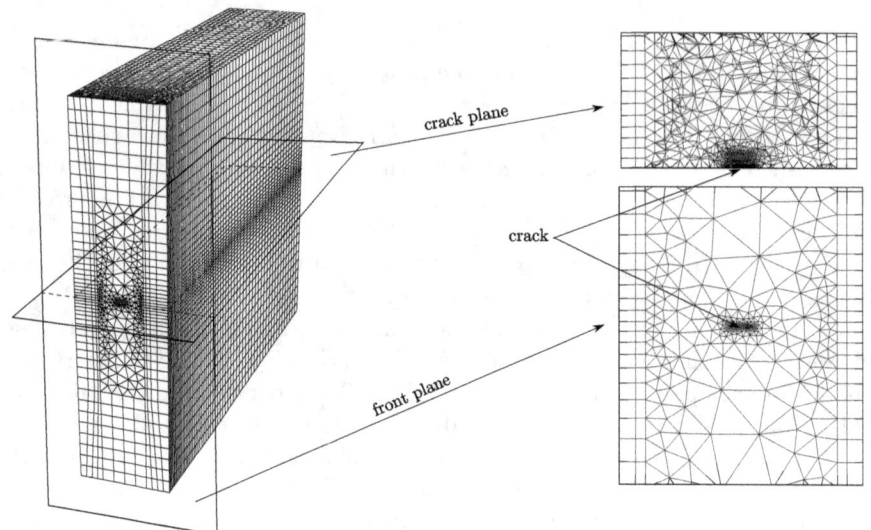

Figure 6.23: Example of mesh for $a = 0.1t_{90}$ and $b = 0.1t_{90}$ generated by FRANC3D.

Figure 6.23 shows an example of mesh corresponding to a model with one of the smallest cracks: $\Delta\hat{a} = 0.1$ and $\Delta\hat{b} = 0.1$. The part which has been remeshed by FRANC3D is clearly distinguishable in the figure. Observe that, for this case, the crack is sufficiently far from the interface and as a consequence a smooth transition is found between the meshes at the two interfaces sides, as clearly showed in the front view. Regarding to the top view, the form of the crack can be clearly observed.

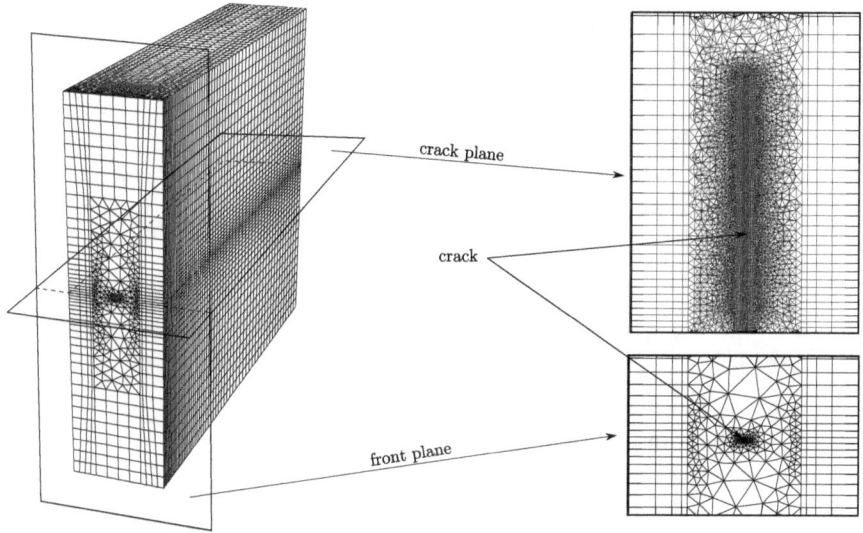

Figure 6.24: Example of mesh for $a = 0.1t_{90}$ and $b = 4.5t_{90}$ generated by FRANC3D.

Figure 6.24 shows an example for an extremal case with a crack very short along the x-direction and very deep along the z-direction corresponding to $\Delta\hat{a} = 0.1$ and $\Delta\hat{b} = 4.5$. Note that, in spite of the extreme aspect ratio of this crack, the mesh is regular enough. In addition, the mesh transition between the two sides of the interface is sufficiently smooth since the crack front is sufficiently far from the interface as in previous example.

Figure 6.25 shows an additional example for an extremal case with a very shallow crack, which is very long along the x-direction, corresponding to $\Delta\hat{a} = 0.95$ and $\Delta\hat{b} = 0.1$. Unlike the previous examples, in this case, the crack front is almost reaching the interface. As a consequence, a strong transition between the meshes at the two sides of the interface can be observed. As a consequence, this model is not able to capture well the weak stress singularity of the crack reaching the interface discussed in Section 6.1.2. A very fine analysis based on remeshing the whole outer plies should be carried if this was the objective of this section. However, for the accuracy of the energy evaluations required here, this model is fine enough.

The models with cracks generated by FRANC3D are subsequently modified by an Abaqus script in order to produce the models necessary for the computation of the released energy for a crack onset in the two different locations studied. Thus, from each model generated by Abaqus, four models are generated:

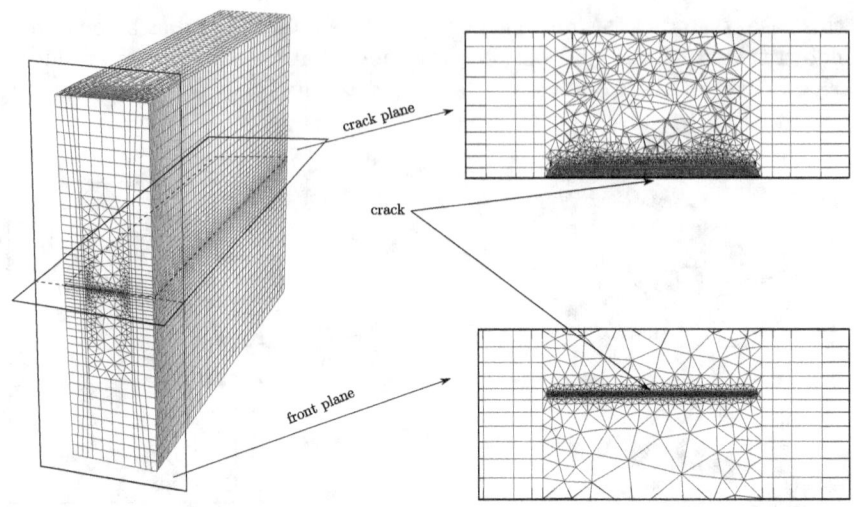

Figure 6.25: Example of the mesh for $a = 0.95t_{90}$ and $b = 0.1t_{90}$ generated by FRANC3D.

- Model with a crack at the free edge: This is the model directly obtained from FRANC3D.

- Model without crack corresponding to the model with a crack at the free edge: This model is obtained by adding a tie condition between the crack surfaces to the previous model. This is equivalent to a laminate without crack and is used to compare the variation of the potential elastic energy.

- Model with crack inside the laminate: This model is obtained by prescribing symmetry conditions along the free edge for $z = 0$, see Figure 6.22(a). Thus, the crack introduced by FRANC3D becomes a crack in the middle of the laminate, far enough from the free edges. This allows studying the two crack locations with the same mesh, which enables to compare both locations without introducing mesh-dependent variations.

- Model without crack corresponding to the model with a crack inside the laminate. This model is generated by prescribing that the crack surfaces of the previous model are tied.

Note that the mesh for the models without cracks is over refined at the zone around the crack since the stress concentration produced by the crack is not longer found after tying its surfaces. Despite it, this strategy is used here because this enables to compute the released energy by comparing magnitudes node to node instead of requiring an additional interpolation, which could introduce inaccuracies.

The four models for each pair of values of $\Delta\hat{a}$ and $\Delta\hat{b}$ listed in Table 6.4 lead to a total amount of 540 models to be solved. The size of the models goes from 809,016 degrees of freedom for the model with the smallest crack to 5,504,094 degrees of freedom for the model with the deepest crack combined with large $\Delta\hat{a}$. These models are solved with Abaqus 6.11.1 Standard.

Once the models have been solved, the released energy can be computed by the different methods which have been described and justified in Section 3.4.1. Two of them have been used here,

- Change of the work of the external forces. This method is applied here by comparing the reaction force given by Abaqus at the nodes situated at the top edge where a vertical displacement was prescribed, see Figure 6.22(a).

- Incremental crack closure technique. This method obtains the change in released energy by computing the work necessary to close the crack. In the present case, considering linear elastic materials and no friction between the crack faces, this work is directly given by the product of the separation of the crack faces when they are not tied and the tractions at the crack faces when they are tied.

Note that both methods require the solution of the models without and with a crack, which justifies the models which have been generated previously.

The percentage difference in the releasd energy computed by the two methods described previously is plotted in Figure 6.26. Observe that for the two hypotheses, with crack inside the laminate and at the free edge, the difference is very low, see Figures 6.26(a) and 6.26(b). It is interesting to observe that the highest differences are found for those models corresponding to small cracks. The reason is that the method based on the change of the external forces is affected by truncation errors since it is based on subtracting two very similar reaction forces with a difference which is lower for smaller cracks. On the contrary, the incremental crack closure method is not affected by these truncation errors because it is based on the product of two magnitudes, thus it is more adequated for smaller cracks. In any case, the percentage difference is sufficiently small in all cases showing an excellent accuracy of the present calculations.

Figure 6.27(a) plots the change in the potential elastic energy $\Delta\Pi$, for cracks situated inside the laminate, as a function of $\Delta\hat{a}$ for several values of $\Delta\hat{b}$. The points represent the values obtained by the numerical analysis described previously and the lines are the result of interpolating these results with cubic splines. This interpolation is necessary for the subsequent application of the energy criterion. Observe that, as expected, the released energy increases with both $\Delta\hat{a}$ and $\Delta\hat{b}$. In particular, for the curves corresponding to large values of $\Delta\hat{b}$, with larger variations, $\Delta\Pi$ clearly increases quadratically with $\Delta\hat{a}$ as occurs typically for cracks in 2D problems in domains without constraining effects. On the contrary, the evolution with $\Delta\hat{b}$ for a fixed $\Delta\hat{a}$ becomes linear for sufficiently large $\Delta\hat{b}$. This can be appreciated in Figure 6.27(a) by comparing the vertical distance between the different points. The difference tends to be the same for fixed increments of $\Delta\hat{b}$. This is due to the tunneling effect which has been widely studied, see

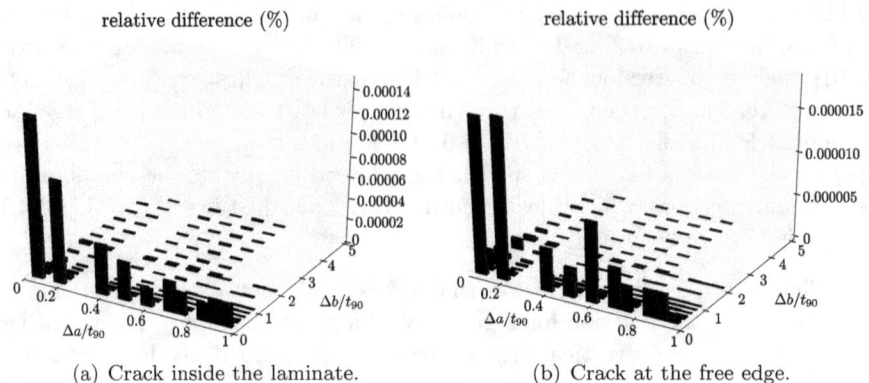

(a) Crack inside the laminate.   (b) Crack at the free edge.

Figure 6.26: Percentage difference in the released energy obtained by the change of external forces and the incremental crack closure technique.

e.g. Hutchinson and Suo (1992). Basically, the reason is that if a crack with a high aspect ratio between the lengths along two directions is enlarged along the large-length direction, the released energy by the opening of a certain increment of length is constrained by the fact that the crack is very short in the other direction. As a consequence the crack surfaces do not separate following the typical form, but the separation is significantly reduced and almost uniform along the large-length direction in the central zone because the crack is relatively short in the other direction. In terms of the energy release rate G, it means that G becomes independent of the length of the crack along this large-length direction. Moreover, in the case of cross-ply laminates, an additional constraining effect arises as a consequence of the high stiffness of the outer plies.

Figure 6.27(b) represents the analogous results for $\Delta\hat{\Pi}$ for the case with the crack at the free edge. The same comments can be given for these results because the results for the two crack locations are qualitatively similar. The percentage difference in $\Delta\hat{\Pi}$ between the crack onset at the free edge and inside the laminate is represented in Figure 6.28 with iso-difference lines. The highest differences are found for the lower values of $\Delta\hat{b}$ which corresponds to the most shallow cracks. The reason is that, when the cracks are deep, the part affected by the free edge is small, because above a certain depth, the constraining effect of the outer plies and the surrounding material affects more strongly than the free edge. Moreover, for $\Delta\hat{b} \gtrsim 3$ the difference is lower than 2.5%, vanishing for large values of $\Delta\hat{b}$. On the other hand, in general, the difference increases slightly with $\Delta\hat{a}$ because a crack larger along this direction increases the influence of the free edge on deeper cracks. For carbon/epoxy and $t_0/t_{90} = 1$, the highest difference found is about 30%, which is a difference which cannot be neglected and could affect the global critical strain obtained by the coupled criterion.

Figure 6.29 shows the released energy, plotted in previous figures, now normalized with the crack depth $\Delta\hat{b}$ for crack onset at the free edge. It is interesting to compare these results with the integration of the value of the ERR $\hat{G}$, see

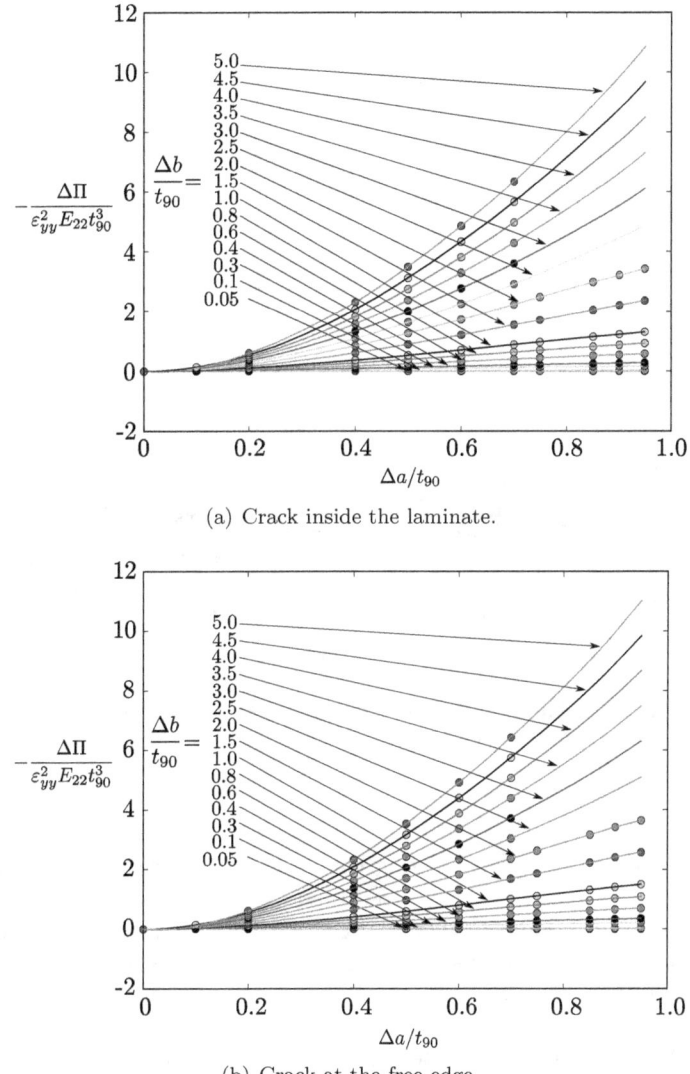

(a) Crack inside the laminate.

(b) Crack at the free edge.

Figure 6.27: Dimensionless released energy $\Delta\hat{\Pi}$ as a function of the crack length $\Delta\hat{a}$ for several values of the crack depth $\Delta\hat{b}$ for carbon/epoxy and $t_0/t_{90} = 1$ inside the laminate.

Figure 6.4, obtained in Section 6.1.2 by means of a Boundary Element code and assuming generalized plane strain. Note that, for moderately large crack depths $\hat{b} \gtrsim 2.5$, the curves are closely approximated by the curve corresponding to generalized plane strain. This is due to the constraining effect for moderately deep cracks discussed above. In fact, the concentration of all curves for $\hat{b} \gtrsim 2.5$ in this plot is equivalent to the observed linear dependence on $\Delta\hat{b}$ when the released

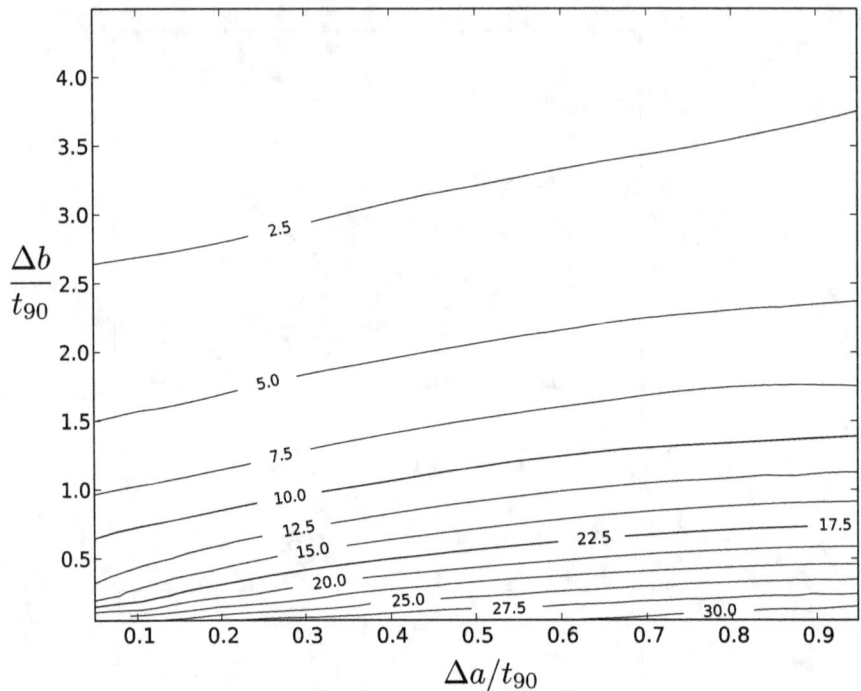

Figure 6.28: Percentage difference between the energy released at the onset of a crack situated at the free edge and inside the laminate.

energy is plotted without normalizing with $\Delta\hat{b}$, see Figure 6.27. The only relevant difference between the generalized plane strain results and the results for large $\Delta\hat{b}$ is found for $\Delta\hat{a} \to 1$. The 2D results show a horizontal tangent at this point, whereas the 3D results seem to predict a finite positive slope tangent for $\Delta\hat{a} \to 1$. In the first case, this is a direct consequence of the fact that the energy release rate $\hat{G}(\hat{a} \to 1) \to 0$, see a detailed discussion in Section 6.1.2. However, the 3D FEM model introduced here is not able to capture accurately such an asymptotic behavior because this would require a much more refined mesh along the crack front including the outer plies. Figure 6.27 shows that, in spite of this detail, the results are approximated enough for the purposes of this work.

Regarding to the main objective of this work, this result showing that the curves from the 3D analysis, for moderate values of the crack depth, approximately collapse to the curve of the generalized plane strain indicates us that, in terms of the released energy, the generalized plane strain analysis can be used, without a relevant loss of accuracy, for sufficiently deep crack onsets. In the case of a crack at the free edge, the 2D model is also very approximated above a threshold depth $\hat{b} \gtrsim 2.5$, which corresponds to a depth very similar to the inner-ply thickness $2t_{90}$. Since the majority of cracks observed in laminates exceed this depth, in terms of the released energy, the generalized plane strain can be

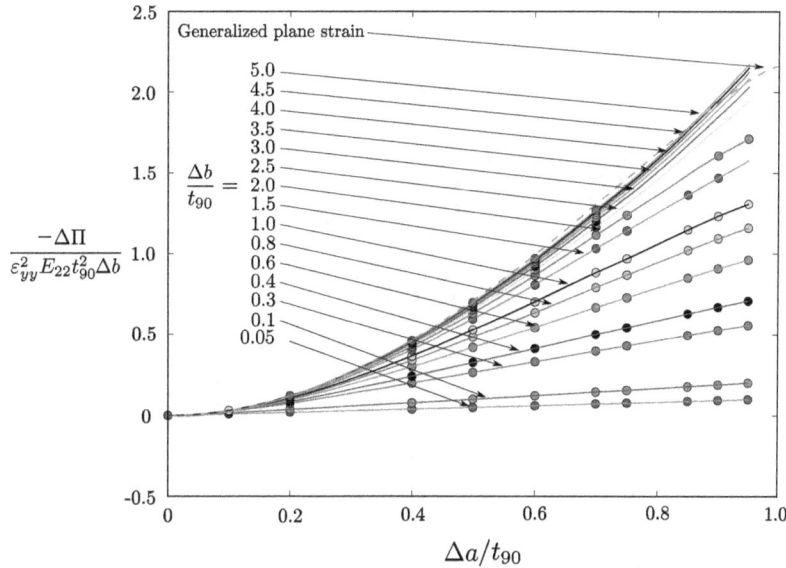

Figure 6.29: Dimensionless released energy $-\Delta\hat{\Pi}$ normalized with the crack depth $\Delta\hat{b}$ as a function of $\Delta\hat{a}$ for carbon/epoxy, $t_0/t_{90} = 1$ and several values of $\Delta\hat{b}$ in comparison with the generalized-plane-strain results.

considered an accurate approximation.

6.2.5 Energy criterion for the 3D model

In view of the results presented in previous section, the energy criterion has to be reformulated with respect to the previous version for the 2D model. Recall the energy condition considered in Section 6.1.2 for the 2D model assuming an initial quasistatic state,

$$- \Delta\Pi \geq \Delta\Gamma, \qquad (6.55)$$

where $\Delta\Pi$ has been obtained in the previous section and $\Delta\Gamma$ is the energy dissipated during the crack onset.

$\Delta\Gamma$ is estimated here in an analogous manner to that used in the 2D model. A uniform fracture energy is assumed and identified with the transverse fracture toughness G_c, see Section 6.1.2 for a discussion about the reasons for this assumption. Thus, $\Delta\Gamma$ is directly calculated by the product of G_c and the area of the surface crack,

$$\Delta\Gamma = G_c t_{90}^2 \Delta\hat{\Gamma}(\Delta\hat{a}, \Delta\hat{b}) = G_c t_{90}^2 \int_{\Delta S(\Delta\hat{a},\Delta\hat{b})} \mathrm{d}\hat{x}\mathrm{d}\hat{z}, \qquad (6.56)$$

where $\Delta S(\Delta\hat{a}, \Delta\hat{b})$ corresponds to the surface of the crack, see Section 6.2.1.

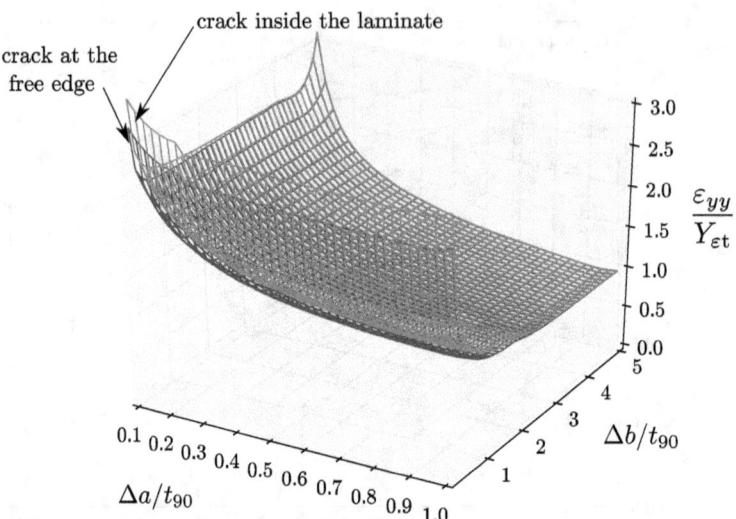

Figure 6.30: Energy criterion for the 3D analysis: value of ε_{yy} required for the crack onset as a function of $\Delta a/t_{90}$ and $\Delta b/t_{90}$ when the crack is situated either at the free edge or inside the laminate for carbon/epoxy, $t_0/t_{90} = 1$ and $\gamma = 1$.

The expression of the energy criterion (6.55) can be rewritten in terms of the strain ε_{yy} by introducing the expressions for the released energy (6.54) and the dissipated energy (6.56) as,

$$\frac{\varepsilon_{yy}}{Y_{\varepsilon t}} \geq \gamma \sqrt{g(\Delta \hat{a}, \Delta \hat{b})}, \qquad (6.57)$$

where γ is the same dimensionless brittleness number defined in (6.19) and $g(\Delta \hat{a}, \Delta \hat{b})$ is the ratio of the dimensionless dissipated to released energy,

$$g(\Delta \hat{a}, \Delta \hat{b}) = \frac{\Delta \hat{\Gamma}(\Delta \hat{a}, \Delta \hat{b})}{\Delta \hat{\Pi}(\Delta \hat{a}, \Delta \hat{b})}. \qquad (6.58)$$

The expression for the energy criterion given by (6.57) for $\gamma = 1$ is represented in Figure 6.30 for the two locations considered for the crack onset. These surfaces show the lowest value required for ε_{yy} by the energy criterion as a function of $\Delta \hat{a}$ and $\Delta \hat{b}$. Note that, analogously to the curve for the 2D model, see Figure 6.6, in this case the strain required for a crack onset of a infinitesimal area, i.e. either $\Delta \hat{a} \to 0$, $\Delta \hat{b} \to 0$ or both, is unbounded, which is the 3D result of the well-known inability of the Griffith criterion to predict crack initiation, see Section 2.1 for a discussion. On the other hand, in this case, a minimum for the energy criterion cannot be observed as for the 2D model. The reason is that this minimum is a direct consequence of the result $\hat{G} \to 0$, see Section 6.E for a proof. Since this result has not been obtained here due to computational limitations, see

Section 6.2.4 for a full discussion, the energy criterion obtained does not present a relative minimum. The effect of this on the accuracy of results in terms of the critical ε_{yy} is negligible as can be deducted from the comparison of the values of the function $g(\Delta\hat{a})$ in Figure 6.7 at the minimum and at $\Delta\hat{a} \to 1$.

The tendency observed in Figure 6.30 is similar to that observed for the 2D analysis. In general, the required value of ε_{yy} for the crack onset decreases with both crack length $\Delta\hat{a}$ and depth $\Delta\hat{b}$, thus, large cracks will be preferential from the energetic point of view.

Regarding to the energy criterion, the differences between the two locations considered for the crack onset can also be observed in Figure 6.30. Both surfaces differ slightly as a direct consequence of the differences for the value of $\Delta\Pi$ described previously. Similarly to $\Delta\Pi$ the difference is from moderate to low for small cracks and almost negligible for large cracks.

6.2.6   Coupled criterion

Once both criteria have been reformulated for the 3D model, it is necessary to study how they are coupled according to Leguillon's hypothesis, see Section 6.1.3 for the analogous section for the 2D model and Section 3.2 for a general discussion about this hypothesis. Thus, the crack onset is given by the minimum value of strain ε_{yy} fulfilling both criteria,

$$\frac{\varepsilon_{yy}^{c}}{Y_{\varepsilon t}} = \min_{\Delta\hat{a},\Delta\hat{b}} \left[\max\left(s\left(\Delta\hat{a}, \Delta\hat{b}\right), \gamma\sqrt{g\left(\Delta\hat{a}, \Delta\hat{b}\right)} \right) \right], \qquad (6.59)$$

where $s\left(\Delta\hat{a}, \Delta\hat{b}\right)$ and $g\left(\Delta\hat{a}, \Delta\hat{b}\right)$ are the functions characterizing the stress and energy criteria, respectively, defined in (6.52) and (6.58).

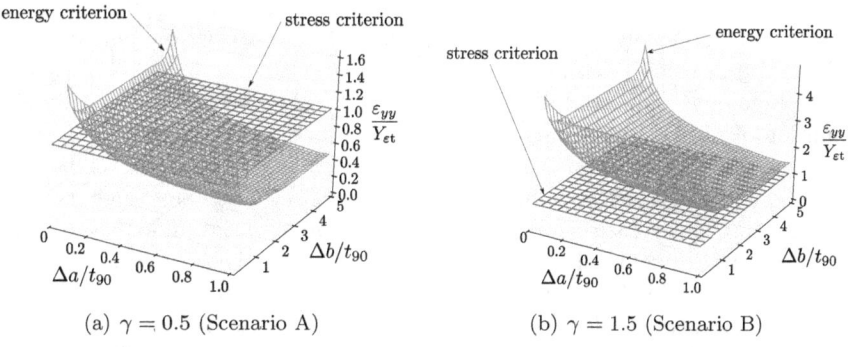

(a) $\gamma = 0.5$ (Scenario A)   (b) $\gamma = 1.5$ (Scenario B)

Figure 6.31: Combination of both stress and energy criteria for two different values of γ.

Figure 6.31 shows two examples of how the curves presented separately above are combined. Analogously to the 2D model, two different scenarios can be defined here depending on the value of γ,

- Scenario A: If the two surfaces intersect, see Figure 6.31(a), the value for the critical strain ε_{yy}^c is given by the lowest point of the curve defined by the intersection of the two criteria surfaces.

- Scenario B: If the surfaces of these two criteria do not intersect, see Figure 6.31(b), the value for the critical strain is given by the minimum of the energy criterion surface, which correspond to the largest crack in the studied domain.

In the context of this work, the results presented in the following section are obtained by a numerical optimization of both criteria in the form expressed in (6.59).

6.2.7 Results and discussion

This section presents the main results obtained from the present 3D model, focusing on the comparison with the 2D model in order to verify the accuracy of this simplified model from the point of view of the 3D analysis.

Figure 6.32 shows the results of the critical crack length $\Delta\hat{a}_c$ and depth $\Delta\hat{b}_c$ predicted by the 3D model as a function of γ for the two locations considered for the crack onset. The results are very similar for both locations. For lower values of γ, the coupled criterion predicts the onset of small cracks. In fact, both $\Delta\hat{a}_c$ and $\Delta\hat{b}_c$ vanish when $\gamma \to 0$. Note that the curve for lower values of γ is not smooth. The reason is that for these values of γ corresponding to scenario A, the values of $\Delta\hat{a}_c$ and $\Delta\hat{b}_c$ are essentially given by the surface corresponding to the stress criterion, which is quite plane. As a consequence, any shallow local minimum, due to, e.g., the interpolation used, can produce these irregularities. At the other extreme for higher values of γ, $\Delta\hat{a}_c$ and $\Delta\hat{b}_c$ tend to the maximum values within the studied domain. Above a certain value of γ, $\Delta\hat{a}_c$ and $\Delta\hat{b}_c$ are fixed to these values. In the case of $\Delta\hat{b}_c$, it is foreseeable that if the analysis included a larger range for $\Delta\hat{b}$, the value predicted for $\Delta\hat{b}_c$ would correspond to the upper bound of this range. In the limit, above a certain value of γ, $\Delta\hat{b}_c$ would span the whole width of the laminate.

The predicted critical strain for the onset ε_{yy}^c is presented in Figure 6.33 as a function of γ for the 2D model and the two crack locations considered in the 3D model. This figure shows that the three curves are very similar in the whole range of γ studied. For low values of γ, the crack onset at the free edge requires a lower ε_{yy}^c, thus this location for the crack onset is preferential. The reason is that, in scenario A, the value of ε_{yy}^c is mainly given by the stress criterion for small cracks which is lower for free-edge cracks, see Figure 6.20, as a consequence of the stress concentration for $\hat{z} \to 0$. On the other extreme, for large γ, the three cases predict a linear relation with a very similar slope. The differences here can be explained by the fact that the computational results obtained for the 3D model overestimate the value of $\Delta\Pi$ when $\Delta\hat{a} \to 0$ as can be observed in Figure 6.29, see also Section 6.2.4 for a discussion. This overestimation produces that the value of the minimum of $g(\Delta\hat{a}, \Delta\hat{b})$ is underestimated, which implies that the slope is underestimated.

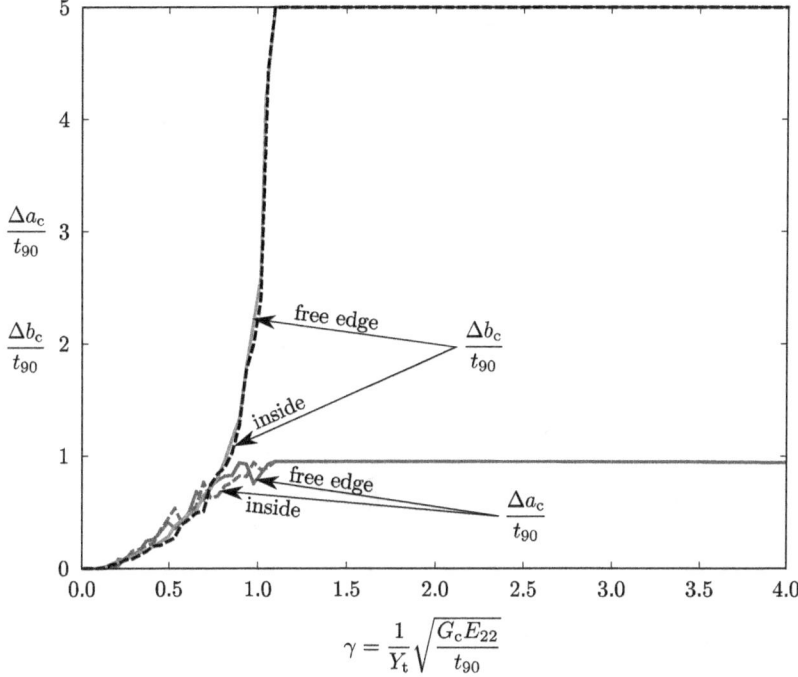

Figure 6.32: Critical crack length $\Delta\hat{a}_c$ and depth $\Delta\hat{b}_c$ predicted by the 3D model as a function of γ when the crack is situated either at the free edge or inside the laminate for carbon/epoxy and $t_0/t_{90} = 1$.

As described in Section 6.1.5.1, the variation with γ can be interpreted as a size effect of the inner-ply thickness. Thus, using the relation (6.36) between γ and t_{90}, the results presented in Figure 6.32 are plotted as a function of t_{90} in Figure 6.34. This figure shows that the size effect predicted by the 3D model is essentially similar to that predicted by the 2D model. The differences are analogous to those described for the results as a function of γ. The main difference is found for thick inner plies which corresponds to $\gamma \to 0$. At this extreme the value of the critical strain ε_{yy}^c in the 2D model is slightly overestimated because the stress concentration at the free edge decreases the value of ε_{yy}^c necessary to fulfill the stress criterion.

The results presented here showed that the 3D version of the FFM analysis confirms that, at least from the FFM perspective, the 2D model is an accurate approximation. This analysis has also shown the drastic increase of the amount and the complexity of the numerical models necessary to carry out the 3D analysis in comparison with the much simpler computational requirements of the 2D model. In fact, the differences between both models are very slight in comparison with typical dispersion in the fracture and strength properties. In view of this outcome, the 2D model will be used in what follows in this chapter.

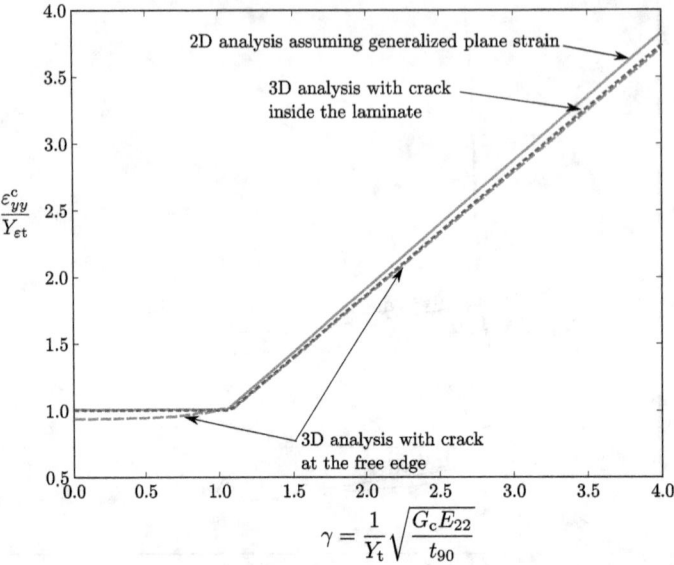

Figure 6.33: Critical strain ε_{yy}^c for the crack onset as a function of γ for the 2D model and the two locations considered in the 3D model for carbon/epoxy and $t_0/t_{90} = 1$.

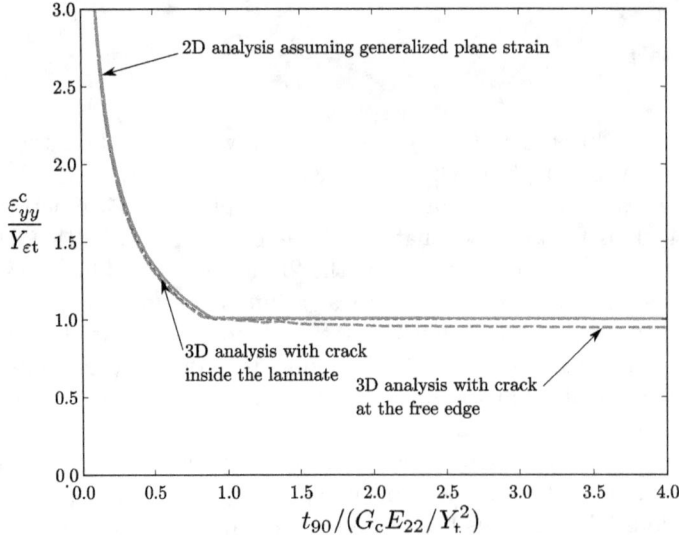

Figure 6.34: Critical strain ε_{yy}^c for the crack onset as a function of the inner-ply thickness t_{90} for the 2D model and the two locations considered in the 3D model for carbon/epoxy and $t_0/t_{90} = 1$.

6.3 2D generalized-plane-strain model considering the influence of residual thermal stresses

Residual thermal stresses have been excluded from the models presented previously. However, it is well known that residual stresses can play an important role in composites with thermosetting resins and coefficient of thermal expansion (CTE) which varies with the orientation of the ply. In these composites, the cooling-down process after the curing of the laminate from the stress-free to the service temperature can cause the presence of residual thermal stresses. For example, they are relevant in laminates formed with unidirectional carbon/epoxy plies due to the high contrast between the longitudinal and transverse CTE caused by the negative CTE of the carbon fibers at their axis-direction. Moreover, this is particularly important in the cases of laminates with a high contrast between the orientations of the different plies, e.g., cross-ply laminates. In view of these facts, this section aims to modify the generalized plane strain model developed in Section 6.1 in order to consider the presence of residual thermal stresses associated to the temperature change from the stress-free to the room temperature $\Delta T \leq 0$.

The presence of residual stresses affects both the stress and energy criteria. On the one hand, the stress criterion is altered since the stresses at the inner ply changes as will be seen in Section 6.3.1. On the other hand, the energy criterion will be modified by the presence of residual stresses because of their effect on the released energy. This is analyzed in Section 6.3.2. Finally, a new expression for the critical strain leading to the first transverse crack onset considering residual thermal stresses is given in Section 6.3.3.

6.3.1 Stress criterion considering residual thermal stresses

The stress criterion applied in the original 2D model developed in Section 6.1 postulates that a crack onset is possible if the inner-ply tension normal to the plane where the crack will appear $\sigma_{yy}^{(90)}$ exceeds or equals the unidirectional transverse strength Y_{t},

$$\sigma_{yy}^{(90)} \geq Y_{\mathrm{t}}. \tag{6.60}$$

Whereas in the original 2D model the left term in the previous condition is directly proportional to the applied strain ε_{yy}, in the presence of residual stresses, this term depends also on the temperature change ΔT.

In the case of combined applied strain and temperature change, the value of $\sigma_{yy}^{(90)}$ can be obtained by the superposition of two partial problems, see Figure 6.35, suitably defined to simplify the analysis:

- A laminate subjected to a uniform strain $\varepsilon_{yy}^{0,\Delta T} + \varepsilon_{yy}$. Taking as reference the state of the laminate prior to the curing process, this strain is the combination of the strain $\varepsilon_{yy}^{0,\Delta T}$ due to the free deformation of the laminate during the cool-down process, and the uniform strain ε_{yy} imposed by the external forces at the service temperature, i.e. the strain measured during a tensile testing at the room temperature. The fact of taking the state previous to the cool-down as reference for the strain is proposed in order to simplify the other partial problem. This strategy will be particularly simplifying in the partial problems proposed for the energy criterion in Section 6.3.2.

- A laminate subjected to a temperature change with vanishing vertical displacements at the top and bottom edges. Note that this is not the actual situation during the cool-down process because the laminate can be freely deformed due to the temperature change. The free deformation of the whole laminate during the cool-down process is taken into account in the first partial problem by the addition of the term $\varepsilon_{yy}^{0,\Delta T}$ as described previously. This separation enables to simplify the analysis of this problem since this is reduced to a temperature change with vanishing displacements, which will be particularly important in the analysis of the energy criterion in Section 6.3.2. Obviously, the stresses $\sigma_{yy}^{(90)}$ physically associated to residual thermal stresses can be obtained by the combination of the stresses given by this partial problem and the stresses corresponding to the strain $\varepsilon_{yy}^{0,\Delta T}$ from the first one. As will be seen later on, both of them depends on ΔT linearly.

The value of $\sigma_{yy}^{(90)}$ in the global problem is given by the sum of the values for the first $\sigma_{yy}^{(90),\varepsilon}$ and second problem $\sigma_{yy}^{(90),\Delta T}$,

$$\sigma_{yy}^{(90)} = \sigma_{yy}^{(90),\varepsilon} + \sigma_{yy}^{(90),\Delta T}. \tag{6.61}$$

The value of $\sigma_{yy}^{(90),\varepsilon}$, corresponding to the first problem, was obtained in Section 6.1.1 by means of the laminate theory, giving,

$$\sigma_{yy}^{(90),\varepsilon} = \tilde{E}_{22}\left(\varepsilon_{yy}^{0,\Delta T} + \varepsilon_{yy}\right), \tag{6.62}$$

where $\tilde{E}_{22}$ is the apparent transverse Young's modulus of the inner 90° ply in the y-direction defined in (6.3). An expression for the mid-plane strain after the cool-down process $\varepsilon_{yy}^{0,\Delta T}$ is obtained by means of the laminate theory in Section 6.B as,

$$\varepsilon_{yy}^{0,\Delta T} = \alpha_2 \Delta T k^{0,\Delta T}\left(\frac{E_{22}}{E_{11}}, \nu_{12}, \frac{t_0}{t_{90}}, \frac{\alpha_1}{\alpha_2}\right), \tag{6.63}$$

where α_1 and α_2 are the CTE along parallel and transverse to the fiber direction, respectively, and $k^{0,\Delta T}$ has been defined in (6.105).

Regarding the value of the tension due to the increment of temperature $\sigma_{yy}^{(90),\Delta T}$ in the second problem, this is obtained using the laminate theory in Section 6.C as,

$$\sigma_{yy}^{(90),\Delta T} = -E_{22}\alpha_2 \Delta T k^{(90),\Delta T}\left(\frac{E_{22}}{E_{11}}, \nu_{12}, \frac{t_0}{t_{90}}, \frac{\alpha_1}{\alpha_2}\right), \tag{6.64}$$

where $k^{(90),\Delta T}$ is an analytical function defined in (6.115).

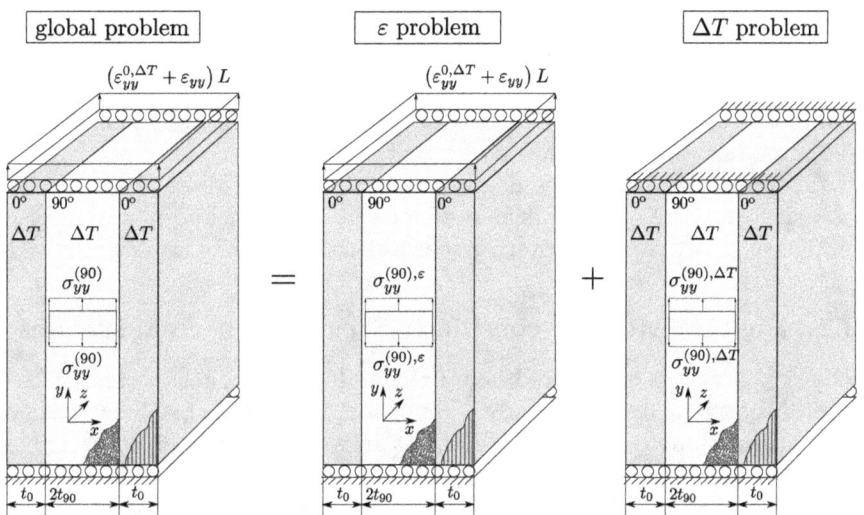

Figure 6.35: Schematic of the superposition employed to obtain the tension normal to the future crack plane, $\sigma_{yy}^{(90)}$, in the global problem through its value $\sigma_{yy}^{(90),\varepsilon}$ and $\sigma_{yy}^{(90),\Delta T}$ at the partial problems ε and ΔT, respectively.

Substituting (6.62) and (6.64) into (6.61), the total value for the normal tension $\sigma_{yy}^{(90)}$ is,

$$\sigma_{yy}^{(90)} = \tilde{E}_{22} \left(\varepsilon_{yy} + \alpha_2 \Delta T k^{0,\Delta T} \right) - E_{22} \alpha_2 \Delta T k^{(90),\Delta T}, \tag{6.65}$$

where the arguments of $k^{0,\Delta T}$ and $k^{(90),\Delta T}$ have not been detailed for the sake of simplicity, analogously to the treatment of $\tilde{E}_{22}$. The expression of $\sigma_{yy}^{(90)}$ is introduced into (6.60) in order to obtain the condition given by the stress criterion in terms of ε_{yy} and ΔT,

$$\tilde{E}_{22} \left(\varepsilon_{yy} + \alpha_2 \Delta T k^{0,\Delta T} \right) - \alpha_2 E_{22} \Delta T k^{(90),\Delta T} \geq Y_t. \tag{6.66}$$

In the case of the problem studied in this chapter, the temperature change ΔT and its associated residual stress is not a magnitude to be varied as ε_{yy} to reach a certain critical value but it is a property of the own manufacturing process of each material. As a consequence, the terms depending on ΔT play here the role of constants. Due to this, the previous expression of the stress criterion is rewritten by isolating ε_{yy} with the objective of defining the condition in analogous manner to that used in the original 2D model in Section 6.1.1, see (6.5),

$$\frac{\varepsilon_{yy}}{Y_{\varepsilon t}} \geq \frac{E_{22}}{\tilde{E}_{22}} + \frac{\alpha_2 \Delta T}{Y_{\varepsilon t}} \left(\frac{E_{22}}{\tilde{E}_{22}} k^{(90),\Delta T} - k^{0,\Delta T} \right). \tag{6.67}$$

This expression replaces the final expression (6.5) of the stress criterion obtained in Section 6.1.1 without considering residual stresses. The only difference with

(6.5) is the second term at the left-hand side of the inequality. This term corresponds to the effect of the residual thermal stresses. Since this appears as an additional term, this can be interpreted as an equivalent residual strain which is added to the critical normalized strain $E_{22}/\tilde{E}_{22}$ reducing or increasing the value of the strain ε_{yy} required for the onset from the stress-criterion perspective. Note that this term does not represent physically any type of residual strain but it is only denoted here as equivalent residual strain for the interpretation of the result, given that the stress criterion is expressed in terms of strains.

6.3.2   Energy criterion considering residual thermal stresses

The energy criterion is revisited here by considering residual thermal stresses. The initial energetic balance proposed in the original model in Section 6.1.2 under the only assumption of quasistatic initial state is expressed in (6.10) as,

$$\int_0^{\Delta a} G(a)\mathrm{d}a \geq G_c\Delta a. \tag{6.68}$$

The left-hand side of the previous expression represents the total released energy due to the crack onset, whereas the right-hand side is the dissipated energy. The presence of residual thermal stresses can modify significantly the total released energy during the crack onset. In contrast, it is not expected to affect to the dissipated energy since the transverse fracture toughness is considered a material property. Thus, the study of the effect of the residual thermal stresses on the energy criterion only requires the study of the variation of the energy release rate G.

The value of G considering residual thermal stresses is obtained in an analogous manner to the procedure used for the stress criterion, i.e. by the superposition of two problems. These problems are the cracked versions of the decomposition proposed previously in Figure 6.35. The two problems are represented in Figure 6.36 along with the global problem. The first problem of the decomposition corresponds to a laminate with a transverse crack of length a in the inner ply, which is subjected to an applied strain $\varepsilon_{yy}^{0,\Delta T} + \varepsilon_{yy}$ through vertical displacements prescribed at its top edge, whereas the second problem correspond to the same laminate with forbidden vertical displacements at these edges and a temperature change in the whole laminate. The introduction of the term $\varepsilon_{yy}^{0,\Delta T}$ into the purely mechanical partial problem is proposed in order to simplify the second partial problem allowing a simpler superposition of G.

The stresses, strains and displacements of the global problem are obtained by summing the corresponding magnitudes of the partial problems. Since the stress intensity factors at the crack tip K_i, corresponding to each mode $i = \mathrm{I, II, III}$, are linear with the stress state, the values of K_i for the global problem is directly the addition of the values of the stress intensity factors corresponding to the first K_i^ε and second $K_i^{\Delta T}$ problems of the decomposition,

$$K_i = K_i^\varepsilon + K_i^{\Delta T}. \tag{6.69}$$

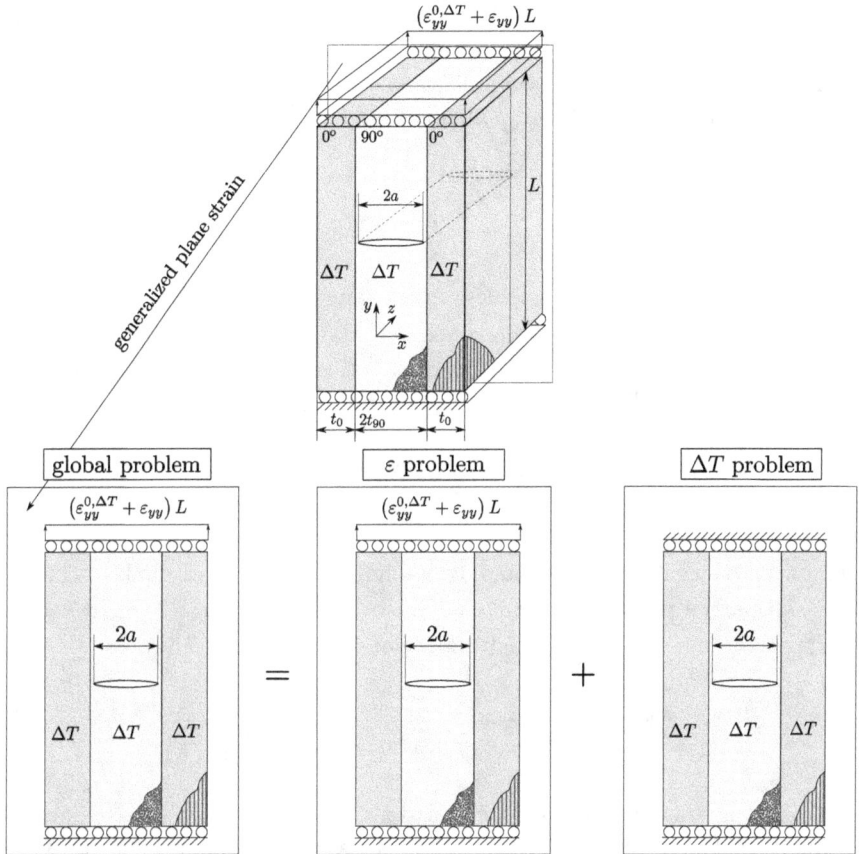

Figure 6.36: Schematic of the superposition employed to obtain the energy release rate of the crack in the global problem through its value G^ε and $G^{\Delta T}$ for the partial problems ε and ΔT, respectively.

Note that this expression cannot be used if a nonlinear behavior is found in some or both of the partial problems.

Due to the symmetry and the loads applied in both partial problems, the transverse crack studied here is opened in pure mode I. Taking into account this fact, the previous superposition (6.69) in terms of K_i can be transformed by using the Irwin's expression relating K_i and G_i as,

$$G(a, \varepsilon_{yy}, \Delta T) = \left(\sqrt{G^\varepsilon \left(a, \varepsilon_{yy}^{0,\Delta T} + \varepsilon_{yy} \right)} + \sqrt{G^{\Delta T}(a, \Delta T)} \right)^2, \qquad (6.70)$$

where G^ε and $G^{\Delta T}$ are the energy release rates corresponding to the partial problems subjected to an applied strain $\varepsilon_{yy}^{0,\Delta T} + \varepsilon_{yy}$ and a temperature change ΔT respectively.

The term $G^\varepsilon(a, \varepsilon_{yy}^{0,\Delta T} + \varepsilon_{yy})$ corresponds to the energy release rate presented in Section 6.1.2 for the problem without considering residual thermal stresses. The dimensionless $\hat{G}^\varepsilon(a)$ proposed in Section 6.D is used here taking into account that the total strain is $\varepsilon_{yy}^{0,\Delta T} + \varepsilon_{yy}$ in this case,

$$\hat{G}^\varepsilon(\hat{a}) = \frac{G^\varepsilon(a, \varepsilon_{yy}^{0,\Delta T} + \varepsilon_{yy})}{E_{22}\left(\varepsilon_{yy}^{0,\Delta T} + \varepsilon_{yy}\right)^2 t_{90}}, \tag{6.71}$$

where $\hat{a} = a/t_{90}$ is the dimensionless crack length. The expression for $\hat{G}^\varepsilon(\hat{a})$ can be manipulated in order to make explicit the normal tension $\sigma_{yy}^{(90)}$ prior to the crack onset. Recall from (6.62) that $\sigma_{yy}^{(90),\varepsilon} = \tilde{E}_{22}(\varepsilon_{yy}^{0,\Delta T} + \varepsilon_{yy})$. Thus the expression (6.71) can be rewritten as,

$$\hat{G}^\varepsilon(\hat{a}) = \frac{G^\varepsilon(a, \varepsilon_{yy}^{0,\Delta T} + \varepsilon_{yy})}{\left(\sigma_{yy}^{(90),\varepsilon}\right)^2 \left(E_{22}/\tilde{E}_{22}\right)^2 t_{90}/E_{22}}. \tag{6.72}$$

Following the same idea, an analogous dimensionless energy release rate for the second partial problem $\hat{G}^{\Delta T}(a)$ is defined substituting $\sigma_{yy}^{(90),\varepsilon}$ in (6.72) with the expression of $\sigma_{yy}^{(90),\Delta T}$ obtained in previous section, see (6.64), as

$$\hat{G}^{\Delta T}(\hat{a}) = \frac{G^{\Delta T}(a, \Delta T)}{E_{22}\left(-\alpha_2 \Delta T k^{(90),\Delta T}\right)^2 \left(E_{22}/\tilde{E}_{22}\right)^2 t_{90}}. \tag{6.73}$$

Introducing the dimensionless expressions for the energy release rates for the first (6.71) and second (6.73) problems into (6.70), and substituting the expression of $\varepsilon_{yy}^{0,\Delta T}$ (6.63), gives,

$$G(a, \varepsilon_{yy}, \Delta T) = E_{22} t_{90} \left(\left(\varepsilon_{yy} + \alpha_2 \Delta T k^{0,\Delta T}\right) \sqrt{\hat{G}^\varepsilon(\hat{a})} - \right.$$
$$\left. - \alpha_2 \Delta T k^{(90),\Delta T} \frac{E_{22}}{\tilde{E}_{22}} \sqrt{\hat{G}^{\Delta T}(\hat{a})} \right)^2. \tag{6.74}$$

In view of this expression, the generalization of the energy criterion to consider residual thermal stresses only requires to compute $\hat{G}^{\Delta T}(\hat{a})$ for a certain temperature change ΔT taken as reference. Given that $\hat{G}^\varepsilon(\hat{a})$ was obtained in Section 6.1.2, once $\hat{G}^{\Delta T}(\hat{a})$ is known, the expression (6.74) enables to obtain directly $G(a, \varepsilon_{yy}, \Delta T)$ for any value of ε_{yy} and ΔT.

For the sake of comparison, $\hat{G}^{\Delta T}(\hat{a})$ is obtained by the means of the boundary element code used in Section 6.1.2 for the same carbon/epoxy laminate taken as reference previously in this chapter with $t_0/t_{90} = 1$ and material properties shown in Table 6.1. The same mesh generated to obtain $\hat{G}^\varepsilon(\hat{a})$ in Section 6.1.2 is used here to avoid mesh-dependent differences. The only difference between the models are the boundary conditions which are defined in accordance with the second partial problem, see Figure 6.36.

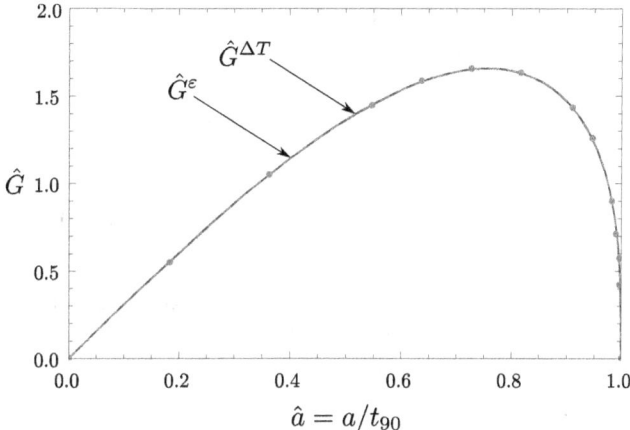

(a) Dimensionless energy release rate of the partial problems of the superposition

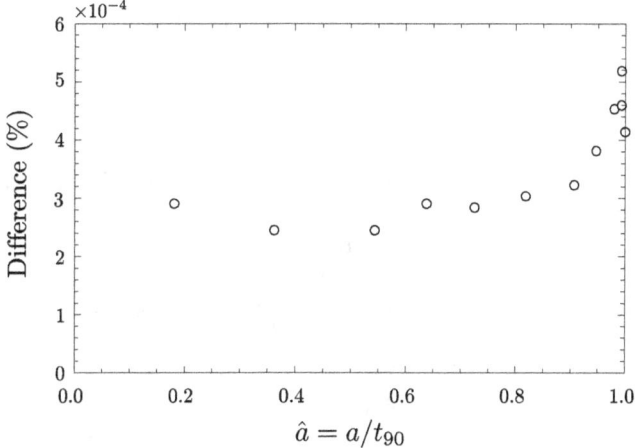

(b) Percentage difference between the energy release rates for the thermal and mechanical partial problems of the superposition

Figure 6.37: Values for the dimensionless energy release rate obtained by a computational analysis employing a boundary element code.

Figure 6.37(a) plots $\hat{G}^{\varepsilon}$ and $\hat{G}^{\Delta T}$ as a function of $\hat{a}$. This figure shows that the numerical results for both functions are apparently identical. The percentage difference between both functions is represented in Figure 6.37(b) as a function of $\hat{a}$. Note that the percentage difference is very low and can be associated to numerical inaccuracies. In fact, following the procedure proposed by Nairn (1997) for a very similar problem, it can be demonstrated that $\hat{G}^{\varepsilon}(\hat{a}) = \hat{G}^{\Delta T}(\hat{a})$. This is a key result because it implies that no thermo-mechanical computation is required to consider residual thermal stresses in the finite fracture mechanics model developed here. For the sake of simplicity, in what follows the same ERR is denoted as $\hat{G}(\hat{a}) = \hat{G}^{\Delta T}(\hat{a}) = \hat{G}^{\varepsilon}(\hat{a})$ in accordance with the notation used in Section 6.1.2.

In view of the previous observation, the expression for $G(a, \varepsilon_{yy}, \Delta T)$ is simplified as,

$$G(a, \varepsilon_{yy}, \Delta T) = E_{22} t_{90} \left(\varepsilon_{yy} + \alpha_2 \Delta T \left(k^{0, \Delta T} - \frac{E_{22}}{\tilde{E}_{22}} k^{(90), \Delta T} \right) \right)^2 \hat{G}(\hat{a}). \quad (6.75)$$

According to this expression, the influence of the residual thermal stresses can be introduced as an additional equivalent strain, which depends on ΔT linearly.

Substituting (6.75) in (6.68) gives,

$$E_{22} t_{90} \left(\varepsilon_{yy} + \alpha_2 \Delta T \left(k^{0, \Delta T} - \frac{E_{22}}{\tilde{E}_{22}} k^{(90), \Delta T} \right) \right)^2 \int_0^{\Delta \hat{a}} \hat{G}(\hat{a}) d\hat{a} \geq G_c \Delta \hat{a}, \quad (6.76)$$

which can be expressed as a condition on the normalized strain $\varepsilon_{yy}/Y_{\varepsilon t}$ in an analogous expression to that presented in Section 6.1.2. First, (6.76) leads to

$$\varepsilon_{yy} \geq \sqrt{\frac{G_c}{E_{22} t_{90}} g(\Delta \hat{a})} + \alpha_2 \Delta T \left(\frac{E_{22}}{\tilde{E}_{22}} k^{(90), \Delta T} - k^{0, \Delta T} \right), \quad (6.77)$$

where $g(\Delta \hat{a})$ is a dimensionless function defined previously in (6.16). This expression for the energy criterion is identical to the expression given for the original 2D model except for the additional term which depends on ΔT linearly.

The previous expression can be normalized with $Y_{\varepsilon t}$, giving a relation for the energy criterion in an analogous form as the expression for the stress criterion,

$$\frac{\varepsilon_{yy}}{Y_{\varepsilon t}} \geq \gamma \sqrt{g(\Delta \hat{a})} + \frac{\alpha_2 \Delta T}{Y_{\varepsilon t}} \left(\frac{E_{22}}{\tilde{E}_{22}} k^{(90), \Delta T} - k^{0, \Delta T} \right), \quad (6.78)$$

where γ has been defined in (6.19). Note that the new thermal term which appears here is identical to the new term in the new expression for the stress criterion (6.67).

6.3.3   Coupled criterion for the thermomechanical finite fracture mechanics model

The application of Leguillon's hypothesis is carried out again with the new expressions for the stress and energy criteria. According to this hypothesis, the critical strain ε_{yy}^c is given by the minimum value of ε_{yy} fulfilling both criteria simultaneously. Thus, combining the new expressions for the stress (6.67) and energy (6.78) criteria, the condition for the onset is,

$$\frac{\varepsilon_{yy}^c}{Y_{\varepsilon t}} = \min_{\Delta \hat{a}} \left[\max \left(\frac{E_{22}}{\tilde{E}_{22}}, \gamma \sqrt{g(\Delta \hat{a})} \right) \right] + \frac{\alpha_2 \Delta T}{Y_{\varepsilon t}} \left(\frac{E_{22}}{\tilde{E}_{22}} k^{(90), \Delta T} - k^{0, \Delta T} \right). \quad (6.79)$$

This expression for the coupled criterion is simplified as a consequence of the discussion of the different scenarios described previously in Section 6.1.3 for the model without considering the residual thermal stresses, which is totally applicable here. Hence, the condition given by the coupled criterion can be rewritten as

$$\frac{\varepsilon^c_{yy}}{Y_{\varepsilon t}} = \max\left(\frac{E_{22}}{\tilde{E}_{22}}, \gamma\sqrt{g(\Delta\hat{a}_{\mathrm{min},g})}\right) + \frac{\varepsilon^{\Delta T}}{Y_{\varepsilon t}}, \tag{6.80}$$

where $\varepsilon^{\Delta T}$ has been introduced as the equivalent strain associated to the residual thermal stresses defined as,

$$\varepsilon^{\Delta T} = \alpha_2 \Delta T \left(\frac{E_{22}}{\tilde{E}_{22}} k^{(90),\Delta T} - k^{0,\Delta T}\right). \tag{6.81}$$

The definition of an equivalent strain $\varepsilon^{\Delta T}$ enables to introduce the effect of the residual thermal stresses on the expression of the critical strain. In spite of being an useful variable for the formulation of the model, it is necessary to remark that it does not correspond to any physical strain in the laminate.

This expression keeps the simplicity of the original expression. Since the new thermal term can be obtained by an analytical expression, only the scalar $g(\Delta\hat{a}_{\mathrm{min},g})$ has to be estimated by computational procedures just as in the model without considering residual thermal stresses. Once this value is obtained, ε^c_{yy} is simply calculated by the sum of $\varepsilon^{\Delta T}$ and the maximum of $E_{22}/\tilde{E}_{22}$ and $\gamma\sqrt{g(\Delta\hat{a}_{\mathrm{min},g})}$.

Figure 6.38 represents the value of $\varepsilon^{\Delta T}$ normalized with $\alpha_2\Delta T$ as a function of t_0/t_{90} for the two bimaterials taken here as reference with elastic properties and CTE shown in Tables 6.1 and 6.5, respectively. The two addends which compose $\varepsilon^{\Delta T}$ are also plotted in this figure. Note that for both bimaterials $\varepsilon^{\Delta T}$ increases with t_0/t_{90}, which means that the effect of the temperature change increases with t_0/t_{90}. On the extreme for $t_0/t_{90} \to 0$, which corresponds to the lack of outer ply, $\varepsilon^{\Delta T} \to 0$ since in the absence of outer ply, the inner ply can contract freely due to the cooling-down process, thus no residual thermal stress is found. The term $k^{0,\Delta T}$ is the mid-plane strain along the $y-$direction during the cooling-down process normalized with $\alpha_2\Delta T$. Regarding to the term $k^{(90),\Delta T} E_{22}/\tilde{E}_{22}$, this can be interpreted as the equivalent residual strain which would appear in the inner ply as a response to a certain stress state if the inner ply could contract freely. The stress state would correspond to a normal tension along the $y-$direction generated when a temperature change is prescribed in the laminate and the displacements at the top and bottom edges are fixed. Thus, in the case of $t_0/t_{90} = 0$, the mid-plane strain after the cooling-down process is given directly by the free expansion $\alpha_2\Delta T$ which corresponds to $k^{0,\Delta T} = 1$. In the case of the term $k^{(90),\Delta T} E_{22}/\tilde{E}_{22}$ for t_0/t_{90}, the equivalent strain which causes the normal tension for the constraint problem is given directly by the strain which the laminate would have without forbidding vertical displacements because in the absence of outer plies, the only constraints are the boundary conditions over the vertical displacements. For $t_0/t_{90} > 0$, the term associated to the mid-plane strains $k^{0,\Delta T}$ decreases because the presence of the outer ply with

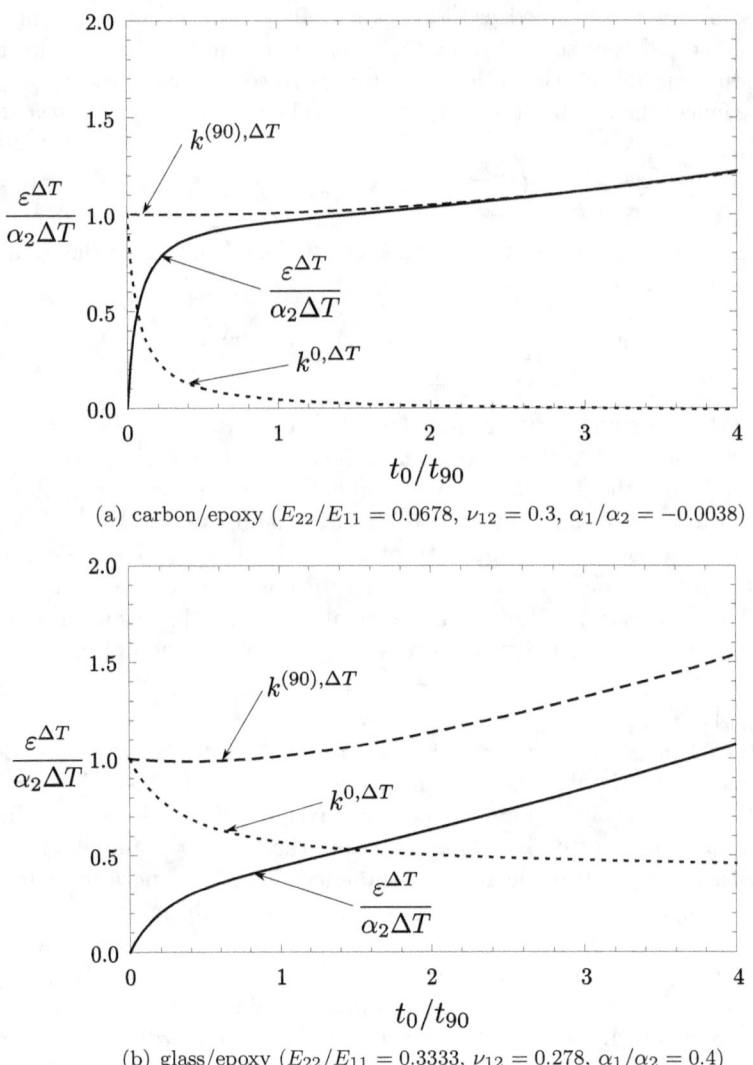

(a) carbon/epoxy ($E_{22}/E_{11} = 0.0678$, $\nu_{12} = 0.3$, $\alpha_1/\alpha_2 = -0.0038$)

(b) glass/epoxy ($E_{22}/E_{11} = 0.3333$, $\nu_{12} = 0.278$, $\alpha_1/\alpha_2 = 0.4$)

Figure 6.38: Equivalent residual thermal stress $\varepsilon^{\Delta T}$ as a function of the ratio of outer- to inner-ply thickness t_0/t_{90} for (a) carbon/epoxy and (b) glass/epoxy

a lower CTE along the y-direction makes more difficult the contraction during the cooling-down. This effect is much more relevant for carbon/epoxy than for glass/epoxy due to the higher contrast between longitudinal and transverse CTE in carbon/epoxy. In fact, for carbon/epoxy, this term vanishes for $t_0/t_{90} \gtrsim 2$. The reason is that, for large values of t_0/t_{90}, the negative value of the longitudinal CTE and the high contrast between the longitudinal and transverse moduli in

	α_1 (°Cμm/m)	α_2 (°Cμm/m)
carbon/epoxy used by París et al. (010a,b)	-1	26
glass/epoxy used by Parvizi et al. (1978)	6	15

Table 6.5: Coefficients of thermal expansion for the two bimaterials employed here as reference.

carbon/epoxy causes that the contraction of the laminate during the cooling-down is almost totally avoided. On the contrary, the term $k^{(90),\Delta T}E_{22}/\tilde{E}_{22}$ increases in general for $t_0/t_{90} > 0$ for both materials very similarly except for a very slight decrease for glass/epoxy and low t_0/t_{90}. Since the evolution of this term is very similar in both bimaterials, the difference is based on the great difference in $k^{0,\Delta T}$, i.e. in the strain after the cooling-down.

According to the expression (6.80), the effect of the residual thermal stresses on the FFM model is given simply by an additional equivalent strain $\varepsilon^{\Delta T}$, which increases or reduces the critical strain of the inner ply. In order to verify which is the sign of the contribution of the residual thermal stresses for the critical strain, it is necessary to study each term in (6.81). For usual composites, $\alpha_2 > 0$ and $\Delta T < 0$. Regarding $(k^{(90),\Delta T}E_{22}/\tilde{E}_{22} - k^{0,\Delta T})$, this is plotted in Figure 6.38, showing that this term is positive in the range of t_0/t_{90} studied for carbon/epoxy and glass/epoxy. As a consequence, the term due to the residual thermal stresses $\varepsilon^{\Delta T}$ in (6.80) is usually negative. From a physical point of view, this result was foreseeable as a consequence of a larger transverse than longitudinal CTE, as usual in the majority of polymeric unidirectional composites, and a negative temperature change due to the curing process at high temperature. The combination of these facts causes the presence of tensions in the inner ply, which reduces the strain necessary for the crack onset.

6.4 Experimental validation through a new set of experiments with geometrically similar laminates

In Section 6.1.6, a preliminary comparison of the model with some experimental results available in the literature has been carried out. This comparison shows a good agreement for the different phenomena predicted. This agreement is particularly good for the size effect, which is the most relevant prediction from an engineering point of view. However, the validation of the FFM model could not be fully accomplished due to some limitations of the experimental results used for this comparison.

First of all, the number of specimens tested by Parvizi et al. (1978) is extremely low for a full validation. This is common with other experimental results presented in the literature focusing on the first transverse crack onset, see e.g. Bader et al. (1980). As described in the corresponding documents, only a single test was carried out for each value of the inner-ply thickness. As a consequence, the presence of outsiders could not be detected and the reliability of the results cannot be evaluated.

Other limiting characteristic of the experimental results by Parvizi et al. (1978) is that they were carried out in the 1970's and 1980's, which is also common to many tests evaluating the first crack onset in the literature. 40 or 30 years is not a long time for the whole evolution of typical metals as steel. However, in the context of composites, this is a long period of time. Composites have been strongly developed during the last decades. On the one hand, from an industrial point of view, it is reasonable to think that the model has to be evaluated with modern composites. It is necessary to dismiss that the agreement with "old" tests is some type of characteristic of the material which is not present in modern composites. On the other hand, the knowledge of the material properties of the "old" materials is much more limited mainly due to two causes: a) testing technology and standards less developed, and b) the impossibility of characterizing with the current means the material used because it is not manufactured anymore.

Parvizi et al. (1978) such as the majority of other authors kept fixed the outer-ply thickness t_0 when they varied the inner-ply thickness t_{90} to study the size effect. However, a dimensional analysis of the problem shows that all the ratios of lengths has to be kept fixed in order to adequately study the size effect. An exception can be assumed with the length L and width W of the specimen when they are large and wide enough compared with t_{90} and under certain conditions, see 6.D. This is not the case of t_0, which is typically of the same order of t_{90} in the experiments in the literature. As a consequence, rigorously the ratio t_0/t_{90} should be kept fixed when the size effect is studied. If it is not, the mixed influence of the size and the variation of t_0/t_{90} is actually obtained.

In view of limiting characteristics of the experimental results in the literature described previously, a new set of tests is carried out in order to evaluate the accuracy of the FFM model. This set is adequately defined in order to overcome the previous limitations. In what follows the details about the specimen design and fabrication are described in Section 6.4.1, justifying the main options chosen. The testing procedure is explained in Section 6.4.2, highlighting the difficulties found, which can be relevant for the interpretation of the results. Finally, the results are compared with the FFM model developed previously and discussed in Figure 6.4.3.

6.4.1   Specimen fabrication

This subsection describes the logical path followed to design the specimen tests in accordance with the reasoning described previously. The first decision to be taken is which are the lay-up configuration tested. The two key conditions to optimize the tests are

- All the specimen tests have to be geometrically similar in order to study the size effect avoiding the effect of other parameters. As discussed previously, it means that t_0/t_{90} is fixed for all the laminates.

- In order to span a range of thickness as wide as possible, the thinnest laminate should have a single layer of prepreg as inner ply. On the other extreme, sufficiently thick laminates should be also tested.

Thus, given that at least a layer at each side has to be placed for the thinnest laminate and that a single layer composes the inner ply, the cheapest laminate in terms of material use is $[0, 90, 0]$. Note that this corresponds to $t_0/t_{90} = 2$. Since this ratio has to be fixed, the lay up configurations are given by $[0_n, 90_n, 0_n]$ where n is the amount of prepreg layers composing the inner ply. According to the experimental observations reported in the literature, the size effect is more relevant for the thinnest laminates, see e.g. Parvizi et al. (1978). Due to this, test specimens were fabricated for $n = 1, 2, 3, 4, 5$. On the other extreme, experiments show a very low size effect for thickest specimens. In addition, these laminates requires much more material. Thus, in order to save material while the experiments span a wide enough range of thickness, three lay up configuration were fabricated for $n = 7, 10, 16$. Based on the experimental results in the literature, it is assumed that the laminates with $n = 16$ are thick enough to capture the tendency for thick laminates.

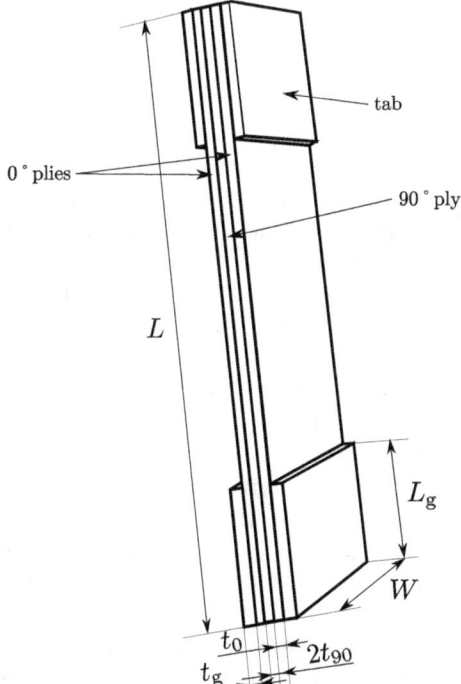

Figure 6.39: Schematic of the test specimens manufactured for the experimental validation of the FFM model, see Table 6.6 for the values of the geometric parameters.

Figure 6.39 shows a schematic of the test specimens fabricated. Observe that this is the typical specimen of tensile testing in composites. It is composed by a rectangle of the laminate with lengths $L \times W$. The values of these parameters are given in Table 6.6. Glass/epoxy tabs were bonded to the extremes of the specimen where the tensile testing machine grips it in order to avoid a premature failure at these zones. These tabs are rectangular with size $L_g \times t_g$. These length are defined with the aim of assuring the correct transmission of the load from the tabs to the laminate without failure of the adhesive. Within a range, a larger adhesion surface is able to transmit higher loads. According to the experimental results found in the literature, the critical strain is of a similar order for all the laminates. Thus, since the thickest laminates requires higher loads to be applied for the same strain level, the tabs were sized to be large for these laminates, see Table 6.6 for details.

n	L (mm)	W (mm)	$2t_{90}$ (mm)	t_0 (mm)	L_g (mm)	t_g (mm)
1	160	25	0.19	0.19	37	1.6
2	160	25	0.38	0.38	37	1.6
3	160	25	0.57	0.57	37	1.6
4	160	25	0.76	0.76	37	1.6
5	160	25	0.95	0.95	40	1.6
7	160	25	1.33	1.33	48	1.6
10	160	25	1.90	1.90	48	1.6
16	160	25	3.04	3.04	56	1.6

Table 6.6: Nominal values for the geometric parameters of the test specimens used in the experiments, see Figure 6.39.

The laminates were made by stacking up unidirectional layers of carbon fiber tape (Hexcel AS-4) preimpregnated in epoxy resin (Hexcel®8852) with a resin volumetric content of $31 - 35\%$, a nominal cured thickness per ply of $0.17 - 0.19$ mm and an areal weight of $268 - 308$ g/m^2. Currently, this prepreg is widely used in structural components in the aeronautical industry.

The longitudinal E_{11} and transverse E_{22} Young's moduli and the major Poisson's ratio ν_{12} for this material obtained by Marlett (2010) following the ASTM Standard D3039 (2006) are taken here as reference. The in-plane shear modulus G_{12} is also taken from the report by Marlett (2010) who measured it using the ASTM Standard D3518 (2007). Unfortunately, no value for the through thickness Poisson's ratio ν_{23} was found in the literature. The value given by Soden et al. (1998) for a similar material is taken as reference. In spite of this value is necessary for the FFM model, the results are not expected to vary significantly in the worst scenario of a large error on the estimation of ν_{23}. Similarly, the values for the longitudinal α_1 and transverse α_2 coefficients of thermal expansion are identified with the values given by Soden et al. (1998) for a similar material.

With regards to the unidirectional tensile transverse strength Y_t, a large difference was found between the value given by the manufacturer Hexply 8852 Product Data (2013) and the value measured by Marlett (2010) under apparently very similar conditions. This variability can be associated to a strong dependence of the fiber-resin bonding quality on the curing process and the state of the resin before it. Due to this, Y_t was measured using unidirectional laminates which were made using prepreg of the same batch and were cured with the main laminates.

The transverse fracture toughness G_c for this material was measured by Renart et al. (2011) with Double Cantilever Beam tests according to the ISO Standard 15024 (2001). The value taken here corresponds to the value obtained for the initiation denoted as 5%MAX. Actually, Renart et al. (2011) gives two values for 5%MAX, either using end blocks bonded to the specimen as indicated in the ISO Standard 15024 (2001) or a novel set of hinges clamped to the end of the specimens. Since the end blocks are currently indicated by the standards, the value given using them is taken here.

Table 6.7 presents the values for the material properties assumed here for the AS-4/8852 and summarizes the references from which these values are extracted.

Property	Value	Reference
E_{11} (GPa)	127.277	Marlett (2010)
E_{22} (GPa)	9.239	Marlett (2010)
ν_{12}	0.302	Marlett (2010)
ν_{23}	0.4	similar material in Soden et al. (1998)
G_{12} (GPa)	4.826	Marlett (2010)
Y_t (MPa)	50.8	own tests
G_c (N/m)	248	value for initiation in Renart et al. (2011)
α_1 (°Cμm/m)	-1	similar material in Soden et al. (1998)
α_2 (°Cμm/m)	26	similar material in Soden et al. (1998)
T_0 (°C)	182	autoclave

Table 6.7: Material properties for carbon/epoxy AS4/8852 necessary for the FFM model.

The entire manufacturing process was carried out at the GERM facilities following the next steps,

1. A laminate was manufactured for each value of n in Table 6.7. The laminates were fabricated by stacking up prepregs layers following the adequate sequence. If the laminate contains more than 5 layers, every 5 layers the partial laminate was subjected to vacuum to avoid the presence of air bubbles. After the full stacking of the laminates, they were introduced in a vacuum bag on an aluminium plate and a set of tests were carried out to detect any vacuum leak.

2. All the laminates were cured simultaneously in an autoclave. In order to avoid air bubbles in the laminate, a controlled level of vacuum was kept at the bag during the curing process, see Figure 6.40(b). In addition, the vessel was subjected to an elevated pressure, which increases the differential level of vacuum inside the bag, see Figure 6.40(c). The temperature evolutes following the curve represented in Figure 6.40(a). First, the temperature is increased up to reach a certain level of temperature which is denoted curing temperature. Once this temperature curing is attained, this is kept

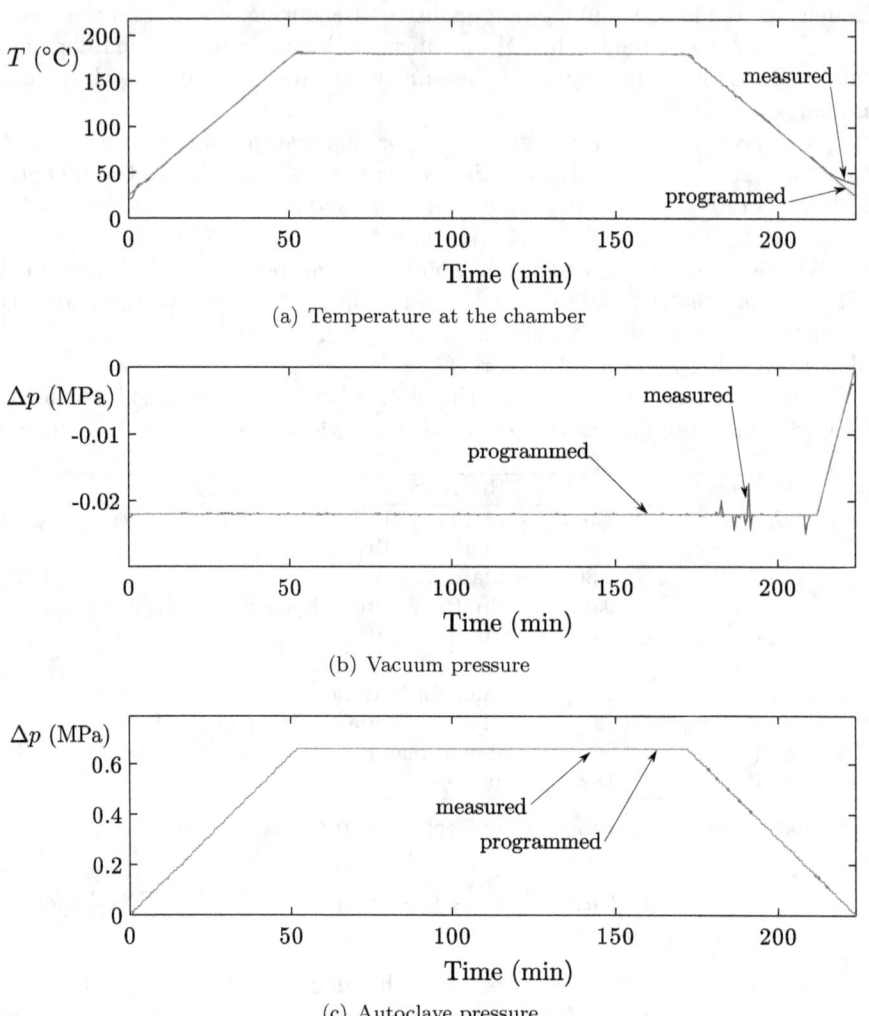

(a) Temperature at the chamber

(b) Vacuum pressure

(c) Autoclave pressure

Figure 6.40: Evolution of the controlled parameters during the curing process of the laminates at the autoclave.

during a sufficiently large period to allow the chemical reactions of curing to occur. Finally, the temperature is decreased up to reach the room level.

3. The tabs, which are made of glass fiber fabric impregnated with epoxy, were cured independently. Once both laminates and tabs were cured and the laminates edges were cut and dismissed, the tabs were bonded to the laminates using an adhesive. The whole set was introduced in a heated press to cure the adhesive.

4. At least 7 specimens were cut of each laminate with the aid of a dia-
   mond cutting disc. Some unexpected crack onsets occurred when cutting
   the thickest laminates ($n = 16$). These cracks appeared at only one of
   the outer plies and along the transverse direction to the fiber axis. This
   is interpreted as a consequence of the residual stresses at the transverse
   direction compared with the critical value Y_t, see Section 6.4.3 for an estimation of the residual stresses for this laminate. Although the residual
   stresses predicted are high, they do not exceed the critical value Y_t. Thus,
   the appearance can be explained by the presence of slight bending promoted for a small deviation of the symmetry of the totally symmetric lay
   up of the laminate. This is coherent with the fact that this occurred only
   for the thickest laminate since the effect of this bending is amplified when
   the thickness increases.

6.4.2 Testing procedure

Once the specimens were manufactured, one of the edges of each one was entirely
sanded and polished to allow post testing observations of the transverse cracks.
The tests were run using an electromechanical testing machine Instron 4483. A
monotonic tensile load was applied in displacement control at a cross-head rate
of 0.5 mm/min. The tests were stopped once a sufficient amount of crack onsets
were recorded and before the global failure of the specimen, the specimens were
unloaded at a cross-head rate of 5 mm/min. Then, the polished edge of each
specimen is inspected at the optical microscope in order to count the number of
cracks and analyze if the recorded events correspond to transverse crack onsets
as will be analyzed later on.

   During the test, a load cell Instron 2525-112 with a maximum capacity of 150
kN is used to record the load at the specimen. The cross-head displacement is
also measured and its value recorded. However this value is not accurate enough
to estimate the strain at the laminate. The cause is that other undesirable
displacements appear at this measure added to the displacements due to the
deformation process of the laminate. The most relevant in this test is the relative
displacement between the tabs and the specimen as a consequence of the own
deformation process of the adhesive. In order to obtain an accurate measure
of the strain, a strain gauge extensometer Instron 2630-112 with 50 mm gauge
length was used to measure the strain through the relative displacement between
two points. It is attached at the center of the specimen at the space without
tabs. Since the length between the tabs is shorter than 50 mm for the thickest
specimens, see Table 6.6, an extensometer 2630-107 with 25 mm gauge length
was used instead for these specimens. For all the specimens, the values of the
three measures were recorded three times every second.

   The onset of cracks on the inner ply of the cross-ply laminates tested does not
entail a significant reduction of the global stiffness of the specimen. This is due
to the very low contribution of the inner ply to the stiffness of the specimen as a
consequence of the high contrast between the longitudinal and transverse Young's
moduli and that the thickness of the plies are similar. As a consequence, the

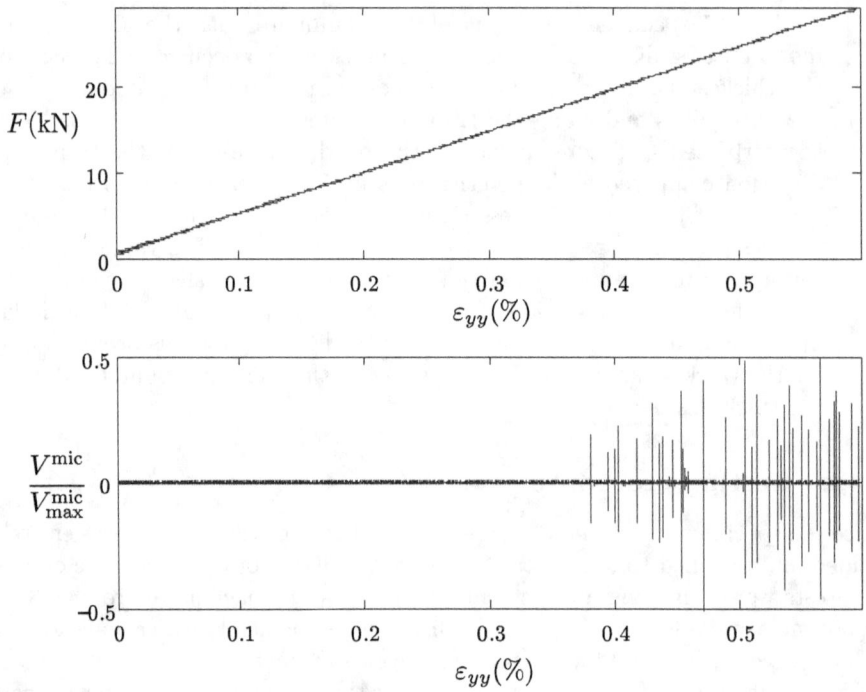

Figure 6.41: Load F and normalized microphone voltage $V^{\mathrm{mic}}/V^{\mathrm{mic}}_{\mathrm{max}}$ as functions of the strain ε_{yy} measured at the extensometer for the specimen 3-1.

onset of a crack in the inner ply cannot be detected by inspecting the load-strain curve. Several alternative methods have been used in the experiments reported in the literature as x-ray in the case of opaque laminates or direct observation for transparent laminates. Another alternative is the use of acoustic emission to detect damage events, which is very common in composites, see e.g. Wevers and Surgeon (2000) for a review. This was the main method implemented in this work. The acoustic emission were measured here by attaching a contact microphone to the specimen. The voltage at the contact microphone V^{mic} is recorded at 44100 Hz enabling a very accurate detection of any event producing an acoustic wave at the specimen.

Figure 6.41 shows the load F and the microphone voltage V^{mic} normalized with the maximum range of values measured $V^{\mathrm{mic}}_{\mathrm{max}}$ as a function of the strain ε_{yy}. Observe that, in spite of the amount of damage events detected by the microphone, the stiffness of the specimen, given by the slope of the F-ε_{yy} curve, remains fixed as described previously. Regarding to the $V^{\mathrm{mic}}/V^{\mathrm{mic}}_{\mathrm{max}}$-$\varepsilon_{yy}$ curve, several peaks are clearly recognized. The peaks corresponding to transverse crack onsets are distinguished by their characteristic spectral content. To confirm the adequate differentiation of the first transverse crack onsets from other damage events, a specimen for each value of n is tested first. These first tests were

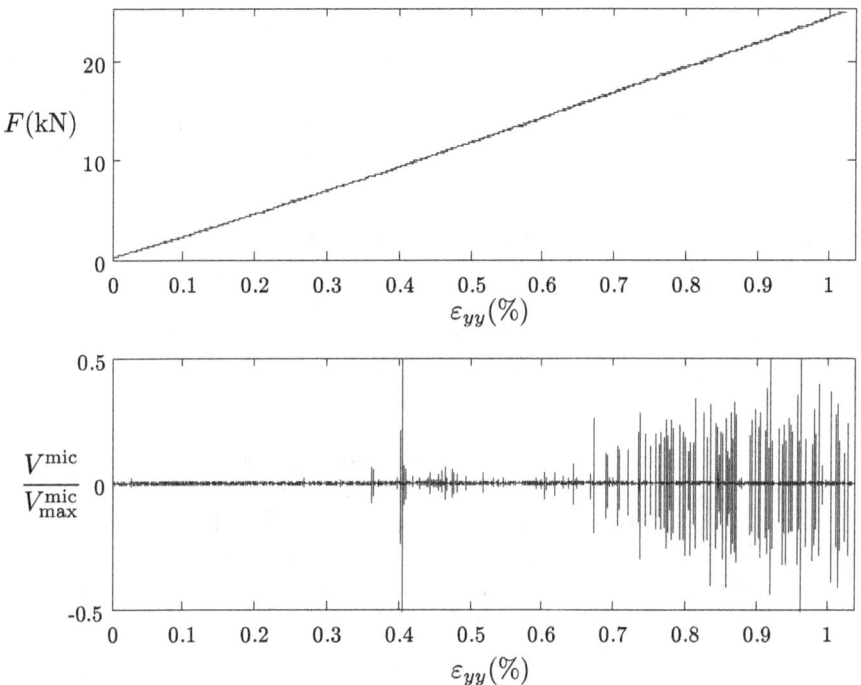

Figure 6.42: Load F and normalized microphone voltage $V^{\mathrm{mic}}/V^{\mathrm{mic}}_{\max}$ as functions of the strain ε_{yy} measured at the extensometer for the specimen 2-3.

stopped after the detection of only a small number of transverse crack onsets to confirm, using the optical microscope, if the events assumed to be a crack onset corresponds actually to visible transverse cracks. The result was successful for all the values of n with the exception of the test for $n = 1$ because the global failure was immediate after the first damage event as will be discussed later on.

In the case of the example shown in Figure 6.41, the critical strain for which the process of transverse cracking begins can be easily recognized because the peaks appears clearly above a certain level of strain. However, in general the critical strain is not as clear as in the previous example. Figure 6.42 shows the same curves corresponding to the specimen 2-3. In this case, a large peak, corresponding to a strong damage event, was recorded for $\varepsilon_{yy} = 0.4040\%$ surrounded by other much softer events and followed by a considerable period without events. For a larger strain: $\varepsilon = 0.6296\%$ a set of contiguous peaks were recorded, much more similar to the behavior observed in previous example.

Given the variability of the results obtained, well exemplified by the cases for the specimens 3-1 and 2-3 plotted, it is necessary to establish a criterion of which is going to be considered the first crack and as consequence which strain is considered as critical. Two different critical value for the strain are taken as reference,

- The critical strain ε_{yy}^{c1} leading to the very first transverse crack onset. This corresponds to the first peak associated to a transverse crack onset without taking into account that the damage event is isolated. This crack onset could correspond to a crack growing from a unusually large flaw. This type of crack onset would be out of the scope of the onsets being claimed to be predicted by the FFM model discussed here.

- The critical strain ε_{yy}^{c2} leading to a transverse crack onset which is almost immediately followed by other crack onsets. Due to the length of the laminate in comparison with the other lengths, it is expected that if the conditions to expect a crack onset are fulfilled, several crack onsets adequately separated along the length are also expected. The crack onsets are not necessarily simultaneous given the slight variability of the material properties along the length. In this case, the difficulty arise to define which is considered simultaneous. On the one hand, it is necessary to define a sufficiently wide range to capture two crack onsets whose difference is based on the material property variability. On the other hand the range has to be narrow enough to reject crack onsets due to unusually large flaws. Thus, based on the observation of the results obtained here, the next conditions are followed to define a crack onset as the first one:

  - A second crack onset followed the first one within the subsequent 5 seconds.

  - A third one within the subsequent 10 seconds.

  - A fourth one within the subsequent 30 seconds.

In what follows the results will show both critical strains because whereas ε_{yy}^{c1} is *a priori* highly affected by the randomness of the presence of a large flaw, this is actually the value for the critical strain more independent of additional assumptions. In addition, it is interesting to observe the difference between both strains in order to evaluate the importance of the flaws in the process.

The post-testing observation at the optical microscope showed some differences on the cracking process between the specimens corresponding to different values of n, i.e. with different inner-ply thickness. In the following, the details found for each n are detailed,

- For $n = 1$, a single strong peak is found, which is followed by the global failure of the specimen. In the majority of the specimens, a large delamination at both interfaces connected by a single transverse crack is found, see Figure 6.43(a) for an example. Since both phenomena occurs almost simultaneously, it is difficult to observe which of them occurs previously. This is basic to discuss which of this damage events promotes the other and as a consequence to know if the failure mode is controlled by transverse cracking. Within this high level of uncertainty some evidences shown in Figure 6.43(a) leads to believe that the first failure for $n = 1$ is the delamination. The main evidence is the angle formed between the transverse crack and

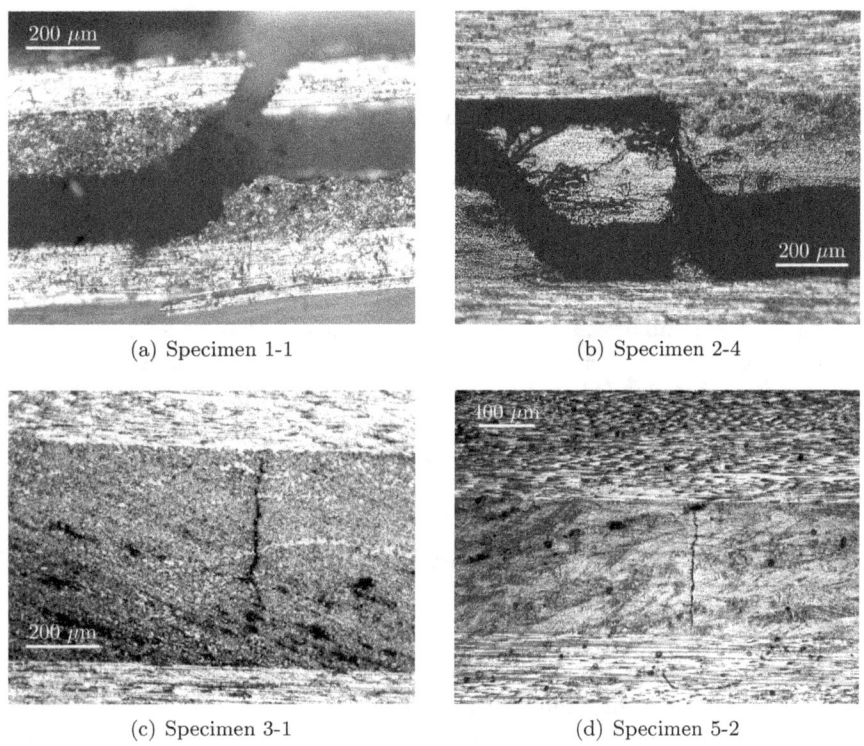

(a) Specimen 1-1   (b) Specimen 2-4

(c) Specimen 3-1   (d) Specimen 5-2

Figure 6.43: Examples of micrographs obtained after the acoustic detection of transverse crack onsets.

the interface. If the interface is not damaged previously, due to the symmetry it is reasonable to think that the transverse crack opens with a certain angle being approximately perpendicular to the load direction. In fact, this is the case for larger values of n, see e.g. Figures 6.43(c) and 6.43(d). However, the case for $n = 1$ in Figure 6.43(a) shows a transverse crack with an angle which is far from being perpendicular to the load. The onset of a transverse crack with this angle could be explained by the bending moment produced when a sudden delamination occurs at one of the interfaces. This delamination would kink out the interface towards the inner ply with an angle which would be affected by the bending moment linked to the lack of symmetry provoked by the delamination. Other evidence supporting this hypothesis is the continuity of the angle of the crack transversing the outer ply. This is coherent with the sequential process of failure described: a delamination which is followed by a transverse crack. The similitude of the crack angles at the inner and outer plies implies that the bending moment promoting this angle was similar before and after the crack going through the interface. As discussed previously the bending moment is given mainly by the presence of the delamination at the interface. Thus, if the bending moment is similar, it implies that the delamination does not change significantly during the process of inner and outer ply cracking. The process would be: delamination, kink out the interface toward the inner ply and cracking of the outer ply. Thus, the failure is not governed by the transverse cracking but it is by the delamination probably associated to the free-edge effect. In this case, the theoretical FFM model to predict the onset of a free-edge delamination proposed by Martin et al. (2010) could be used.

- According to the behavior found for $n = 2$, its corresponding thickness can be considered a transition between the behavior observed for $n = 1$ and for thicker laminates. Whereas some specimens showed the same behavior that for $n = 1$, others showed the usual transverse cracking observed for larger values of n. Moreover, in some cases, see e.g. Figure 6.43(b), both behaviors seem to appear in the same specimen. Note that two transverse cracks are visible at the micrograph. Whereas the transverse crack on the left has a similar angle that the transverse crack described for $n = 1$, the transverse crack on the right is perpendicular to the load in the majority of its length. In addition two delaminations appears being connected by both transverse cracks. This complex behavior can be explained by the next sequence,

  1. A transverse crack onset occurs which corresponds to the crack on the right shown in Figure 6.43(b).

  2. The onset of a delamination occurs probably associated to the free-edge effect.

  3. The delamination kinks out the interface toward the inner ply, which corresponds to the transverse crack on the left.

  4. A new delamination occurs at the other interface due to the proximity of the new transverse crack

  5. The new delamination reaches the first transverse crack connecting both transverse cracks.

- For larger values of $n = 3, 4, 5, 7, 10, 16$, the observation at optical microscope shows clearly a set of transverse cracks perpendicular to the load, see e.g. Figures 6.43(c) and 6.43(d). These transverse cracks are sometimes accompanied by small delaminations at the interface near the crack tip very probably promoted by the stress concentration at the interface when a transverse crack approaches it, see París et al. (010a,b) for a detailed explanation of this phenomenon. Excluding the critical values of strain studied in next section, the only relevant difference between the behavior observed for the specimens corresponding to these values of n is the frequency of the crack onsets. It is observed that the frequency decreases for thicker laminates. The cause is that for these laminates, the specimens are shorter in dimensionless terms since the length has not been scaled with the thickness. As a consequence, the presence of a transverse crack affects a large part of the length for large n making more difficult the subsequent transverse crack onsets.

In spite of the different behavior described here for different values of n, the critical values of the strain were recorded in all the cases for the sake of comparison.

6.4.3   Results and discussion

The experimental results recorded for the two critical values of the strain are presented and discussed here. The experimental results are compared with the critical strain predicted by the FFM 2D model developed in Section 6.1 combined with the correction proposed in Section 6.3 to take into account the residual stresses.

Table 6.8 shows the values obtained for the two critical strains ε_{yy}^{c1} and ε_{yy}^{c2} defined previously for each specimen. In addition, the inner-ply thickness of each specimen defined as the mean of the measured thickness at three distant points with the aid of the optical microscope.

Values for ε_{yy}^{c1} and ε_{yy}^{c2} were obtained for the majority of the specimens. However some problems arose for certain values of n leading to a lack of these values for some specimen as can be observed in Table 6.8. In the case of the specimens corresponding to $n = 1$, the lack of values for ε_{yy}^{c2} is common to all of them. The cause for this lack is based on the failure sequence described in the previous section for $n = 1$. These specimens only show a single peak corresponding to a transverse crack before the global failure. Moreover this transverse crack is probably associated to a previous delamination as discussed previously. The fact that only a transverse crack onset is detected makes impossible to fulfill the conditions defined for ε_{yy}^{c2}. The critical value could not be obtained for some specimens corresponding to $n = 10$ and $n = 16$. In these cases, the cause for the lack is based on the premature failure of the adhesive which bonds the tabs and the laminate due to the high loads necessary to provide a similar level of strain for thick specimens. Note that unlike the case for $n = 1$, the lack of ε_{yy}^{c2} for thick specimens is not based on a different behavior of the failure process but in a phenomenon associated to the testing procedure.

In general, a decreasing tendency of both critical strains when the laminates are thicker can be observed in the results in Table 6.8. This tendency is clearer for the thinnest specimens in spite of for these thickness a larger dispersion is found. For the thick specimens the critical strain seems to tend to a fixed value around $0.1 - 0.3\%$. Note that this value is significantly lower that the unidirectional critical transverse strain $Y_{\varepsilon t} = Y_t/E_{22} = 0.55\%$ of the lamina, see Table 6.7. This is caused by the presence of high residual stresses at the inner ply due to the cooling-down from the curing temperature as will be seen in the following.

The experimental results obtained here are compared with the FFM 2D model developed in Section 6.1 corrected to take into account the residual stresses in Section 6.3. Recall that the value predicted for the critical strain ε_{yy}^{c} when the

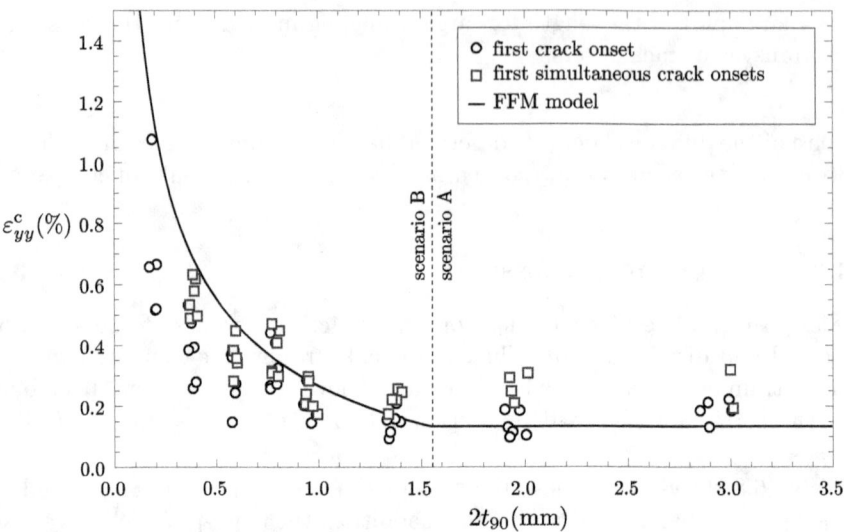

Figure 6.44: Comparison between the experimental results obtained using laminates with $t_0/t_{90} = 2$ of carbon/epoxy AS4/8852 and the theoretical model based on the coupled criterion of the finite fracture mechanics.

residual stresses are taken into account is given by the expression (6.80),

$$\frac{\varepsilon_{yy}^c}{Y_{\varepsilon t}} = \max\left(\frac{E_{22}}{\tilde{E}_{22}}, \gamma\sqrt{g(\Delta\hat{a}_{\min,g})}\right) + \frac{\varepsilon_{\Delta T}}{Y_{\varepsilon t}}, \tag{6.82}$$

where the ratio $E_{22}/\tilde{E}_{22}$ is a function of the thickness ratio t_0/t_{90} and the elastic properties of the lamina, see (6.3). For the material and laminate used here

$$E_{22}/\tilde{E}_{22} = 1.01091. \tag{6.83}$$

The brittleness number γ was defined previously in (6.19) as,

$$\gamma = \frac{1}{Y_t}\sqrt{\frac{G_c E_{22}}{t_{90}}}, \tag{6.84}$$

where the value of all the parameters is known, see Tables 6.7 and 6.8.

Regarding to the value of $g(\Delta\hat{a}_{\min,g})$, this is given by the energy release rate G as a function of the crack length, see Section 6.1.2 for a detailed analysis. This value, which depends only on the elastic properties and t_0/t_{90}, is obtained here by using the BEM code developed by Blázquez et al. (1998) as carried out in Section 6.1.2. For the elastic properties shown in Table 6.7 and $t_0/t_{90} = 2$, the value obtained is

$$g(\Delta\hat{a}_{\min,g}) = 0.887603, \tag{6.85}$$

which corresponds to

$$\Delta\hat{a}_{\min,g} = 0.966162. \tag{6.86}$$

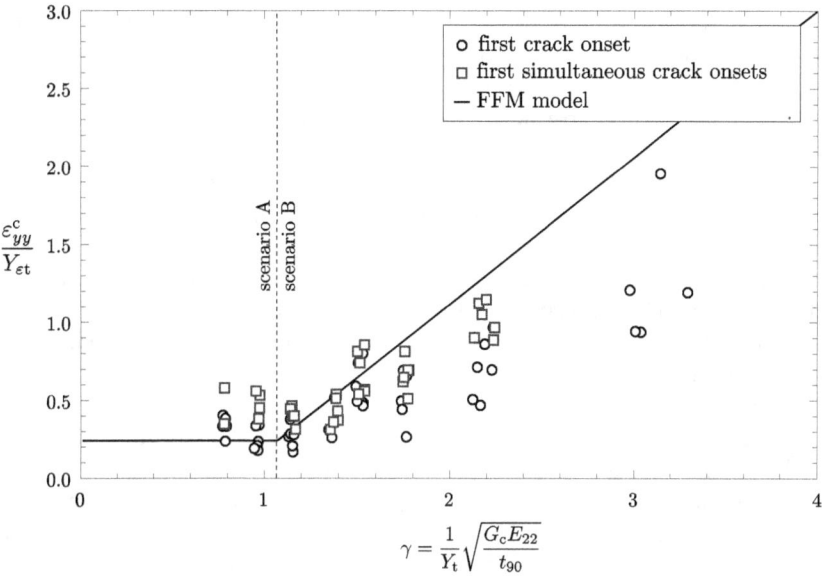

Figure 6.45: Comparison between the experimental results obtained using laminates with $t_0/t_{90} = 2$ of carbon/epoxy AS4/8852 and the theoretical model based on the coupled criterion of the finite fracture mechanics in dimensionless terms.

The value $\varepsilon^{\Delta T}$ necessary to introduce the influence of the residual stresses is calculated by using the expression (6.81) as a function of the thermoelastic material properties and t_0/t_{90} and the difference between the curing temperature, see Table 6.7, and the room temperature $T_0 = 23°C$. Thus the next value is obtained,

$$\varepsilon^{\Delta T} = 0.422901\%. \tag{6.87}$$

Note that this does not correspond to the residual strain at the laminate but that it is the equivalent strain which would be generated by the residual stresses actually existing in the laminate, see Section 6.3 for a full description. This was defined with the objective of introducing the residual stresses in the 2D model, which is defined in term of strains. It is interesting to discuss the value of $\varepsilon^{\Delta T}$ normalized with the unidirectional critical value for the transverse strain $Y_{\varepsilon t}$ of the lamina as appears in (6.82),

$$\frac{\varepsilon^{\Delta T}}{Y_{\varepsilon t}} = 0.76913, \tag{6.88}$$

which means that the residual stress exceed about 75 % of the nominal transverse strength of the lamina.

Once all the parameters necessary for the model have been obtained, the predicted value for ε^c_{yy} as a function of the inner-ply thickness $2t_{90}$ is plotted in Figure 6.44 along with the experimental results. Both the decreasing tendency with $2t_{90}$ and the minimum value for the laminates are well predicted by the FFM model. In addition, the transition between the scenario A governed by the stress criterion and the scenario B governed by the energy criterion predicted by the FFM model, see Section 6.1.3, agrees with the effect observed in the experimental results. Regarding to the value of the critical strain for the thickest laminates, it is well approximated within the level of uncertainty found at the experimental results. At the other extreme, for the thinnest laminate, the tendency is well predicted but the model overestimates the strength of the inner-ply thickness. This can be due to the fact that the material utilized for the specimen fabrication had expired, which could reduce the fracture toughness with respect to the value used here. As a consequence, this would decrease the theoretical curve for the thinnest laminates. On the other hand the results for $n = 1$ corresponding to the values $2t_{90} \approx 0.19$ have to be excluded from this comparison because these values correspond likely to other type of failure as discussed in Section 6.4.2.

The previous comparison can also be represented in dimensionless terms using the key parameters for the model. Figure 6.45 shows the normalized critical strain $\varepsilon_{yy}^{c}/Y_{\varepsilon t}$ as a function of γ. The result of the comparison is independent of the terms chosen to represent it but this figure highlights some of the results discussed previously. The agreement between experiments and the model at the transition between scenarios is more clearly observed here. On the other hand, this representation shows sharply that the model overestimates the critical strain for great values of γ which correspond to thinner laminates.

n_{90°	specimen	t_{90} (mm)	$\varepsilon_{yy}^{c1}(\%)$	$\varepsilon_{yy}^{c2}(\%)$
1	1-1	0.201	0.6766	-
1	1-2	0.164	0.6684	-
1	1-3	0.193	0.5276	-
1	1-4	0.180	1.0872	-
1	1-5	0.197	0.5302	-
2	2-1	0.357	0.5436	0.5436
2	2-2	0.380	0.2698	0.5892
2	2-3	0.386	0.4040	0.6296
2	2-4	0.372	0.4846	0.6430
2	2-5	0.359	0.3946	0.4994
2	2-6	0.395	0.2900	0.5074
3	3-1	0.570	0.3798	0.3932
3	3-2	0.591	0.2846	0.3530
3	3-3	0.573	0.3772	0.3932
3	3-4	0.588	0.2550	0.3678
3	3-5	0.584	0.3920	0.4590
3	3-6	0.573	0.1584	0.2926
4	4-1	0.761	0.2778	0.3248
4	4-2	0.762	0.2684	0.3194
4	4-3	0.794	0.2832	0.3088
4	4-4	0.801	0.3356	0.4590
4	4-5	0.788	0.4188	0.4188
4	4-6	0.763	0.4524	0.4818
5	5-1	0.940	0.2966	0.3074
5	5-2	0.985	0.1838	0.1838
5	5-3	0.925	0.2174	0.2174
5	5-4	0.942	0.2120	0.2940
5	5-5	0.929	0.2134	0.2496
5	5-6	0.961	0.1558	0.2108
7	7-1	1.375	0.2202	0.2672
7	7-2	1.395	0.1598	0.2564
7	7-3	1.369	0.1692	0.2308
7	7-4	1.339	0.1046	-
7	7-5	1.346	0.1262	0.2322
7	7-6	1.329	0.1664	0.1852
10	10-1	1.899	0.2000	-
10	10-2	1.915	0.1422	0.3046
10	10-3	1.937	0.1276	0.2214
10	10-4	1.923	0.1100	0.2604
10	10-5	2.002	0.1168	0.3194
10	10-6	1.97	0.1974	-
16	16-1	2.892	0.1424	-
16	16-2	2.988	0.2336	0.3300
16	16-3	3.001	0.1960	0.2040
16	16-4	2.886	0.2228	-
16	16-5	2.847	0.1960	-

Table 6.8: Experimental results obtained for the new set of tests of geometrically similar laminates of carbon/epoxy AS4/8852 with $t_0/t_{90} = 2$.

6.5 Concluding remarks

A novel theoretical model has been developed to predict the first transverse crack onset in cross-ply laminates. Actually, this also refers to the sufficiently distant transverse crack initiations which appear almost simultaneously with the first one.

The combination of the coupled criterion and a generalized-plane-strain analysis has allowed a semianalytical expression for the prediction of the critical strain that originates the first crack to be obtained. This expression depends on the computational results only through a scalar value, which is a function of the elastic properties of the lamina and the geometric parameter t_0/t_{90}.

The theoretical model developed can also predict the final length of the crack achieved after the onset, an arrest length of the crack very close to the inner-ply thickness. This fact is coherent with the experimental results found in the bibliography. A lower bound is also predicted for the arrest length of the crack. This bound is a very accurate estimation of the final crack length because it is very close to the interface (96.4 % for carbon/epoxy and 99.86 % for glass/epoxy) and a more accurate approximation is not plausible due to the non validity of homogeneous material hypothesis for these distances. Surprisingly, this lower bound value has been demonstrated, in accordance with the model described, to be independent of the strength and fracture properties Y_t and G_c.

A size effect is predicted for the critical strain originating a crack as a function of the inner-ply thickness t_{90}. Apparent strength of the inner ply is predicted to increase for thinner plies. This size effect is a direct consequence of the energy criterion. A simple expression of this size effect for geometrically similar laminates has been obtained as a function of the structural parameter γ, brittleness number. The resulting expression is composed by a constant and a linear function of γ depending on a scalar value $g(\Delta \hat{a}_{\min,g})$.

Several authors have highlighted the importance of the phenomena occurring at the transverse direction to the plane taken for the analysis with generalized plane strain. In order to evaluate the validity and accuracy of the 2D model, a 3D FFM model has been developed. The comparison of the results obtained by this 3D model and the previous 2D model shows that the 2D model is an accurate approximation. Some slight differences are found for the most brittle configurations which corresponds to the thickest laminates. On this extreme, the crack onset is predicted to occur in the vicinity of the free edge due to the stress concentration. The difference in terms of critical strain is considered insignificant in comparison with the typical value of dispersion obtained in experiments. The combination of this fact and the high complexity of the 3D analysis with respect to the 2D model makes this last one recommendable in the majority of applications.

The influence of the residual thermal stresses is included in the model by studying their influence on the stress and energy criteria separately. This analysis enables to obtain an expression for the 2D model considering these residual stresses, without adding complexity to the application of the model. The new semianalytical expression obtained depends on the same scalar value from the computational results and no thermal computations analysis is required. In fact, it is demonstrated that the influence of the residual thermal stresses is equivalent to add an equivalent strain to the expression of the critical value $\varepsilon^{\Delta T}$, which is obtained by the means of the laminate theory leading to an explicit expression.

Experimental evidence from Parvizi et al. (1978) is reanalyzed in view of the present theoretical developments. The present theoretical results for the size effect have been confirmed by these experimental results. This theoretical model agrees with the observed threshold value of the inner-ply thickness where the critical strain changes abruptly its behavior and the predictions for both scenarios with a good accuracy.

A new set of experiments to validate the model have been carried out. The main characteristic of this new set is that the laminates are geometrically similar, i.e. t_0/t_{90} is kept fixed. In addition at least 5 specimens are tested for each value of the inner-ply thickness. The model agrees reasonably well with the experimental results obtained within the dispersion found in the tests.

A physical interpretation of the size effect has been introduced. It is based on the different geometric dimensions associated to dissipation and release of energy at the crack onset. The released energy by the crack onset is taken from the potential energy stored in a volume. In contrast, the energy is dissipated along a surface. This difference of dimensions causes that a higher strain is necessary to allow energetically a crack onset to appear when the inner ply is thinner.

As also described by París et al. (010a) and confirmed experimentally by París et al. (010b), the study of crack onset is only the first step in the description of this failure mechanism. The next step corresponds to a further increase of load and the appearing of a debond along the interface between both plies. The present approach applied to the problem of the crack initiation can be used to develop a similar theoretical model and to predict the critical parameters for this failure as well. A similar problem has been studied profusely in the last decade, see e.g. Martin et al. (2008); Leguillon and Martin (2013a).

An indirect procedure to obtain the transverse fracture toughness is proposed using the present theoretical model. Simplicity of the test and the specimen to be carried out is the key advantage of this procedure. A first test of the procedure has been carried out for the experimental results by Parvizi et al. (1978).

Appendices to Chapter 6

6.A   An expression for $E_{22}/\tilde{E}_{22}$

Laminate theory determines that for a symmetric and orthogonal laminate, efforts can be expressed as a function of the mid-plane strains by the following equation:

$$
\begin{bmatrix} N_y \\ N_z \\ N_{yz} \end{bmatrix} = \begin{bmatrix} A_{11} & A_{12} & 0 \\ A_{12} & A_{22} & 0 \\ 0 & 0 & A_{66} \end{bmatrix} \begin{bmatrix} \varepsilon^0_{yy} \\ \varepsilon^0_{zz} \\ \gamma^0_{yz} \end{bmatrix}, \tag{6.89}
$$

where N_y, N_z and N_{yz} are the efforts, ε^0_{yy}, ε^0_{zz} and γ^0_{yz} are the mid-plane strains. In general, the apparent laminate stiffness matrix A_{ij} is defined as

$$
A_{ij} = \sum_{k=1}^{3} Q_{ij}^{(k)}(x_k - x_{k-1}), \tag{6.90}
$$

where Q_{ij}^k is the stiffness matrix of the k-th ply and x_k is the position of the interface between the k-th and $(k+1)$-th ply. Then, in view of the expression (6.89), taking into account that all efforts applied are expected to be parallel to the laminate plane and that the studied laminate has a symmetric configuration, it can be assumed that

$$
\varepsilon_{yy}^{(k)} = \varepsilon^0_{yy}, \; k = 1, \ldots, 3, \tag{6.91a}
$$

$$
\varepsilon_{zz}^{(k)} = \varepsilon^0_{zz}, \; k = 1, \ldots, 3. \tag{6.91b}
$$

$$
\gamma_{yz}^{(k)} = \gamma^0_{yz} = 0, \; k = 1, \ldots, 3. \tag{6.91c}
$$

In order to obtain the relation between the material transverse Young modulus and the apparent Young modulus of the inner ply, it is sufficient to use the first two scalar equations in (6.89)

$$
N_y = A_{11}\varepsilon^0_{yy} + A_{12}\varepsilon^0_{zz}, \tag{6.92a}
$$

$$
N_z = A_{12}\varepsilon^0_{yy} + A_{22}\varepsilon^0_{zz}. \tag{6.92b}
$$

As $N_z = 0$, the next expression is obtained from (6.92b)

$$\varepsilon_{zz}^0 = -\frac{A_{12}}{A_{22}}\varepsilon_{yy}^0. \tag{6.93}$$

On the other hand, it is known from the laminate theory that the value of $\sigma_{yy}^{(k)}$ is

$$\sigma_{yy}^{(k)} = Q_{11}^{(k)}\varepsilon_{yy}^{(k)} + Q_{12}^{(k)}\varepsilon_{zz}^{(k)}. \tag{6.94}$$

Taking into account the expressions of $Q_{ij}^{(0)}$ and $Q_{ij}^{(90)}$ for $i,j = 1,2$

$$Q_{ij}^{(0)} = \frac{1}{1 - \nu_{12}\nu_{21}}\begin{bmatrix} E_{11} & \nu_{12}E_{22} \\ \nu_{12}E_{22} & E_{22} \end{bmatrix} \tag{6.95}$$

and

$$Q_{ij}^{(90)} = \frac{1}{1 - \nu_{12}\nu_{21}}\begin{bmatrix} E_{22} & \nu_{12}E_{22} \\ \nu_{12}E_{22} & E_{11} \end{bmatrix}, \tag{6.96}$$

and using these stiffness matrices and the expression of A_{ij} from (6.90),

$$A_{ij} = 2t_{90}\begin{bmatrix} \frac{E_{11}\frac{t_0}{t_{90}} + E_{22}}{1 - \nu_{12}\nu_{21}} & \frac{\nu_{12}E_{22}\left(1 + \frac{t_0}{t_{90}}\right)}{1 - \nu_{12}\nu_{21}} \\ \frac{\nu_{12}E_{22}\left(1 + \frac{t_0}{t_{90}}\right)}{1 - \nu_{12}\nu_{21}} & \frac{E_{22}\frac{t_0}{t_{90}} + E_{11}}{1 - \nu_{12}\nu_{21}} \end{bmatrix}. \tag{6.97}$$

By combining the above expressions and (6.93), (6.94) and (6.91), the relation between $\sigma_{yy}^{(90)}$ and the longitudinal applied strain ε_{yy}^0 is found as

$$\sigma_{yy}^{(90)} = \frac{E_{22}}{1 - \nu_{12}\nu_{21}}\left(1 - \frac{\nu_{12}\nu_{21}\left(1 + \frac{t_0}{t_{90}}\right)}{1 + \frac{E_{22}}{E_{11}}\frac{t_0}{t_{90}}}\right)\varepsilon_{yy}^0, \tag{6.98}$$

then, defining $\tilde{E}_{22}$ as $\sigma_{yy}^{(90)}/\varepsilon_{yy}^0$, the desired expression is obtained,

$$\frac{E_{22}}{\tilde{E}_{22}} = \frac{1 - \nu_{12}\nu_{21}}{1 - \nu_{12}\nu_{21}\frac{1 + \frac{t_0}{t_{90}}}{1 + \frac{E_{22}}{E_{11}}\frac{t_0}{t_{90}}}}. \tag{6.99}$$

6.B   An expression for $\varepsilon_{yy}^{0,\Delta T}$

Laminate theory determines that for a symmetric and orthogonal laminate subjected to an increment of temperature, the mid-plane strain $\varepsilon_{yy}^{t_0}$ after the curing can be obtained by solving the next system of equations,

$$\begin{bmatrix} N_y \\ N_z \end{bmatrix} = \begin{bmatrix} A_{11} & A_{12} \\ A_{12} & A_{22} \end{bmatrix}\begin{bmatrix} \varepsilon_{yy}^{t_0} \\ \varepsilon_{zz}^{t_0} \end{bmatrix} - \begin{bmatrix} N_y^T \\ N_z^T \end{bmatrix}, \tag{6.100}$$

where A_{ij} is the apparent laminate stiffness matrix defined in (6.90). N_y, N_z are the mechanical efforts and N_y^T, N_z^T are the thermal efforts defined as,

$$\begin{bmatrix} N_y^T \\ N_z^T \end{bmatrix} = \int_k \begin{bmatrix} \bar{Q}_{yy} & \bar{Q}_{yz} \\ \bar{Q}_{yz} & \bar{Q}_{zz} \end{bmatrix}^{(k)}\begin{bmatrix} \alpha_y \\ \alpha_z \end{bmatrix}^{(k)}\Delta T\mathrm{d}x, \tag{6.101}$$

where $\bar{Q}_{ij}^{(k)}$ corresponds to the stiffness matrix of each ply k adequately orientated, i.e. the expression given in (6.95) (6.96) for the outer and inner plies respectively. Analogously, the CTE $\alpha_y^{(k)}$ and $\alpha_z^{(k)}$ corresponds to the adequately orientated values of the CTE of each ply, i.e. $\alpha_y^{(0)} = \alpha_z^{(90)} = \alpha_1$ and $\alpha_y^{(90)} = \alpha_z^{(0)} = \alpha_2$.

Thus, taking into account that during the cooling after the curing process $N_y = N_z = 0$, the previous system (6.100) can be rewritten as,

$$\begin{bmatrix} N_y^T \\ N_z^T \end{bmatrix} = \begin{bmatrix} A_{11} & A_{12} \\ A_{12} & A_{22} \end{bmatrix} \begin{bmatrix} \varepsilon_{yy}^{to} \\ \varepsilon_{zz}^{to} \end{bmatrix}. \tag{6.102}$$

Solving this system of equations for ε_{yy}^{to}, the next expression is obtained,

$$\varepsilon_{yy}^{to} = \frac{N_y^T - \frac{A_{12}}{A_{22}} N_z^T}{A_{11} - \frac{A_{12}^2}{A_{22}}}. \tag{6.103}$$

Substituting (6.90), (6.95), (6.95) and (6.101) in (6.103) gives

$$\varepsilon_{yy}^{to} = \alpha_2 \Delta T k^{0,\Delta T} \left(\frac{E_{22}}{E_{11}}, \nu_{12}, \frac{t_0}{t_{90}}, \frac{\alpha_1}{\alpha_2} \right), \tag{6.104}$$

where k_{t_0} is

$$k^{0,\Delta T} \left(\frac{E_{22}}{E_{11}}, \nu_{12}, \frac{t_0}{t_{90}}, \frac{\alpha_1}{\alpha_2} \right) = \frac{k_1 \left(\frac{E_{22}}{E_{11}}, \nu_{12}, \frac{t_0}{t_{90}}, \frac{\alpha_1}{\alpha_2} \right) - k_2 \left(\frac{E_{22}}{E_{11}}, \nu_{12}, \frac{t_0}{t_{90}}, \frac{\alpha_1}{\alpha_2} \right)}{k_3 \left(\frac{E_{22}}{E_{11}}, \nu_{12}, \frac{t_0}{t_{90}}, \frac{\alpha_1}{\alpha_2} \right)}, \tag{6.105}$$

where

$$k_1 = \frac{t_0}{t_{90}} \frac{\alpha_1}{\alpha_2} + \frac{E_{22}}{E_{11}} \left(1 + \left(\frac{t_0}{t_{90}} \right)^2 \frac{\alpha_1}{\alpha_2} + \nu_{12} \frac{t_0}{t_{90}} \left(1 - \frac{\alpha_1}{\alpha_2} \right) \right), \tag{6.106a}$$

$$k_2 = \left(\frac{E_{22}}{E_{11}} \right)^2 \left(\frac{t_0}{t_{90}} \left(\nu_{12} \left(1 - \frac{\alpha_1}{\alpha_2} \right) - 1 \right) + \nu_{12}^2 \left(1 + \frac{t_0}{t_{90}} \right) \left(1 + \frac{t_0}{t_{90}} \frac{\alpha_1}{\alpha_2} \right) \right), \tag{6.106b}$$

$$k_3 = \frac{t_0}{t_{90}} + \frac{E_{22}}{E_{11}} \left(1 + \left(\frac{t_0}{t_{90}} \right)^2 \right) - \left(\frac{E_{22}}{E_{11}} \right)^2 \left(\nu_{12}^2 \left(1 + \frac{t_0}{t_{90}} \right)^2 - \frac{t_0}{t_{90}} \right). \tag{6.106c}$$

6.C An expression for $k^{(90),\Delta T}$

Laminate theory determines that for a symmetric and orthogonal laminate subjected to an increment of temperature, the value of the tension $\sigma_{yy}^{(90),\Delta T}$ at the inner ply can be obtained by the next expression,

$$\sigma_{yy}^{(90),\Delta T} = Q_{22}(\varepsilon_{yy}^0 - \alpha_2 \Delta T) + Q_{12}(\varepsilon_{zz}^0 - \alpha_1 \Delta T), \tag{6.107}$$

where

$$Q_{ij} = \frac{1}{1 - \nu_{12}\nu_{21}} \begin{bmatrix} E_{11} & \nu_{12}E_{22} \\ \nu_{12}E_{22} & E_{22} \end{bmatrix}, \tag{6.108}$$

ε_{yy}^0 and ε_{zz}^0 are the mid-plane strain along the $y-$ and $z-$directions respectively. α_1 and α_2 are, respectively, the longitudinal and transverse coefficients of thermal expansion (CTE) and ΔT is the uniform temperature change.

According to the boundary conditions, see Figure 6.35, $\varepsilon_{yy}^0 = 0$. Thus, the only unknown in (6.107) is ε_{zz}^0. This can be obtained from the global system of equation which relates the mechanical and thermals efforts and the mid-plane strains,

$$\begin{bmatrix} N_y \\ N_z \end{bmatrix} = \begin{bmatrix} A_{11} & A_{12} \\ A_{12} & A_{22} \end{bmatrix} \begin{bmatrix} \varepsilon_{yy}^0 \\ \varepsilon_{zz}^0 \end{bmatrix} - \begin{bmatrix} N_y^T \\ N_z^T \end{bmatrix}, \tag{6.109}$$

where A_{ij} is the apparent laminate stiffness matrix defined in (6.90). N_y, N_z are the mechanical efforts and N_y^T, N_z^T are the thermal efforts defined as,

$$\begin{bmatrix} N_y^T \\ N_z^T \end{bmatrix} = \int_k \begin{bmatrix} \bar{Q}_{yy} & \bar{Q}_{yz} \\ \bar{Q}_{yz} & \bar{Q}_{zz} \end{bmatrix}^{(k)} \begin{bmatrix} \alpha_y \\ \alpha_z \end{bmatrix}^{(k)} \Delta T dx, \tag{6.110}$$

where $\bar{Q}_{ij}^{(k)}$ corresponds to the stiffness matrix of each ply k adequately orientated, i.e. the expression given in (6.95) (6.96) for the outer and inner plies respectively. Analogously, the CTE $\alpha_y^{(k)}$ and $\alpha_z^{(k)}$ corresponds to the adequately orientated values of the CTE of each ply, i.e. $\alpha_y^{(0)} = \alpha_z^{(90)} = \alpha_1$ and $\alpha_y^{(90)} = \alpha_z^{(0)} = \alpha_2$.

Thus, taking into account $N_z = 0$, the value of ε_z^0 can be extracted from (6.109) as,

$$\varepsilon_{zz}^0 = \frac{N_z^T}{A_{22}}, \tag{6.111}$$

, where N_z^T is obtained from the second equation in (6.110),

$$N_z^T = t_0 \Delta T (Q_{12}\alpha_1 + Q_{22}\alpha_2) + t_{90}\Delta T (Q_{12}\alpha_2 + Q_{11}\alpha_1) \tag{6.112}$$

and A_{22} is given by (6.90) as,

$$A_{22} = t_0 Q_{22} + t_{90} Q_{11}. \tag{6.113}$$

Substituting (6.112) and (6.113) into (6.111) and the result into (6.107) and after some simplifications gives,

$$\sigma_{yy}^{(90),\Delta T} = -\alpha_2 E_{22} \Delta T k^{\Delta T} \left(\frac{E_{22}}{E_{11}}, \nu_{12}, \frac{t_0}{t_{90}}, \frac{\alpha_1}{\alpha_2} \right), \tag{6.114}$$

where $k^{\Delta T}$ is

$$k^{(90),\Delta T} \left(\frac{E_{22}}{E_{11}}, \nu_{12}, \frac{t_0}{t_{90}}, \frac{\alpha_1}{\alpha_2} \right) =$$

$$= \frac{1 - \frac{E_{22}}{E_{11}} \left(\frac{t_0}{t_{90}} \left(\nu_{12} \left(1 - \frac{\alpha_1}{\alpha_2} \right) - 1 \right) + \nu_{12}^2 \left(1 + \frac{t_0}{t_{90}} \frac{\alpha_1}{\alpha_2} \right) \right)}{\left(1 - \frac{E_{22}}{E_{11}} \nu_{12}^2 \right) \left(\frac{E_{22}}{E_{11}} \frac{t_0}{t_{90}} + 1 \right)}. \tag{6.115}$$

6.D Dimensional analysis of the energy release rate assuming generalized plane strain

Estimation of the value of ERR for a cracked specimen can be calculated directly if the stress and displacement distribution in the whole laminate is known. Thus, the value of ERR G depends on the stress tensor and the displacement vector in the laminate and the length of the crack (which is a geometric parameter)

$$G = f(\sigma_{ij}(x, y, z), u_i(x, y, z)). \tag{6.116}$$

The solution of stresses and displacements depends on the problem geometry (crack length included), mechanical properties of material, boundary conditions and volume loads. Therefore, the stresses in the present laminate can be expressed as

$$\sigma_{ij} = \sigma_{ij}(a, t_{90}, t_0, H, W, E_{11}, E_{22}, \nu_{12}, \nu_{23}, G_{12}, \varepsilon_{yy}, x, y, z), \tag{6.117}$$

where a is the semilength of the transverse crack and ε_{yy} represents the external applied strain. Other parameters have been defined in Section 6.1.1. The displacement vector u_i depends generally on the same parameters as σ_{ij},

$$u_i = u_i(a, t_{90}, t_0, H, W, E_{11}, E_{22}, \nu_{12}, \nu_{23}, G_{12}, \varepsilon_{yy}, x, y, z). \tag{6.118}$$

For a fixed geometry (crack length a included), this problem is linear, thus the values of σ_{ij} and u_i are directly proportional to the unique external solicitation, the applied strain component ε_{yy}. Consequently, the above expressions of σ_{ij} and u_i leads to

$$\sigma_{ij} = \varepsilon_{yy}\sigma_{ij}(\cdots, \varepsilon_{yy} = 1, \cdots) \tag{6.119a}$$

and

$$u_i = \varepsilon_{yy}u_i(\cdots, \varepsilon_{yy} = 1, \cdots), \tag{6.119b}$$

where $\varepsilon_{yy} = 1$ does not refer to its physical sense but mathematical one. It is known from the LEFM theory that the value of ERR is a homogeneous and linear function of the product of displacement and stresses. Considering this result, the expression of the ERR G leads to

$$G = \varepsilon_{yy}^2 G(a, t_{90}, t_0, H, W, E_{11}, E_{22}, \nu_{12}, \nu_{23}, G_{12}, \varepsilon_{yy} = 1). \tag{6.120}$$

In order to obtain the dimensionless parameters that govern the present value of G, two[1] independent dimensional parameters must be selected to normalize this expression.

A good choice is to take the semilength of the inner-ply t_{90} and the transverse Young's modulus E_{22}. Then, using the Buckingham Π theorem, see e.g. Barenblatt (1996), the above expression can be rewritten in terms of a smaller number of variables in the following special form:

---

[1]A system of units consisting of two units (a unit for the measurement of length and a unit for the measurement of force) is sufficient for measuring the properties of a static mechanical problem

$$G = E_{22}\varepsilon_{yy}^2 t_{90}\hat{G}\left(\frac{a}{t_{90}}, \frac{t_0}{t_{90}}, \frac{W}{t_{90}}, \frac{H}{t_{90}}, \frac{E_{11}}{E_{22}}, \frac{G_{12}}{E_{22}}, \nu_{12}, \nu_{23}\right), \tag{6.121}$$

where a/t_{90} is the ratio of the half-crack length to the thickness t_{90} of the inner-ply and t_0/t_{90}, W/t_{90}, H/t_{90} are dimensionless geometric parameters and $\hat{G}$ is the dimensionless ERR. E_{11}/E_{22}, G_{12}/E_{22}, ν_{12} and ν_{23}, are dimensionless parameters which depend on the material properties. These parameters are denoted as: "E.P." in the following to simplify the expressions. Geometrical parameters $W/t_{90} \gg 1$ and $H/t_{90} \gg 1$ typically in laminates, and $G(W/t_{90} \to \infty) = $ cte and $G(H/t_{90} \to \infty) = $ const. Therefore, as shown by Barenblatt (1996), H/t_{90} and W/t_{90} can be assumed not essential and the above expression can be consequently simplified for sufficiently large H/t_{90} and W/t_{90} as

$$G = E_{22}\varepsilon_{yy}^2 t_{90}\hat{G}\left(\frac{a}{t_{90}}, \frac{t_0}{t_{90}}, \text{E.P.}\right), \tag{6.122}$$

where t_0/t_{90} is the ratio of the thicknesses of the 0° lamina to the 90° lamina. This expression enables to to generalize a computational result for G to any values of E_{22}, ε_{yy} and t_{90}.

6.E   Proof of that the function g has a minimum strictly between the location of the maximum of $\hat{G}$ and the interface

The following basic facts obtained from the definition of $\hat{G}$ assuming $t_0/t_{90} > 0$, and the relations (6.12) and (6.13) are used in this proof:

1. The continuity of $\hat{G}$ in the interval $[0, 1]$.

2. $\hat{G}(\hat{a}) > 0$ in the interval $(0, 1)$.

3. $\hat{G}(0) = \hat{G}(1) = 0$

Proposition: There exists a minimum of the function $g(\Delta\hat{a})$ for a point $0 < \Delta\hat{a} < 1$.

Proof:
Denoting $I_{\hat{G}}(\Delta\hat{a}) = \int_0^{\Delta\hat{a}} \hat{G}(\hat{a})d\hat{a}$, the definition of g in (6.16) writes as

$$g(\Delta\hat{a}) = \frac{\Delta\hat{a}}{I_{\hat{G}}(\Delta\hat{a})}, \tag{6.123}$$

The derivative of the function g is

$$g'(\Delta\hat{a}) = \frac{I_{\hat{G}}(\Delta\hat{a}) - \Delta\hat{a}I'_{\hat{G}}(\Delta\hat{a})}{I_{\hat{G}}^2(\Delta\hat{a})}. \tag{6.124}$$

The sign of this derivative at the extremes of the interval $[0, 1]$ can be calculated as follows. At the extreme $\Delta\hat{a} = 0$, $\hat{G}$ can be approximated as

$$\hat{G}(\hat{a} \to 0^+) \cong \pi\hat{a}. \tag{6.125}$$

Thus,

$$I_{\hat{G}}(\Delta\hat{a} \to 0^+) \cong \int_0^{\Delta\hat{a}} \pi\hat{a}d\hat{a} = \pi\frac{\Delta\hat{a}^2}{2}. \tag{6.126}$$

Introducing this approximation in the expression of g', it is obtained that

$$g'(\Delta\hat{a} \to 0^+) \cong \frac{\pi\frac{\Delta\hat{a}^2}{2} - \pi\Delta\hat{a}^2}{\pi^2\frac{\Delta\hat{a}^4}{4}} = -\frac{2}{\pi\Delta\hat{a}^2} < 0, \tag{6.127}$$

which implies that g is a decreasing function for small $\Delta\hat{a} \gtrsim 0$.

At the other extreme $\Delta\hat{a} = 1$, the expression of g' in (6.124) can be simplified by applying $I'_{\hat{G}} = \hat{G}$. According to the discussion in Section 6.1.2, $\hat{G}(\hat{a} \to 1^-) = \hat{G}(1) = 0$. Introducing this result in the expression of g' in (6.124) leads to

$$g'(\Delta\hat{a} \to 1^-) = \frac{I_{\hat{G}}(\Delta\hat{a})}{I_{\hat{G}}^2\Delta\hat{a}} = \frac{1}{I_{\hat{G}}(\Delta\hat{a})}. \tag{6.128}$$

As the dimensionless ERR $\hat{G}$ is non-negative, $\hat{G} \geq 0$,

$$I_{\hat{G}}(\Delta\hat{a}) = \int_0^{\Delta\hat{a}} \hat{G}(\hat{a})d\hat{a} > 0, \quad \text{for } \Delta\hat{a} > 0 \tag{6.129}$$

which implies that

$$g'(\Delta\hat{a} \to 1^-) = \frac{1}{I_{\hat{G}}(\Delta\hat{a})} > 0. \tag{6.130}$$

The continuity of the derivative of g' is assured by the continuity of $\hat{G}$ except possibly for the points where the denominator $I_{\hat{G}} = 0$ in (6.124) vanishes. However, according to (6.129) this happens only for $\Delta\hat{a} = 0$. Therefore in the interval $\Delta\hat{a} \in (0, 1]$ the derivative g' is continuous. Combining this result with the facts that $g'(\Delta\hat{a}) < 0$ for $\Delta\hat{a} \to 0^+$ and $g'(\Delta\hat{a}) > 0$ for $\Delta\hat{a} \to 1^-$, the function g has at least one minimum point at $\Delta\hat{a}_{min,g} \in (0, 1)$.

Proposition: The (first) minimum $\Delta\hat{a}_{min,g}$ of g is situated closer to the interface than the (first) maximum of $\hat{G}$, $\Delta\hat{a}_{min,g} > \hat{a}_{max,\hat{G}}$.

Proof:

Since $\hat{G}(0) = \hat{G}(1) = 0$, $\hat{G} > 0$ in $(0, 1)$, and $\hat{G}$ is continuous in $[0, 1]$, $\hat{G}$ has at least one maximum in an interior point of the interval $(0, 1)$.

The second derivative of g can be calculated from the first one, obtaining,

$$g'' = \frac{\left(I_{\hat{G}} - \Delta\hat{a}I'_{\hat{G}}\right)' I_{\hat{G}}^2 - 2I_{\hat{G}}I'_{\hat{G}}\left(I_{\hat{G}} - \Delta\hat{a}I'_{\hat{G}}\right)}{I_{\hat{G}}^4}, \tag{6.131}$$

$$\left(I_{\hat{G}} - \Delta \hat{a} I'_{\hat{G}}\right) = 0 \text{ for } \Delta \hat{a} = \Delta \hat{a}_{\min,g} \text{ because } g'(\Delta \hat{a}_{\min,g}) = 0. \text{ Then}$$

$$g''\left(\Delta \hat{a}_{\min,g}\right) = \frac{\left(I_{\hat{G}} - \Delta \hat{a} I'_{\hat{G}}\right)'}{I_{\hat{G}}^2}, \qquad (6.132)$$

and knowing that $I_{\hat{G}}^2 > 0$, the sign of the second derivative is reduced to

$$\text{sign}\left(g''\left(\Delta \hat{a}_{\min,g}\right)\right) = -\text{sign}\left(G'\left(\Delta \hat{a}_{\min,g}\right)\right), \qquad (6.133)$$

where the fundamental theorem of calculus has again been applied.

This relation implies that the (first) minimum point of g (where $g''(\Delta \hat{a}_{\min,g}) > 0$) is situated in the region where $\hat{G}$ is a decreasing function ($\hat{G}' < 0$), i.e. after the (first) maximum of $\hat{G}$,

$$\Delta \hat{a}_{\min,g} > \hat{a}_{\max,\hat{G}}, \qquad (6.134)$$

where $\hat{a}_{\max,\hat{G}}$ corresponds to the (first) maximum of $\hat{G}$.

6.F   Proof of that $\Delta \hat{a}_{\min,g}$ is a lower bound for the arrest length

Proposition: $\hat{a}_{\mathrm{a}} \geq \Delta \hat{a}_{\min,g}$

Proof: In scenario B, the proof has been given in Section 6.1.4. In scenario A, let $\Delta \hat{a}_{\mathrm{c}}^{\alpha}$ denote the length of the crack at the onset and $\varepsilon_{yy}^{\mathrm{c},\alpha}$ the critical strain of a determinated configuration α. Let ε_{yy}^{β} denote the minimum strain required by the energy criterion for a given value of $\Delta \hat{a}^{\beta} \in (\Delta \hat{a}_{\mathrm{c}}^{\alpha}, \Delta \hat{a}_{\min,g})$, it means the value of the right-hand side of (6.17) for $\Delta \hat{a} = \Delta \hat{a}^{\beta}$. Taking into account that $g(\Delta \hat{a})$ is a decreasing function for $\Delta \hat{a} < \Delta \hat{a}_{\min,g}$,

$$\varepsilon_{yy}^{\mathrm{c},\alpha}(\Delta \hat{a}_{\mathrm{c}}^{\alpha}) > \varepsilon_{yy}^{\beta}(\Delta \hat{a}^{\beta}), \qquad (6.135)$$

and introducing from (6.120) the dependence of G on ε_{yy},

$$G\left(\hat{a}, \varepsilon_{yy}^{\mathrm{c},\alpha}(\Delta \hat{a}_{\mathrm{c}}^{\alpha})\right) > G\left(\hat{a}, \varepsilon_{yy}^{\beta}(\Delta \hat{a}^{\beta})\right), \ \forall \hat{a} \in (0,1) \qquad (6.136)$$

and from (6.30) that $G\left(\Delta \hat{a}^{\beta}, \varepsilon_{yy}^{\beta}(\Delta \hat{a}^{\beta})\right) > G_{\mathrm{c}}$ for $\Delta \hat{a}^{\beta} < \Delta \hat{a}_{\min,g}$ implies that

$$G\left(\Delta \hat{a}^{\beta}, \varepsilon_{yy}^{\mathrm{c},\alpha}(\Delta \hat{a}_{\mathrm{c}}^{\alpha})\right) > G_{\mathrm{c}}, \text{ for } \Delta \hat{a}_{\mathrm{c}}^{\alpha} < \Delta \hat{a}^{\beta} < \Delta \hat{a}_{\min,g}. \qquad (6.137)$$

Then, the condition to arrest the crack (6.33) is not fulfilled for values of the length lower than $\Delta \hat{a}_{\min,g}$. Therefore, $\Delta \hat{a}_{\min,g}$, which is independent of the fracture and strength properties, is a lower bound of the arrest length.

CHAPTER 7

Crack initiation at a weak surface in a V-notch

Crack initiation at stress singular points, e.g. corner singularities, has been studied for decades and is still an open subject. They are preferential failure points and their presence has a strong effect on the global strength of structures. Despite this importance, the problem of predicting the critical load originating the failure and its direction is still an open problem, the scientific debate is not closed on this point, especially for complex loadings. One of the reasons is that the linear elastic fracture mechanics (LEFM) itself is not able to predict the onset of a crack at such a location. Several criteria have been developed based on very different approaches. They mainly consider that the generalized stress intensity factor (GSIF) is the critical parameter to predict failure at corner singularities. This idea is based on many experimental evidences, see e.g. Reedy Jr and Guess (1993); Seweryn (1994); Dunn et al. (1997).

In the context of the finite fracture mechanics theory (FFM), a new model based on the coupled criterion was proposed by Leguillon (2002) to predict the onset of a crack at corner singularities. It enables to obtain the critical value of the GSIF under symmetric loading as shown in previous reference or by other authors, see e.g. Carpinteri et al. (2008). The predictions agree quite well with the experimental results. Moreover, one of the main advantages of this new criterion is that the critical GSIF can be expressed as a function of two commonly used material properties: the tensile strength σ_c and the fracture toughness G_c of the material.

Yosibash et al. (2006) and Priel et al. (2008) generalized this criterion to corner singularities under non-symmetric loads. In this case the two first singular terms of Williams' expansion must be accounted for. The leading one corresponds to a symmetric mode and the second one corresponds to an antisymmetric mode which is activated only if the loading is not symmetric. Then, the generalization developed here allows obtaining the critical load, the critical length and the

247

angle of the new created crack, which is not situated necessarily symmetric (i.e. along the bisector). Despite the non-symmetry of the problem the influence of the shear stresses in the stress criterion and the effect of the stress mixity at the crack tip in the energy criterion are neglected in these approaches. This is because the new crack is assumed to grow in pure mode 1 (or very close to, depending on the choice of the criterion, see Erdogan and Sih (1963)). This angle is very close to that where the normal stresses and the energy release rate reach their maximum value. Moreover, shear stresses at this angle vanish as can be derived from the equilibrium equations. This hypothesis is consistent with the criterion of local symmetry, see Barenblatt and Cherepanov (1961); Erdogan and Sih (1963), usually employed to predict the crack path in LEFM. This is formulated for homogeneous solids having homogeneous properties of fracture and strength and it gives a very good agreement with experimental results for corner points as have been shown by Yosibash et al. (2006) and Priel et al. (2008).

However, the hypothesis of pure mode 1 after the crack onset is not longer valid in the presence of a preferential crack path imposed by an interface or a weak surface with weaker fracture and strength properties. The most simple example of such a preferential direction is the presence of a weak adhesive joint between similar or dissimilar materials. In this case, the joint is generally a preferential path for the crack despite the facts that in general the crack does not grow in pure mode 1 and that the shear stresses do not vanish. Thus, both shear stresses and the variation of the fracture toughness with the stress mixity can play a crucial role. Recent experiments (Toda et al. (2001); Ogihara and Koyanagi (2010)) have shown the importance of the shear stresses in failure of adhesive joints. As a consequence, in the cases where a high ratio of shear to normal stresses is expected, shear stresses should not be neglected. Similarly, the so called mode-mixity cannot be neglected in the energy criterion because the fracture toughness can depend strongly on the mixity as shown by several experiments, see e.g. Hutchinson and Suo (1992) or Banks-Sills et al. (2000). Therefore, it is necessary to generalize the criterion presented by Yosibash et al. (2006) including the shear stresses and mode-dependent fracture toughness in order to study a crack initiating and growing along the weak surface. The crack initiation at a corner in the presence of an interface is a common problem in composites and can be associated with the first step of some failure mechanisms. A well studied example is the failure initiation at the multimaterial corner which can be found in double lap joints, see Barroso (2007); Vicentini (2012). The model developed in this chapter can be easily extended to account for the study of this particular case.

Regarding the computational methods, the influence of shear stresses and fracture mode-mixity has been implemented in some models. A general law taking into account these dependencies for cohesive zone models (CZM) for interfaces is presented in Tvergaard and Hutchinson (1993). In this model, the normal and shear stress laws depend on the normal and tangential separations. In the limit where the tangential (or normal) separation vanishes, the normal (or shear) stress peak is the tensile (or shear) strength and the work of separation is the fracture toughness in pure mode 1 (or 2). In the general case the proposed model couples both effects. Based on this pioneering work, several cohesive laws have been proposed and implemented during the last years, among them it is worth mentioning the work by Freed and Banks-Sills (2008) which proposed a new CZM which coincides with the well known law of Hutchinson and Suo (1992) if the mixity is defined as the ratio of tangential to normal separation. In the similar way, Távara et al. (2010) proposes a linear elastic brittle interface model (LEBIM). In the context of the coupled criterion criterion of the FFM, Mantič (2009) implements a mode-dependent toughness in order to apply the criterion to particular problems of interfaces where a high mixity exists and proposes also as a future development the introduction of Mohr-Coulomb type law in order to take into consideration the influence of the shear stresses on the failure.

This chapter proposes the generalization of Yosibash et al. (2006) and Priel et al. (2008) for the case where a weak surface exists on a V-notched specimen under non-symmetric loading. The singular stress field at a V-notch is shortly described in Section 7.1. Stress and energy criteria considering the shear stresses and mode-dependent fracture toughness are implemented to the present problem in Sections 7.2 and 7.3 respectively. Both conditions are combined in Section 7.4 where the first general expressions are obtained. In Section 7.5, these expressions are applied to a particular problem, the simulation of the Arcan test on V-notched specimens embedding a weak surface starting from the notch root. The solution for the crack onset is a competition between two possible paths: the crack is initiated along the weak surface if it is sufficiently weak or in the bulk at an optimum angle calculated by Yosibash et al. (2006). Finally, both possibilities are studied in Sections 7.6.1 and 7.6.2 and compared for a particular case of material and weak-surface properties in Section 7.6.3.

7.1 The singular stress field at a V-notch

Stress singularities around a corner point are nowadays a well-studied problem. The V-notch corner induces a stress singularity in its vicinity, whose level depends strongly on the V-notch opening angle ω. With stress-free boundary condition along the two notch faces, it is a weaker singularity than that of a crack (corresponding to $\omega = 0°$). In plane elasticity and under general mixed modes loading conditions, the displacement field around this point (Wieghardt (1907); Williams (1952)) can be expressed as

$$\underline{U}(r,\theta) = \underline{U}(0) + k_1 r^{\lambda_1} \underline{u}_1(\theta) + k_2 r^{\lambda_2} \underline{u}_2(\theta) + \cdots \qquad (7.1)$$

where $\underline{U}(r,\theta)$ is the displacement vector at the polar coordinates (r,θ) with the origin at the corner root O and $\theta = 0°$ corresponding to the bisector (see Figure 7.1). The term $\underline{U}(0)$ holds for the displacement of the corner root and is constant. This term is irrelevant and does not have any importance in the subsequent developments but is present for consistency. The second term $k_1 r^{\lambda_1} \underline{u}_1(\theta)$ is the contribution to the displacement field due to the symmetric loads. The factor k_1 is the corresponding generalized stress intensity factor (GSIF). It is proportional to the remote load, units are MPa.m$^{1-\lambda_1}$. The singularity exponent λ_1 lies in the domain $1/2 \leq \lambda_1 \leq 1$ and depends strongly on the V-notch opening ω. It is $1/2$ and 1 respectively for a crack ($\omega = 0°$) and a straight edge ($\omega = 180°$). The vector $\underline{u}_1(\theta)$ is a symmetric angular shape function with dimensions MPa^{-1}. This may seem surprising, this is due to the fact that the elastic material properties are implicitly embedded in the denominator of $\underline{u}_1(\theta)$. The third term $k_2 r^{\lambda_2} \underline{u}_2(\theta)$ corresponds to the antisymmetric mode. Analogously, k_2, λ_2 and $\underline{u}_2(\theta)$ are the GSIF, singularity exponent and angular shape function of the antisymmetric mode. Properties of these parameters are similar to that described for the symmetric case except that λ_2 can be larger than 1 roughly for $\omega \geq 105°$.

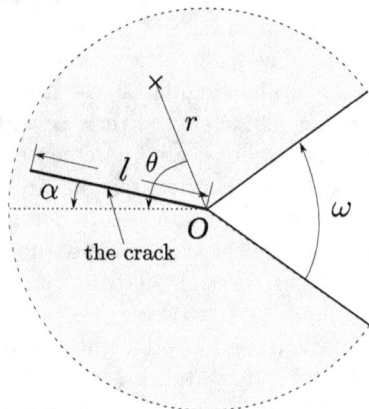

Figure 7.1: Focus on the vicinity of the corner point, the shaded side corresponds to the solid part. The dashed line is the bisector of this region.

Applying the constitutive elastic law provides an analogous expansion of the stress field in the vicinity of the origin

$$\underline{\underline{\sigma}}(r,\theta) = k_1 r^{\lambda_1 - 1} \underline{\underline{s}}_1(\theta) + k_2 r^{\lambda_2 - 1} \underline{\underline{s}}_2(\theta) + \cdots \qquad (7.2)$$

where $\underline{\underline{\sigma}}$ is the stress tensor at point (r,θ). The functions $\underline{\underline{s}}_1(\theta)$ and $\underline{\underline{s}}_2(\theta)$ are the dimensionless associated angular functions derived from (7.1). As a consequence the GSIF units are MPa.m$^{1-\lambda_i}$, they vary with the exponent value. This also explains the previous remark on the unit of vectors $\underline{u}_1(\theta)$ and $\underline{u}_2(\theta)$.

Note that both terms are unbounded for $r \to 0$ if the exponent $\lambda_i < 1$ ($i = 1, 2$). As mentioned above, this is true for the first exponent λ_1 for any opening angle smaller than $180°$, whereas the second exponent λ_2 can be higher than 1 for widely opened corners (roughly $\omega > 105°$). As a consequence, in this special case, stresses in the vicinity of the corner can be non singular under a pure antisymmetric loading (since $k_1 = 0$ in this case).

The exponents λ_1, λ_2 and the angular functions $\underline{u}_1(\theta)$, $\underline{u}_2(\theta)$ are solutions to an eigenvalue problem (Leguillon and Sanchez-Palencia (1987)), see Table 7.1 for some values of exponents.

The values of the GSIF k_1 and k_2 are proportional to the remote load and can be extracted, in general, from a numerical analysis of the global elastic solution in the vicinity of the corner, see e.g. Leguillon and Sanchez-Palencia (1987); Labossiere and Dunn (1999).

ω ($^\circ$)	0	30	60	90	120	150	180
λ_1	0.5	0.502	0.512	0.545	0.616	0.752	1
λ_2	0.5	0.598	0.730	0.906	1.149	1.484	2

Table 7.1: Exponents λ_1 and λ_2 for different values of the notch opening ω

As in LEFM, a combined equivalent GSIF is proposed here as

$$k_{\text{eq}}^2(r) = \underbrace{k_1^2 r^{2\lambda_1-1}}_{(k_1^*)^2} + k_2^2 \underbrace{r^{2\lambda_2-1}}_{(k_2^*)^2} \tag{7.3}$$

This equivalent parameter k_{eq} has the same units as the usual SIF of a crack, e.g. MPa.mm$^{1/2}$. It enables to define new dimensionless parameters $\hat{a}$ and $\hat{b}$ as

$$\hat{a} = \frac{k_1}{k_{\text{eq}}} r^{\lambda_1-1/2} \tag{7.4a}$$

$$\hat{b} = \frac{k_2}{k_{\text{eq}}} r^{\lambda_2-1/2} \tag{7.4b}$$

Note that $\hat{a}$ and $\hat{b}$ can be related by the equation (7.3). Combining (7.3) and (7.4), next relation is obtained:

$$\hat{a}^2 + \hat{b}^2 = 1 \tag{7.5}$$

As a consequence the pure modes can be easily defined:

- In pure mode 1: $\hat{a} = 1 \Rightarrow \hat{b} = 0$ and $k_2 = 0$.

- In pure mode 2: $\hat{b} = 1 \Rightarrow \hat{a} = 0$ and $k_1 = 0$.

One can rewrite the expansions (7.1) and (7.2) functions of these new parameters

$$\underline{U}(r,\theta) = \underline{U}(0) + k_{\text{eq}} r^{1/2} \left(\hat{a}\underline{u}_1(\theta) + \hat{b}\underline{u}_2(\theta) \right) + \cdots \tag{7.6}$$

and

$$\underline{\sigma}(r,\theta) = k_{\text{eq}} r^{-1/2} \left(\hat{a}\underline{s}_1(\theta) + \hat{b}\underline{s}_2(\theta) \right) + \cdots \tag{7.7}$$

These two relations are apparently similar to the usual expansions for a crack simply modified by the parameters $\hat{a}$ and $\hat{b}$. However, the dependence on λ_1 and λ_2 has not disappeared, it is implicitly embedded in $\hat{a}$ and $\hat{b}$ which depend also on the distance r to the corner point.

These expressions should allow defining a mixed-mode criterion based on a critical value for the equivalent SIF k_{eq} with an expression which is totally symmetric with respect to the two modes. It is necessary to remark the symmetric aspect of this formulation which differs from what is done usually. Pure symmetric mode is traditionally characterized by a vanishing mixity parameter whereas the pure antisymmetric mode is obtained when this parameter tends to infinity (a convenient way of presenting this property is to consider an angle varying from $0°$ to $90°$ and its tangent, see below in Section 7.3.2). Another new expression was presented recently by Hills and Dini (2011) as a function of a length with a very interesting physical meaning: a point closer or farther to the corner than this reference length will have a stress dominated either by the symmetric or the antisymmetric modes respectively. However this length is defined by normalizing with k_1 and this may cause some inaccuracy when studying quasi antisymmetric modes, what it is sought to avoid.

7.2 Stress criterion

The stress criterion is based on the assumption of a critical value of the tensile or shear stress all along the line where the crack is expected to appear. Originally Leguillon (2002) proposed a stress criterion based on the assumption of a critical value σ_c so-called tensile strength for the normal tractions. An abrupt onset is possible over a finite extension l if for all points x in this extension

$$\sigma(r) \geq \sigma_c, \text{ for } 0 < r \leq l \tag{7.8}$$

where $\sigma(r)$ corresponds to the normal traction acting to the plane of the new crack prior to its onset.

Later, other authors have proposed different definitions of the stress criterion as Cornetti et al. (2006) who proposed the use of the averaged value of the normal traction over the finite crack extension and comparing it to the critical value σ_c. All these definitions are based on the hypothesis of pure or nearly mode 1.

In the mixed-mode case, it is necessary to take into account the shear stresses $\tau(r)$ along this line; whereas, none of the above definitions included this possible influence. To this aim, Mantič (2009) proposed to use the Mohr-Coulomb criterion, which is intensively employed in soils mechanics, to predict interface debonding in composite materials. Experimental evidences of the influence of τ have been presented in Toda et al. (2001) and Ogihara and Koyanagi (2010) for glass/epoxy interfaces. In the mixed-mode case, it is necessary to take into account both σ and τ. To generalize the Mohr-Coulomb criterion, the following relationship between σ and τ is classically used

$$\xi = \left(\frac{\sigma(r)}{\sigma_c}\right)^p + \left(\frac{|\tau(r)|}{\tau_c}\right)^p \geq 1 \tag{7.9}$$

where $\sigma = \sigma_{\theta\theta}$ and $\tau = \tau_{r\theta}$ can be extracted from (7.7). σ_c and τ_c denote respectively the tensile and the shear strengths respectively. The shear component τ is involved regardless of its sign. The parameter p is related to the shape of the failure curve as shown in Figure 7.2. Thanks to this definition, the main failure curves used in the literature can be obtained for different values of p (Figure

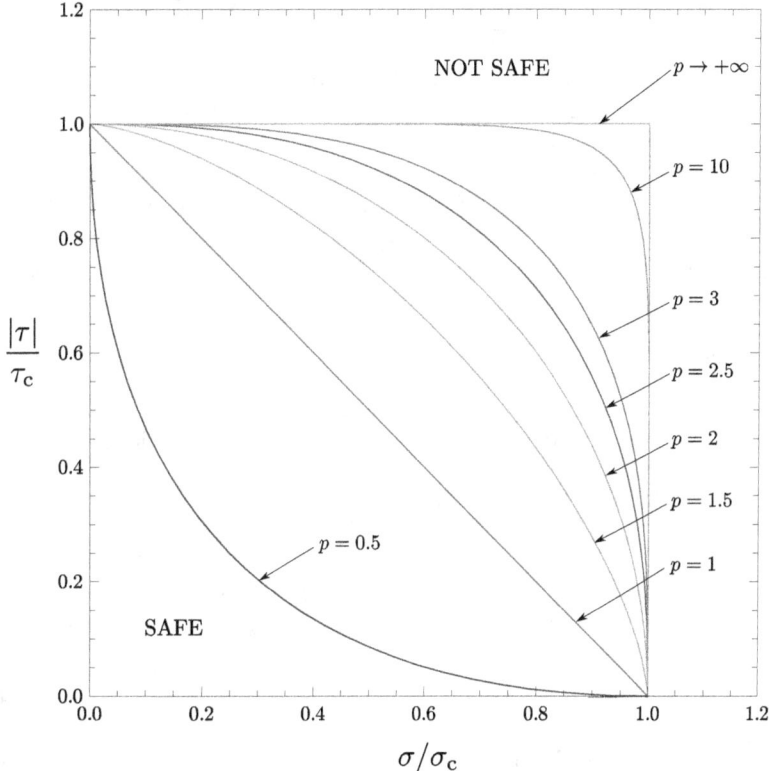

Figure 7.2: The stress criterion evaluated for several values of the parameter p.

7.3). This failure curves are concave for $p \geq 1$. In particular, failure curve for $p = 1$ and $p = 2$ corresponds respectively to the Mohr-Coulomb criterion and an elliptical failure curve. Finally $p \to \infty$ gives a rectangular safe region which corresponds to impose an uncoupled fulfillment of the pure mode stress criteria to failure: $\sigma \geq \sigma_c$ or $\tau \geq \tau_c$. The choice of the power law (7.9) is not essential. It is convenient for analytical calculations, it may be noted that simply considering to different exponents for the tension and the shear terms complicated greatly the calculations. Of course, the exponent p needs to be identified, 2 is often selected, see Seweryn et al. (1997).

Under the form (7.9), the criterion corresponds to a tension dominated situation. However, with a slight change in (7.9), the model could be obviously extended by considering that a compression may act on the failure zone, see Section 3.3.

Transcribing (7.7) into the components of interest of the stress tensor leads to

$$\sigma(r, \theta) = k_{eq} r^{-1/2} \left(\hat{a} s_1(\theta) + \hat{b} s_2(\theta) \right) + \cdots \tag{7.10a}$$

$$\tau(r,\theta) = k_{\text{eq}} r^{-1/2} \left(\hat{a} t_1(\theta) + \hat{b} t_2(\theta) \right) + \cdots \qquad (7.10\text{b})$$

where s_1 and s_2 (respectively t_1 and t_2) hold for the dimensionless normal (respectively tangent) angular functions extracted directly from $\underline{s}_1$ and $\underline{s}_2$. Figure 7.3 shows the values of these parameters as functions of the polar angle θ for several values of the opening angle ω in the half domain $\theta > 0°$. Clearly s_1 and t_2 are decreasing functions of θ because the maximum value of the normal stresses under a symmetric loading and shear stresses under an antisymmetric loading are expected along the bisector. Moreover, it is foreseeable that:

- $s_2(\theta = 0) = 0$ and $t_1(\theta = 0) = 0$ because of the symmetry and antisymmetry, respectively.

- $s_1(\theta = 180° - \omega/2) = s_2(\theta = 180° - \omega/2) = t_1(\theta = 180° - \omega/2) = t_2(\theta = 180° - \omega/2) = 0$, due to the stress-free boundary conditions (Figure 7.3).

Moreover, the eigenvectors $\underline{u}_1(\theta)$ and $\underline{u}_w(\theta)$ are normalized in order to have $s_1(\theta = 0°) = 1$ and $t_2(\theta = 0°) = 1$. Note that, for a crack, these last equalities slightly differ from the classical notation by a multiplicative factor $\sqrt{2\pi}$.

The combination of relations in (7.10) together with (7.9) amounts to a new way to express the stress condition

$$k_{\text{eq}} r^{-1/2} \left[\left(\hat{a} s_1(\theta) + \hat{b} s_2(\theta) \right)^p \tau_c^p + |\hat{a} t_1(\theta) + \hat{b} t_2(\theta)|^p \sigma_c^p \right]^{1/p} \geq \tau_c \sigma_c \qquad (7.11)$$

Crack onset over a length l is allowed if the above inequality holds true for all the values of $r \leq l$. If both $\sigma(r,\theta)$ and $|\tau(r,\theta)|$ are decreasing functions of r, it suffices to fulfill this condition at $r = l$. As already mentioned, this assumption is not valid for τ if $\omega > 105°$, because $\lambda_2 > 1$ and then τ now increases with r and the condition $\tau \geq \tau_c$ can never be satisfied all along the crack path: failure cannot occur under a pure shearing mode if $\omega > 105°$. Thus the final formulation splits into two parts, below and above the transition value $\omega = 105°$ (see Table 7.1), below both tension and shear are used whereas only tension is involved above.

In conclusion, the stress condition imposes, in general, a maximum value of the crack onset length l as a function of the load (implicit through the k_i's and then k_{eq}). When the remote load increases, k_{eq} increases, then l increases expanding the range of allowed lengths of the crack onset according to this criterion.

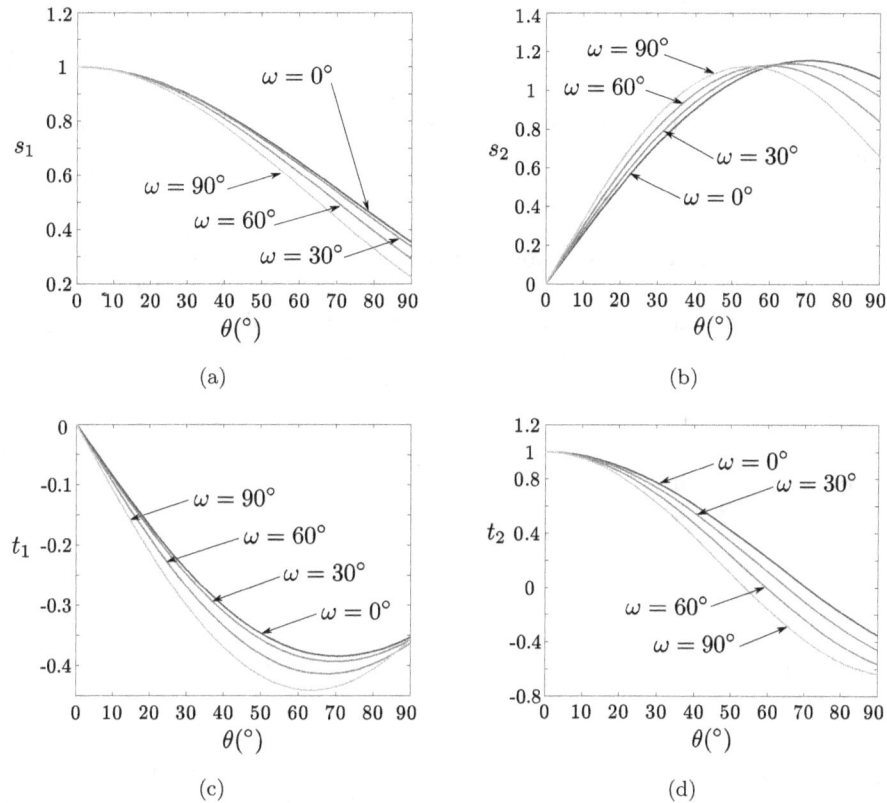

Figure 7.3: Dimensionless angular functions (a) s_1, (b) s_2, (c) t_1, and (d) t_2 for several values of the opening angle $\omega = 0°, 30°, 60°, 90°$. The functions s_1 and t_2 (resp. s_2 and t_1) are symmetric (resp. antisymmetric) with respect to $\theta = 0°$. Results courtesy of Professor D. Leguillon published by García and Leguillon (2012).

7.3 Energy criterion

The Griffith criterion is originally based on an energy balance between an initial equilibrium state and another one with an infinitesimal crack extension,

$$\Delta\Pi + \Delta E_k + \Delta\Gamma = 0 \tag{7.12}$$

where $\Delta\Pi$ and ΔE_k, are the changes in potential and kinetic energy and where $\Delta\Gamma$ denotes the energy dissipated in the process of fracture. Assuming a quasi-static initial state, $\Delta E_k \geq 0$, the above balance is rewritten as,

$$-\Delta\Pi \geq \Delta\Gamma \tag{7.13}$$

where the terms $\Delta\Pi$ and $\Delta\Gamma$ are developed in the following.

7.3.1 Released energy

According to the procedure described in Leguillon (1989) (see also Yosibash et al. (2006)), the asymptote of the potential energy change prior to and after the short crack onset in the direction $\theta = \alpha$ can be expressed, under plane strain assumption as

$$- \Delta\Pi = \left(A_1(\alpha)k_1^2 l^{2\lambda_1} + A_{12}(\alpha)k_1 k_2 l^{\lambda_1+\lambda_2} + A_2(\alpha)k_2^2 l^{2\lambda_2} \right) e \qquad (7.14)$$

where e and l hold respectively for the thickness of the specimen and the length of the new crack after the onset. The terms A_1, A_{12} and A_2 (MPa^{-1}) are real coefficients which depend on the angle of fracture α. They can be efficiently computed as described in Yosibash et al. (2006). Figure 7.4 shows the corresponding dimensionless coefficients A_1^*, A_{12}^* and A_2^*, obtained multiplying A_{ij} by the effective Young modulus $E^* = E/(1-\nu^2)$, as functions of α for several values of the opening angle ω. Introducing $\hat{a}$ and $\hat{b}$ (7.4) in (7.14), the released energy can be rewritten as,

$$- \Delta\Pi = k_{eq}^2 l \left(\hat{a}^2 A_1(\alpha) + \hat{a}\hat{b} A_{12}(\alpha) + \hat{b}^2 A_2(\alpha) \right) e + \cdots \qquad (7.15)$$

7.3.2 Dissipated energy

In pure opening mode 1, the dissipated energy is approximated by the linear relationship

$$\Delta\Gamma = G_{1c} l e \qquad (7.16)$$

where G_{1c} is the fracture toughness in mode 1 normally employed in LEFM. In the present case, under a mixed-mode loading, the influence of the mode mixity on the fracture toughness and the dissipated energy can be relevant. In order to take into account this dependence, the dissipated energy can be calculated by integrating $G_c(\psi)$ which depends on the mixity, denoted ψ, as proposed in Mantič (2009) (a more precise definition of ψ will be provided lately)

$$\Delta\Gamma = e \int_0^l G_c(\psi(r)) dr = e l \bar{G}_c(l) \ \text{ with } \ \bar{G}_c(l) = \frac{1}{l} \int_0^l G_c(\psi(r)) dr \qquad (7.17)$$

The parameter G_c can be approximated for instance by a phenomenological law proposed in Hutchinson and Suo (1992), however, any regular monotonic curve connecting G_{1c} and G_{2c} would give neighboring results

$$G_c = G_{1c} \left(1 + \tan^2[(1-\Lambda)\psi] \right) \qquad (7.18)$$

where Λ is a dimensionless parameter that weights the influence of the mixity on the fracture toughness. If $\Lambda = 1$, then $G_c = G_{1c}$, the fracture toughness is independent of the mode mixity. On the other extreme, if $\Lambda = 0$, the fracture toughness is infinite for pure mode 2. In general, Λ can be calculated if the values of G_{1c} and G_{2c} (the fracture toughness in pure mode 1 and 2) are known

$$\Lambda = 1 - \frac{2}{\pi} \arctan \sqrt{\frac{G_{2c}}{G_{1c}} - 1} \qquad (7.19)$$

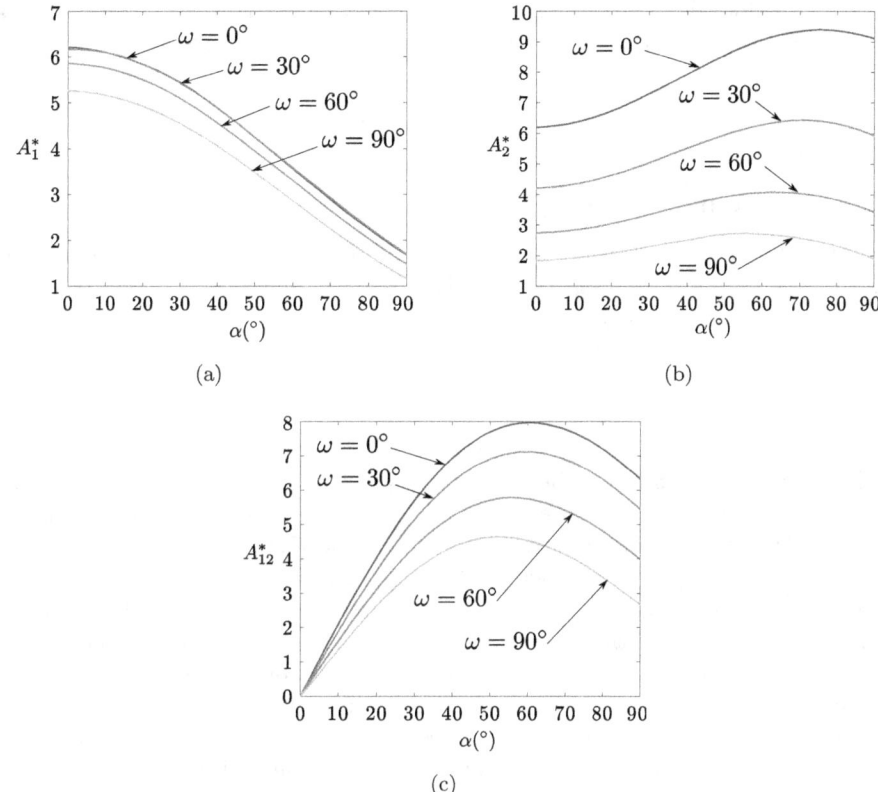

Figure 7.4: Dimensionless parameters (a) $A_1^* = A_1 E^*$; (b) $A_2^* = A_2 E^*$; (c) $A_{12}^* = A_{12} E^*$ for several values of the opening angle ω. The coefficients A_1^* and A_2^* are symmetric with respect to $\alpha = 0°$ whereas A_{12}^* is antisymmetric. The horizontal axis is limited here to 90° because failures in directions making an angle greater than this value are unlikely to occur. Results courtesy of Professor D. Leguillon published by García and Leguillon (2012).

The mode-mixity ψ is usually defined as the angle between the resultant of the traction vector acting onto a plane and the normal to this plane ahead the crack tip, assuming slow crack growth. Nevertheless, this approach may present difficulties as highlighted in Mantič (2009) for interface cracks. An alternative is based on employing τ and σ along the presupposed crack path prior to the crack onset. It has some advantages as has been detailed in Section 3.4.2. Thus, the mode mixity ψ in (7.18) is defined in what follows as

$$\psi(r) = \arctan\left(\frac{\tau(r)}{\sigma(r)}\right) = \arctan\left(\frac{\hat{a}t_1 + \hat{b}t_2}{\hat{a}s_1 + \hat{b}s_2}\right) \qquad (7.20)$$

Both expressions depend implicitly on r through $\hat{a}$ and $\hat{b}$. The average toughness $\bar{G}_c$ can be also derived from (7.17) using (7.20), (7.18) and $\Delta S = l.e$

$$\bar{G}_c = \frac{1}{l} \int_0^l G_c(r) \mathrm{d}r \qquad (7.21)$$

7.3.3   Incremental energy criterion. Final expression

The initial incremental energy balance can be rewritten by taking into account the expressions developed for $\Delta\Gamma$ and $\Delta\Pi$. Then, combining (7.13), (7.15), (7.17) and (7.21) the energy criterion

$$G^{\mathrm{inc}} = k_{\mathrm{eq}}^2 \left(\hat{a}^2 A_1(\alpha) + \hat{a}\hat{b} A_{12}(\alpha) + \hat{b}^2 A_2(\alpha) \right) \geq \bar{G}_c(l) \qquad (7.22)$$

where the incremental energy release rate is denoted G^{inc}. The coefficients $\hat{a}$ and $\hat{b}$ depend on r and must be taken at $r = l$. It is interesting to observe that for $\hat{a} = 0$ and $\hat{b} = 0$ this expression reduces to the well known expressions for pure mode 1 and 2 respectively. Whatever l, the right-hand side of (7.22) is bounded $G_{1c} \leq \bar{G}_c(l) \leq G_{2c}$ whereas the incremental energy G^{inc} grows at least like $l^{2\lambda_1 - 1}$. Thus, inequality (7.22) has always solutions, except in the extreme (and somewhat unrealistic) case where G_{2c} is infinite, but it means that failure is inhibited in pure mode 2. However the situation is not as straightforward as in the single mode case, some difficulties could occur with multiple crossing points in some very particular cases. Nevertheless, $\bar{G}_c(l)$ increases monotonically from G_{1c} to G_{2c} and the working range for l is small as numerically checked in the examples, thus $\bar{G}_c(l)$ is almost constant within this range and difficulties vanish. Otherwise it should be necessary to have a discussion like in Martin et al. (2008) although the discussion arises for completely different reasons. Therefore, this criterion imposes a minimum value of the loading (through k_{eq}) for the crack onset which depends on the unknown (up to now) increment l.

7.4   Coupled criterion

The twofold criterion proposed in Leguillon (2002) is based on the coupling of two usual conditions in stress and energy. In the previous sections, both expressions have been developed using special functions measuring the mode mixity $\hat{a}$ and $\hat{b}$, they involve the crack onset length l, the equivalent combined SIF k_{eq} defined in (7.3), some angular functions s_1, s_2, t_1, t_2, A_1, A_{12} and A_2 and material properties like fracture toughness in pure mode 1: G_{1c} and 2: G_{2c} and tensile σ_c and shear τ_c strength. Values of these material properties are *a priori* uncorrelated and can be estimated experimentally in an independent manner. However some correlations between them can be found in the literature for some specific materials or interfaces, e.g. Zhang (2002); Kamp et al. (2002). These relations cannot be generalized because they are based on the microstructure or on other material-dependent properties. This section aims at coupling necessary fracture conditions in stress and energy to obtain a condition for crack onset which is assumed to be sufficient. Expressions for both criteria can be summarized in the next two inequalities:

$$k_{eq}^2 \left(\hat{a}^2 A_1(\alpha) + \hat{a}\hat{b} A_{12}(\alpha) + \hat{b}^2 A_2(\alpha) \right) \geq \bar{G}_c(l) \tag{7.23a}$$

$$k_{eq} r^{-1/2} \left[\left(\hat{a} s_1(\theta) + \hat{b} s_2(\theta) \right)^p \tau_c^p + |\hat{a} t_1(\theta) + \hat{b} t_2(\theta)|^p \sigma_c^p \right]^{1/p} \geq \tau_c \sigma_c \text{ for } 0 \leq r \leq l \tag{7.23b}$$

As discussed above for $\lambda_1 \leq 1$ and $\lambda_2 \leq 1$, it is enough to verify the second condition for $r = l$ because $\sigma(r)$ and $\tau(r)$ are decreasing functions. Under this assumption the two inequalities are contradictory for a small loading intensity, the compatibility is only obtained when the load is sufficiently increased. Then combining the above inequalities, the critical length l_c corresponding to the crack onset can be derived by solving the next equation:

$$l = \frac{\bar{G}_c}{\sigma_c^2 \tau_c^2} \frac{\left((\hat{a} s_1 + \hat{b} s_2)^p \tau_c^p + |\hat{a} t_1 + \hat{b} t_2|^p \sigma_c^p \right)^{2/p}}{\hat{a}^2 A_1 + \hat{a}\hat{b} A_{12} + \hat{b}^2 A_2}. \tag{7.24}$$

This is an implicit equation in l because the right-hand side of the equation depends on l through $\hat{a}$, $\hat{b}$ and $\bar{G}_c$. This equation can be solved by a numerical algorithm providing the critical crack length at onset l_c. It is the initial length of the crack immediately after the onset but it is not necessarily the arrest length, the crack can go on growing in an unstable manner beyond this length. It is more precisely the smallest that can be observed, a smaller one being incompatible with the energy balance.

Introducing the expression of l_c solution to (7.24) in (7.23b) gives the critical value of the equivalent combined SIF

$$k_{eq} \geq k_{eq}^f = \sqrt{\frac{\bar{G}_c}{\hat{a}^2 A_1 + \hat{a}\hat{b} A_{12} + \hat{b}^2 A_2}} \tag{7.25}$$

The critical value k_{eq}^f, i.e. k_{eq} at failure, depends implicitly on l_c, σ_c and τ_c. It is important to note that this relation is totally symmetric with respect to the two modes and has the form of an Irwin-like criterion (i.e. involving the SIF). Critical k_{eq} depends also on the crack initiation angle α through the angular functions s_i, t_i $(i = 1, 2)$ and the A_j's $(j = 1, 2, 12)$.

Starting from the critical value of k_{eq}^f, the critical values for k_1 and k_2 can be directly derived by using (7.4)

$$k_1^f = \hat{a} k_{eq}^f l^{1/2-\lambda_1} = \hat{a} \left(\frac{\bar{G}_c}{\hat{a}^2 A_1 + \hat{a}\hat{b} A_{12} + \hat{b}^2 A_2} \right)^{1-\lambda_1} \times$$

$$\times \left(\frac{\sigma_c \tau_c}{\left((\hat{a} s_1 + \hat{b} s_2)^p \tau_c^p + |\hat{a} t_1 + \hat{b} t_2|^p \sigma_c^p \right)^{1/p}} \right)^{2\lambda_1 - 1} \tag{7.26a}$$

$$k_2^{\mathrm{f}} = \hat{b} k_{\mathrm{eq}}^{\mathrm{f}} l^{1/2-\lambda_2} = \hat{b} \left(\frac{\bar{G}_c}{\hat{a}^2 A_1 + \hat{a}\hat{b} A_{12} + \hat{b}^2 A_2} \right)^{1-\lambda_2} \times$$

$$\times \left(\frac{\sigma_c \tau_c}{\left((\hat{a} s_1 + \hat{b} s_2)^p \tau_c^p + |\hat{a} t_1 + \hat{b} t_2|^p \sigma_c^p \right)^{1/p}} \right)^{2\lambda_2 - 1} \quad (7.26\mathrm{b})$$

For the particular case of $\alpha = 0$, by symmetry $A_{12} = 0$, $s_2 = t_1 = 0$ and by an appropriate normalization $s_1 = t_2 = 1$. Then (7.26) leads to:

$$k_1^{\mathrm{f}} = \hat{a} k_{\mathrm{eq}}^{\mathrm{f}} l^{1/2-\lambda_1} = \hat{a} \left(\frac{\bar{G}_c}{\hat{a}^2 A_1 + \hat{b}^2 A_2} \right)^{1-\lambda_1} \left(\frac{\sigma_c \tau_c}{\left((\hat{a} s_1 \tau_c)^p + |\hat{b} t_2 \sigma_c|^p \right)^{1/p}} \right)^{2\lambda_1 - 1}$$
$$(7.27\mathrm{a})$$

$$k_2^{\mathrm{f}} = \hat{b} k_{\mathrm{eq}}^{\mathrm{f}} l^{1/2-\lambda_2} = \hat{b} \left(\frac{\bar{G}_c}{\hat{a}^2 A_1 + \hat{b}^2 A_2} \right)^{1-\lambda_2} \left(\frac{\sigma_c \tau_c}{\left((\hat{a} s_1 \tau_c)^p + |\hat{b} t_2 \sigma_c|^p \right)^{1/p}} \right)^{2\lambda_2 - 1}$$
$$(7.27\mathrm{b})$$

Note that the above relations reduce to simplified forms in case of pure modes 1 or 2:

- Pure mode 1: As discussed in Section 7.1, pure mode 1 corresponds to $\hat{a} = 1$ and $\hat{b} = 0$. Introducing these values in (7.26) leads to:

$$k_1^{\mathrm{f}} = \hat{a} k_{\mathrm{eq}}^{\mathrm{f}} l^{1/2-\lambda_1} = \left(\frac{G_{1c}}{A_1} \right)^{1-\lambda_1} \left(\frac{\sigma_c}{s_1} \right)^{2\lambda_1 - 1} \qquad (7.28\mathrm{a})$$

$$k_2^{\mathrm{f}} = \hat{b} k_{\mathrm{eq}}^{\mathrm{f}} l^{1/2-\lambda_2} = 0 \qquad (7.28\mathrm{b})$$

  The first expression matches with that presented in Leguillon (2002) for pure mode 1.

- Pure mode 2 corresponds to $\hat{a} = 0$ and $\hat{b} = 1$. Introducing these values in (7.26) leads to:

$$k_1^{\mathrm{f}} = \hat{a} k_{\mathrm{eq}}^{\mathrm{f}} l^{1/2-\lambda_1} = 0 \qquad (7.29\mathrm{a})$$

$$k_2^{\mathrm{f}} = \hat{b} k_{\mathrm{eq}}^{\mathrm{f}} l^{1/2-\lambda_2} = \left(\frac{G_{2c}}{A_2} \right)^{1-\lambda_2} \left(\frac{\tau_c}{t_2} \right)^{2\lambda_2 - 1} \qquad (7.29\mathrm{b})$$

  This last expression, dedicated to mode 2, is clearly the analogous to (7.28) for pure mode 1.

7.5   Analysis of a weak surface at $\alpha = 0°$ in a V-notch

The model developed in the previous sections can be directly applied to the problem of a corner at a plate made from a homogeneous material. The plate is cut into two parts along a certain plane which contains the corner to be lately bonded. Thus, a weak surface between similar materials is emanating from the corner with a certain angle α. For simplicity, it is assumed that this weak surface is situated at $\alpha = 0$ but the reasoning holds true regardless of its position. This weak surface is assumed to have weaker failure properties than the surrounding material and often crack initiates within it. The thickness of this surface is assumed to be vanishing, therefore it is out of the scope of this chapter the discussion between adhesive and cohesive failure, see e.g. Gersappe and Robbins (1999). More precisely, the problem is formulated at a larger scale, i.e. the competition between failure in either the weak surface or in the bulk material as a function of the properties in the two directions. Since the weak-surface thickness is assumed vanishing, the functions s_i, t_i and A_{ij} remain the same than for the case of homogeneous material without a weak surface, but one has to consider now tensile and shear strengths, fracture mode 1 and mode 2 toughness both for the weak surface and the bulk material. In the following the properties of the weak surface will be denoted by the superindex "ad" and the properties of the homogeneous material by the superindex "h". It is summarized below

$$\sigma_c(\alpha), \tau_c(\alpha), G_{1c}(\alpha), G_{2c}(\alpha) = \begin{cases} \sigma_c^{ad}, \tau_c^{ad}, G_{1c}^{ad}, G_{2c}^{ad} \text{ for } \alpha = 0° \\ \sigma_c^{h}, \tau_c^{h}, G_{1c}^{h}, G_{2c}^{h} \text{ for } \alpha \neq 0° \end{cases} \qquad (7.30)$$

In view of these definitions, two different mechanisms of failure should be studied in order to compare the corresponding critical loads

- Crack initiation at the weak surface $\alpha = 0°$.

- Crack initiation at the optimum angle $\alpha \neq 0°$ in the bulk material. This optimum minimizes the critical load taking into account strength and toughness of the homogeneous material.

In a first stage, the aim is to obtain the critical load triggering the crack onset as predicted by each mechanism of failure. For an arbitrary load intensity represented by a force R^0 the values of k_1^0 and k_2^0 can be calculated by a direct and linear numerical analysis using for example the finite element method (FEM). Due to the linearity the critical force R^f causing the failure can be calculated as

$$R^f = \frac{k_1^f}{k_1^0} R^0 = \frac{k_2^f}{k_2^0} R^0 = \frac{k_{eq}^f}{k_{eq}^0} R^0 \qquad (7.31)$$

where the last expression is the most adequate because it is valid for pure modes. In the following, the application of the model to both mechanisms is described.

7.5.1 Crack initiation at the weak surface

The crack initiation at the weak surface $\alpha = 0°$ is governed by relations (7.27). As previously discussed, this case corresponds to $s_2(0) = t_1(0) = 0$, $A_{12}(0) = 0$

and $s_1(0) = t_2(0) = 1$. Thus, the equation (7.24) for l can be simplified leading to

$$l = \frac{\bar{G}_c^{\text{ad}}}{(\sigma_c^{\text{ad}} \tau_c^{\text{ad}})^2} \frac{\left((\hat{a}\tau_c^{\text{ad}})^p + |\hat{b}\sigma_c^{\text{ad}}|^p\right)^{2/p}}{\hat{a}^2 A_1(0) + \hat{b}^2 A_2(0)}, \tag{7.32}$$

and the expression of the critical GSIF k_{eq} is reduced to

$$k_{\text{eq}}^{\text{f}} = \sqrt{\frac{\bar{G}_c^{\text{ad}}}{\hat{a}^2 A_1(0) + \hat{b}^2 A_2(0)}}, \tag{7.33}$$

Finally k_1^{f} and k_2^{f} write

$$k_1^{\text{f}} = \hat{a} \left(\frac{\bar{G}_c^{\text{ad}}}{\hat{a}^2 A_1(0) + \hat{b}^2 A_2(0)}\right)^{1-\lambda_1} \left(\frac{\sigma_c^{\text{ad}} \tau_c^{\text{ad}}}{\left((\hat{a}\tau_c^{\text{ad}})^p + |\hat{b}\sigma_c^{\text{ad}}|^p\right)^{1/p}}\right)^{2\lambda_1-1}, \tag{7.34a}$$

$$k_2^{\text{f}} = \hat{b} \left(\frac{\bar{G}_c^{\text{ad}}}{\hat{a}^2 A_1(0) + \hat{b}^2 A_2(0)}\right)^{1-\lambda_2} \left(\frac{\sigma_c^{\text{ad}} \tau_c^{\text{ad}}}{\left((\hat{a}\tau_c^{\text{ad}})^p + |\hat{b}\sigma_c^{\text{ad}}|^p\right)^{1/p}}\right)^{2\lambda_2-1}. \tag{7.34b}$$

The corresponding load intensity can be calculated by means of (7.31).

The computation of these values for a particular problem are described in Section 7.6. This will point out the importance of taking into account the mixture of modes.

7.5.2   Crack initiation in the bulk at the optimum angle

In this case the crack initiation angle α is *a priori* unknown. The critical GSIF depends on α by (7.25) through the A_j's and $\hat{a}$ and $\hat{b}$

$$k_{\text{eq}}^{\text{f}} = \sqrt{\frac{\bar{G}_c^{\text{h}}}{\hat{a}^2 A_1(\alpha) + \hat{a}\hat{b} A_{12}(\alpha) + \hat{b}^2 A_2(\alpha)}}, \tag{7.35}$$

where $\hat{a}$ and $\hat{b}$ depends on l which varies itself on α through the following equation:

$$l = \frac{\bar{G}_c^{\text{h}}}{(\sigma_c^{\text{h}} \tau_c^{\text{h}})^2} \frac{\left((\hat{a}^2 s_1(\alpha) + \hat{b} s_2(\alpha))^p (\tau_c^{\text{h}})^p + |\hat{a} t_1(\alpha) + \hat{b} t_2(\alpha)|^p (\sigma_c^{\text{h}})^p\right)^{2/p}}{\hat{a}^2 A_1(\alpha) + \hat{a}\hat{b} A_{12}(\alpha) + \hat{b}^2 A_2(\alpha)}. \tag{7.36}$$

The actual direction of fracture is obtained by minimizing the value of the critical force (7.31) with respect to α. According to the expression which uses k_1^{f} and k_2^{f}, the optimization of the critical load R^{f} is equivalent to minimize k_1^{f} or k_2^{f} because k_1^0 and k_2^0 does not depend on l. Unfortunately, both methodologies meet a difficulty, the optimization is ill-conditioned in case of near pure modes.

One or the other expression is unusable and it is not possible to build a general formula valid in all cases. In order to keep the formulation general, the third equality providing R^f in (7.31) will be employed. In this case it is necessary to note that the optimization of R^f is not equivalent to minimize k_{eq}^f. The reason is that k_{eq}^0 does depend on the critical length l_c which can differ for different values of α. Therefore, it is necessary to minimize the ratio k_{eq}^f/k_{eq}^0

$$\frac{k_{eq}^f}{k_{eq}^0} = \sqrt{\frac{\bar{G}_c^h(\alpha,l)}{\left((k_1^0)^2 l^{2\lambda_1-1} + (k_2^0)^2 l^{2\lambda_2-1}\right)\left(\hat{a}^2 A_1(\alpha) + \hat{a}\hat{b} A_{12}(\alpha) + \hat{b}^2 A_2(\alpha)\right)}}. \quad (7.37)$$

Then, the following numerical procedure should be implemented to obtain the optimum angle α and the minimum value of the critical force R^f:

- Compute by a numerical code, e.g. linear FEM, the problem of interest assuming an arbitrary load intensity R^0 without crack,

- Extract the GSIF k_1^0 and k_2^0 from the solution

- Obtain the angular functions s_i, t_i ($i = 1, 2$) solving the eigenvalue problem and the A_j's ($j = 1, 2, 12$) from a numerical computation (see e.g. Yosibash et al. (2006)),

- Minimize the expression (7.37), l being obtained solving the nonlinear equation (7.36). An iterative algorithm should be implemented to obtain α and the ratio (7.37),

- Compute the critical resultant force using (7.31).

7.6 Predictions of the critical load at failure in an Arcan test

By way of example of application, the procedure developed in the above Sections 7.5.1 and 7.5.2 is applied to the prediction of the failure load in compact shear tension (CTS) V-notched specimens. Figure 7.5 shows the mixed-mode loading device proposed by Richard and Benitz (1983). The CTS specimen is assumed to be made of PMMA and the jaws are from steel. Properties of both materials are listed in Table 7.6, they are extracted from Leguillon et al. (2009) and Tran et al. (2010). The mode mixity is varied by modifying the loading angle β of the applied remote force R from a pure symmetric mode for $\beta = 0°$ to a pure antisymmetric one for $\beta = 90°$. Note that the angles β and the previously mentioned ψ (7.20) measured at the distance l are likely closed to each other but there is no reason that they coincide except for $0°$ and $90°$.

In order to obtain the parameters k_1 and k_2 corresponding to given R^0 and β a numerical analysis is carried out using FEM. It is quite tricky because the notched specimens are not symmetric with respect to the load direction and any applied force creates a bending moment. For this reason, during the tests

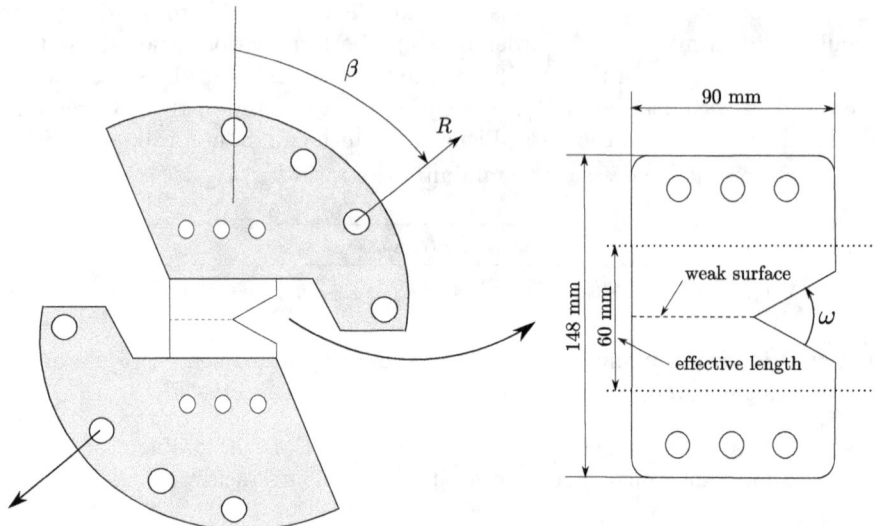

Figure 7.5: Arcan experimental setup and specimen

Property	E (GPa)	ν	σ_c (MPa)	τ_c (MPa)	G_{1c} (MPa.mm)	G_{2c}/G_{1c}
PMMA	3.250	0.3	75	225	0.350	5
Adhesive	-	-	36.5	16.5	0.200	2
Steel	200	0.3	-	-	-	-

Table 7.2: Material properties

the jaws are allowed to rotate around a horizontal axis. Therefore, meshing only the CTS specimen with prescribed boundary conditions along the top and bottom faces is not sufficient (Figure 7.6 middle). In view of this conclusion, the entire setup (CTS and jaws) is modeled using FEM (Figure 7.6 left). A linear elastic simulation assuming perfect bonding (i.e. continuity of displacement and stress vectors) is sufficient to extract k_1 and k_2 using path independent integrals (Leguillon and Sanchez-Palencia (1987); Labossiere and Dunn (1999)).

The main geometric characteristics of the simulated specimen are shown in Figure 7.5. The mesh employed is represented in Figure 7.6. Triangular linear elements are used, their maximum and minimum diameters are respectively 11 mm and 40 μm (near the corner, Figure 7.6 right). A 1 mm displacement is prescribed at the loading points situated along the outer part of the jaws. Finally, k_1 and k_2 are extracted from the FE solution, see Table 7.3 for several values of the opening of the corner ω ant the load angles β. The contrast between k_1 and k_2 depends strongly on β proving the efficiency of this type of test to experiment specimens under mixed-mode loading.

The data needed to apply the model developed in Section 7.5 are now avail-

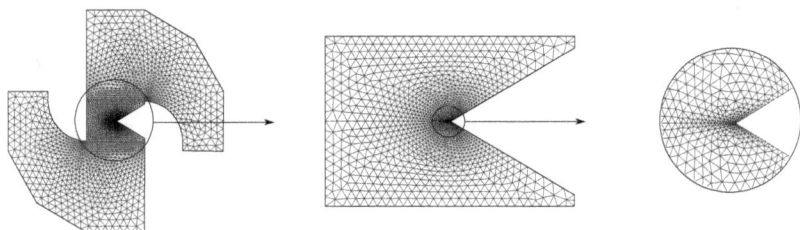

Figure 7.6: The finite element mesh of the entire structure

β (°)	$\omega = 0°$		$\omega = 30°$		$\omega = 60°$		$\omega = 90°$	
	k_1	k_2	k_1	k_2	k_1	k_2	k_1	k_2
0	0.0971	0	0.0974	0	0.0978	0	0.0959	0
30	0.0827	0.0223	0.0821	0.0210	0.0809	0.0172	0.0774	0.0119
60	0.0455	0.0369	0.0443	0.0339	0.0423	0.0269	0.0389	0.0179
75	0.0228	0.0399	0.0220	0.0363	0.0208	0.0287	0.0190	0.0189
83	0.0115	0.0402	0.0111	0.0366	0.0105	0.0288	0.0095	0.0189
86	0.0058	0.0401	0.0055	0.0364	0.0052	0.0287	0.0047	0.0188
90	0	0.0397	0	0.0361	0	0.0285	0	0.0187

Table 7.3: GSIF values extracted from the numerical results in Modulef for several values of β and ω. The values of k_i are expressed in MPa.mm$^{(1-\lambda_i)}$

able. The parameter p which defines the stress criterion (7.9), will be set to $p = 2$. As discussed in the previous section, the critical load for either a failure at the weak surface or in the bulk material are separately determined. A subsequent comparison between these two critical loads will give the critical load at failure.

7.6.1 Crack initiation at the weak surface

Results presented in the following are computed using the procedure described in Section 7.5.1. Figure 7.7 shows the safe and failure regions as a function of the GSIF values k_1 and k_2, safe is below and unsafe above the curves. The different subfigures correspond to 4 different values of the opening angle ω. Different values of ω lead to different exponents and thus different units for k_1^f and k_2^f. As a consequence the different failure curves cannot be plotted together. The coupling between k_1^f and k_2^f decreases when the value of the opening angle ω increases. In the limit, when $\lambda_2 > 1$ for an opening angle $\omega \approx 105°$, the shear stresses at the weak surface (where $t_1 = 0$) is not longer singular at the corner point and the influence of mode 2 vanishes. Hence k_1^f become totally independent of k_2^f.

Figure 7.8 shows the values of critical loads as a function of the load angle β. Different subfigures correspond to the different values of ω. The minimum value corresponds to the symmetric load $\beta = 0°$ when crack initiates at the weak surface in pure mode 1. For $\beta > 0°$ the mixture of modes causes an increase of

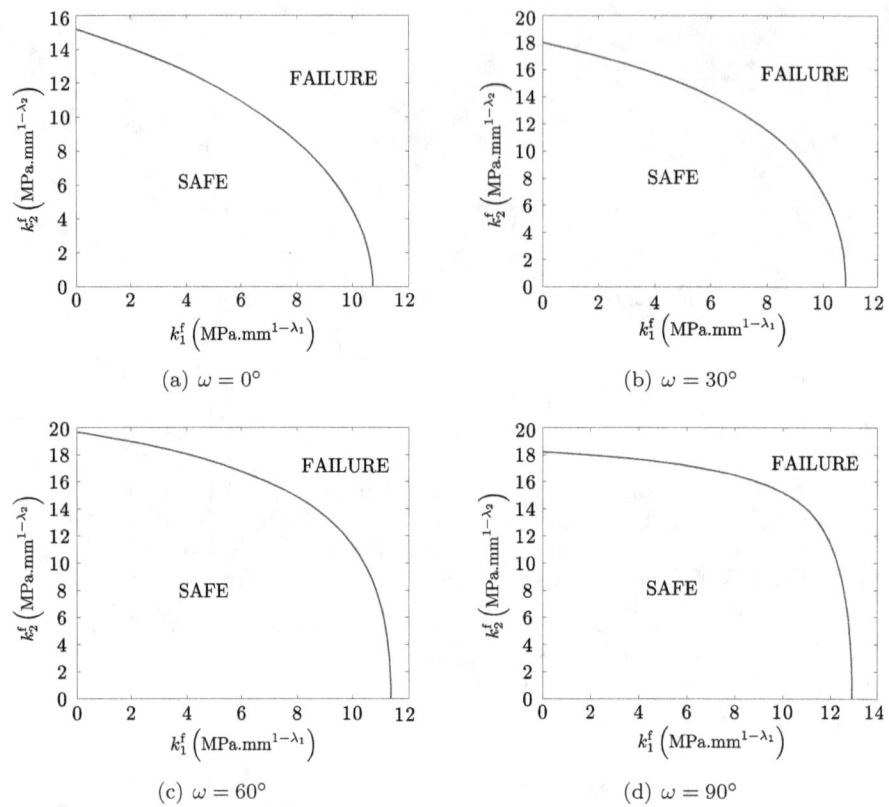

Figure 7.7: Failure (above) and safe (below) regions as a function of the GSIF values for $p = 2$ and (a) $\omega = 0°$, (b) $\omega = 30°$, (c) $\omega = 60°$, (d) $\omega = 90°$.

the critical load. This figure shows also the comparison between the critical load obtained by the model presented and other simplified models. It is worth noting the moderate consequences of ignoring mode 2 and the shear stresses, except, as expected, for high load angles where mode 2 and the shear stresses become predominant. The simplified models seem to have an admissible accuracy, say up to $\beta \simeq 50°$, their predictions are no longer conservative above this value and even meaningless in pure mode 2. In Figure 7.8, the reference load R_f^* is the remote load R_f normalized to 1 for $\omega = 0°$ and $\beta = 0°$ (a crack in pure mode 1).

7.6.2   Crack initiation in the bulk at the optimum angle

Under the assumption of crack initiation in the bulk material, the angle of the crack onset α is *a priori* not known and is calculated by means of the optimization scheme described in Section 7.5.2. Following this scheme, the angle α, the critical load R_f and the critical GSIF k_1^f and k_2^f are computed.

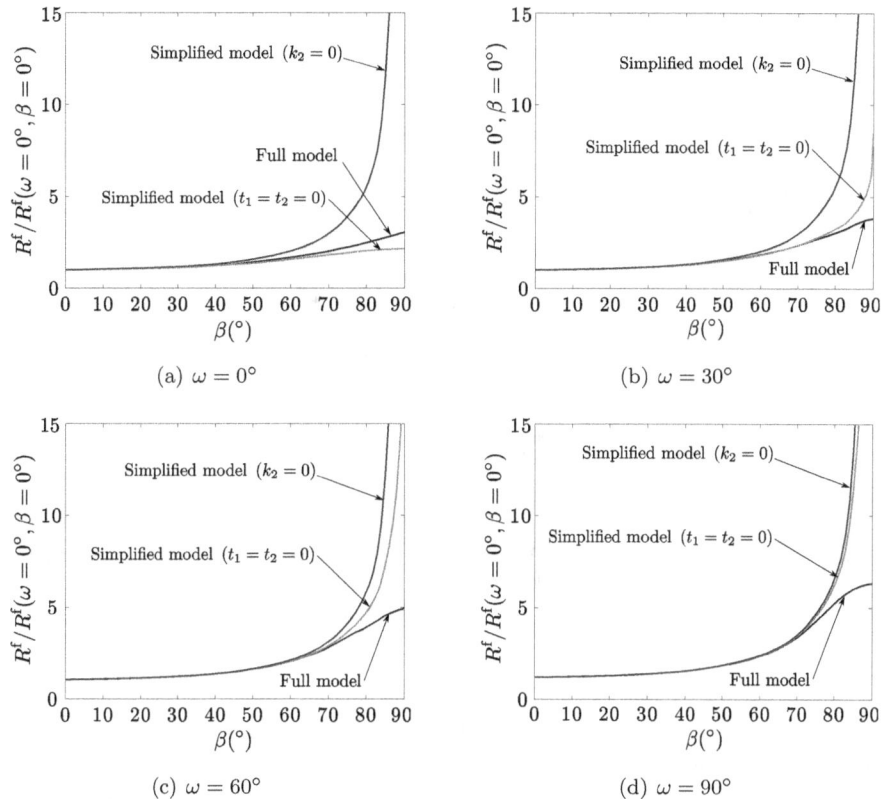

Figure 7.8: Critical loading as a function of the loading angle β for (a) $\omega = 0°$, (b) $\omega = 30°$, (c) $\omega = 60°$ and (d) $\omega = 90°$. Simplified model ($k_2 = 0$) corresponds to the initial criterion taking into account the single mode 1 and tensile stress (Leguillon (2002)); simplified model ($t_1 = t_2 = 0$) to the improved model proposed by Yosibash et al. (2006) taking into account two modes but neglecting the influence of the shear stresses.

Figure 7.9 shows the optimum angle α as a function of the load angle β for several values of the opening angle ω. This is equivalent to the principle of local symmetry (Barenblatt and Cherepanov (1961) and Erdogan and Sih (1963)). The value calculated here is compared to the predictions by Yosibash et al. (2006) (dashed line) which corresponds to neglecting the shear stresses, i.e. taking $t_1 = t_2 = 0$. The results show that this generalization of the model gives similar values of the optimum angle but it is strongly dependent on the material properties used (namely in the present case $\tau_c^{PMMA} \gg \sigma_c^{PMMA}$ and $G_{c2}^{PMMA} \gg G_{c1}^{PMMA}$).

Then, in the case of the full model, according to the results shown in Figure 7.9, the optimum angle is similar to the one predicted by the model when neglecting the shear stresses. This means that, in this case, the full model predicts

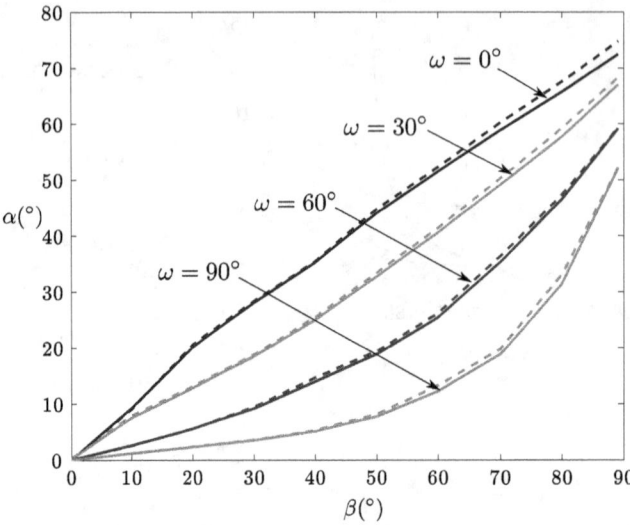

Figure 7.9: Predictions of the crack initiation angle α as a function of the load angle β for several values of the opening angle ω. Full model: solid line; simplified model ($t_1 = t_2 = 0$): dashed line.

a nearly pure mode 1 crack onset. Thus, as described above, this angle corresponds to a maximum in normal tractions and vanishing shear stresses. Hence, the value estimated of the critical resultant by the full model will be very similar to the results presented in Yosibash et al. (2006). For brevity these results are not presented in this subsection but in the next one where the comparison between the two hypotheses is carried out.

7.6.3 Global prediction of failure load

The global prediction of the failure load can be obtained by comparing the critical loads under the two hypotheses. Figure 7.10 shows this critical load as a function of the opening ω and the load β angles. It shows that the failure occurs in general at the weak surface except for the combination of low values of ω and high values of β where the antisymmetric mode is predominant. In this case, the competition is between a failure in pure mode 1 at the optimum angle and a nearly pure mode 2 at the weak surface. As already mentioned $\omega = 0°$ corresponds to a situation where the failure is governed by the energy criterion, in this case, due to the similarity between the values of G_{c1}^{h} and G_{c2}^{ad} (see Table 7.6), the crack onset occurs in pure mode 1 at the optimum angle. On the contrary, $\omega = 90°$ corresponds to a situation much more governed by the stress condition and the failure is initiated at the weak surface. This is why the crack onset occurs at the weak surface, the tensile strength of the PMMA being higher than the shear strength of the weak surface. In other cases with low or moderate values of β the

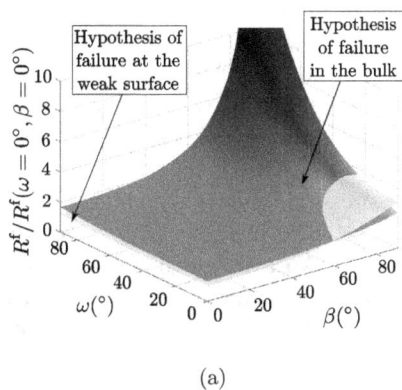
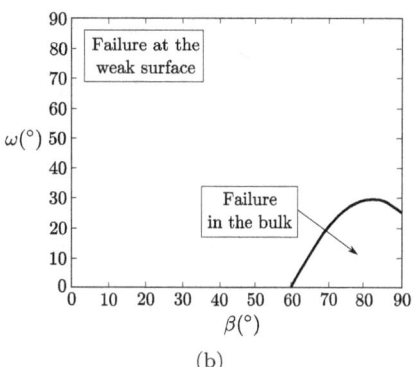

(a)   (b)

Figure 7.10: (a) Comparison of the critical load obtained by the two hypotheses: Crack initiation either at the weak surface for $\alpha = 0°$ or in the bulk. (b) Values of β and ω for which the failure is predicted in accordance with each hypothesis.

mode 1 contribution at the weak surface is sufficient to ensure the onset at the weak surface due to its weaker strength and fracture properties as can be seen in Figure 7.10.

These results confirm previous ones, particularly the principle of local symmetry, but they are obviously closely related to the fracture properties of the material in question, i.e. higher constants in shear than in tension. It seems that they could be questioned in the opposite case of a hypothetical material with failure properties significantly lower in shear than in tension.

7.7 Concluding remarks

A semi-analytical coupled stress and energy criterion has been developed in order to predict the failure of V-notched specimens containing weak surfaces between similar materials under complex loading. The procedure introduced requires first to solve a nonlinear equation for the crack length at onset. Then, the knowledge of this length enables to calculate the critical load to failure using a straightforward expression. In fact, this criterion goes much further and can be used in all situations where there is a competition between tension and shear.

Two dimensionless parameters have been defined which measure the contribution of each mode as a function of the length of the crack onset. The obtained expression is totally symmetric with respect to the two modes and reduces as expected to the known expressions in the pure mode cases. An equivalent combined GSIF which depends on the distance to the corner root has been defined. This has allowed defining the failure as a critical value of this parameter in a similar way to the Irwin criterion in LEFM.

A new expression of the stress condition has been introduced in order to consider the effect of the shear stresses. This is generated by a general law which couples the critical normal and shear stresses. This law has a free parameter p. It gives the relationship between tensile and shear stress at failure defining a curve in the plane $\sigma - \tau$. Some typical curves can be generated for different values of p as the Mohr-Coulomb or the elliptical failure curves. In the studied examples, a value of $p = 2$ has been used but the expression of the criterion remains of course valid for any other value of p.

Regarding to the energy condition, a new expression based on the phenomenological law of Hutchinson and Suo has been introduced in order to consider the influence of the mixity on the fracture toughness. An average toughness is defined as the mean dissipated energy per unit of surface along the line defined by the crack at onset. This mean value is necessary because the mode mixity depends on the distance to the corner root and therefore it is the same for the fracture toughness.

The criterion has been applied to a particular example: The simulation of an Arcan test on a V-notched specimen in PMMA with a weak surface under several loading angles. In the case of the weak surface, the coupling between the critical GSIF of mode 1 and that of mode 2 decreases when the opening angle of the notch increases. The reason is that the difference between the singularity orders of the two modes increases when the opening angle increases and mode 2 is becoming less important.

The competition between the crack onset in the homogeneous material and at the weak surface has been analyzed for the particular values of strength and fracture properties considered here. The results show that the onset is generally promoted at the weak surface except if two combined conditions hold: the solution is governed by the energy condition (low values of the opening angle) and the onset at the weak surface is in almost pure mode 2 whereas it is almost in pure mode 1 in the homogeneous material. The reason is that in cases governed by the energy condition the value of G_{2c} of the weak surface can be close to the value of the G_{1c} in the homogeneous material. In conclusion, the failure is predicted in general at the weak surface except for the cases that combine high mixed-mode loading and low values of the opening angle.

A comparison has been carried out between the results of this criterion and simplified versions which have been presented in the literature. This demonstrates that even in the case of moderate mixed-mode loading, mode 2 must not be neglected because the predictions overestimates strongly the critical load to failure. If shear stresses are neglected but still two modes are taken into account, the difference between the predictions is less pronounced but it cannot be assumed in general. However, no additional comparison is proposed herein with other approaches such as the averaged stress criterion or CZM, it is beyond the scope of this work. A preliminary analysis was given by Leguillon and Yosibash (2003) in the symmetric case. The mixed criterion (Leguillon (2002)) was compared with point stress criterion of McClintock (1958) and the average stress criterion of Novozhilov (1969). The main difference concerns the characteristic length that is used. Nevertheless the criteria are not highly sensitive of this length which anyway are of the same order of magnitude, thus differences in the failure load prediction remain small: about 15 % for $\omega = 120°$ and less for smaller opening angles. Indeed the difference lies in the fact that these authors assume the characteristic lengths to be intrinsic and related to the microstructure, reasoning which is refuted by the coupled criterion at the present work.

CHAPTER 8

Conclusions and future developments

8.1 Conluding remarks

In this work, the application of the coupled stress and energy criterion of the finite fracture mechanics to the modeling of failure initiation in composites has been investigated. This criterion has been shown to be a powerful tool for characterizing the crack initiation at different scales from a practical point of view. In what follows, the main conclusions of this work are discussed from the perspective of the three main objectives established in Section 1.3.

The first objective is the contribution to the development of the coupled criterion. Given the context of this thesis, this objective focuses partially on those developments required by the specific characteristics of failure initiation in composites but formulated in general way. Chapters 2 and 3 summarize the main developments associated to this objective within a more general analysis of the state-of-art.

Given that the concept of crack initiation, especially under static loading, is controversial, Chapter 2 is devoted to introduce this concept in the context of the thesis. In particular, the physical interpretation under which this concept is understood in the rest of the thesis is established. In addition, the main techniques employed in the literature for the prediction of crack initiation are reviewed to put into context the technique used in this thesis. A special attention is paid to the finite fracture mechanics because it is the framework in which the coupled criterion was proposed.

In Chapter 3, the coupled criterion is reviewed from its first historical evidences to the most recent generalizations, analyses and methods developed within its framework. The most relevant modifications of the original criterion are discussed and presented within a general formulation, with a special attention to the modifications particularly relevant for the application of the coupled criterion to the problems of crack initiation in composites studied in this thesis. Some of these contributions can be considered partially as a result of this thesis and more particularly as a derivation of the first objective. These contributions are presented in an integrated form along with the whole formulation of the coupled criterion in order to give to Chapter 3 more of a reviewing structure. In the next paragraphs, the main contributions to the development of the coupled criterion are described.

A general formulation of the coupled criterion has been proposed in order to address the different problems following a common structure of the analysis. This common structure is expressed as a combination of two dimensionless functions and one dimensionless parameter which governs the problem. These functions and parameters are particularized for each problem, whereas the expression of the coupled criterion in terms of these functions and parameters is generalized. The coupled criterion is expressed as the minimization on an explicit function over the possible crack geometries. The minimized function depends on a dimensionless parameter γ, which modulates the coupling between the two stress and energy criteria, and two dimensionless functions s and g corresponding to these criteria respectively. Thus, the analysis of any problem is simplified to obtain the value of γ and the particular expressions for s and g corresponding to the problem.

The parameter γ, introduced by Mantič (2009) for the problem studied in Chapter 4, is generalized by the formulation presented here to any problem analyzed by means of the finite fracture mechanics. For $\gamma \to 0^+$, the critical load is exclusively given by the stress criterion whereas for $\gamma \to +\infty$ it is given by the energy criterion. On the other hand, s and g are dimensionless functions measuring the "efficacy" of a certain crack surface for crack onset according to the stress and energy criteria, respectively. Thus, s is calculated as the dimensionless ratio of the effective strength to the driven stress for a certain crack surface. Analogously, g is the dimensionless ratio of dissipated to released energy for a certain crack surface.

Initially the coupled criterion was formulated neglecting the influence of shear stresses at the failure plane. Following the first proposal by Mantič (2009), this work has expanded to take into account this influence, which is particularly relevant for the study of the crack initiation at interfaces, typically found in failure mechanisms in composites. As discussed in Chapter 3, interfaces are preferential crack paths requiring consideration of the fracture mode mixity. The reason is that the preference of a certain path avoids the crack deviating to follow a pure (or nearly pure) opening mode. In the context of the coupled criterion, different modifications have been proposed to introduce the influence of mode mixity in both the stress and energy criteria.

Regarding the stress criterion, the influence of the shear stresses associated to the future crack surface has been introduced by generalizing the stress criteria initially employed in the coupled criterion. A general polynomial expression has been proposed to define a family of critical curves in the normal-shear stresses plane. The polynomial degree p modulating these curves is considered a material-dependent parameter. The expression proposed can take the form of the different stress criteria, which have been found in the literature, to be employed in the coupled criterion, e.g., the Mohr-Coulomb criterion ($p = 1$) and the quadratic criterion ($p = 2$). In addition, this expression has been generalized for cases with compressions and high shear stresses.

Regarding the energy criterion, the influence of the fracture mode-mixity can affect the expression of the dissipated energy. It is well known that in the problem of crack propagation studied by the linear elastic fracture mechanics, the fracture toughness is typically assumed to vary with the fracture mode mixity. The generalization of this result in the case of the crack onset predicted by the coupled criterion is controversial and has been discussed here. In the physical sense of generalizing this influence in the case of a finite length is discussed, and the different forms of approximating the energy dissipated in the presence of fracture mode mixity are described highlighting the reasonings behind them.

The second general objective of this work is to show that the coupled criterion is a suitable technique to be employed in the study of the failure initiation in composites. This objective has been developed in the whole thesis as a result of analyses of different cases. The main contributions resulting from this work are highlighted in what follows.

The general formulation for the coupled criterion proposed in Chapter 3 has been applied to several problems of a very different nature and scale. These problems differ also in the manner in which the elastic solutions required for the application of the coupled criterion have been obtained or approximated. In spite of this, the formulation can be employed similarly. Moreover, the relevance of the brittleness number γ has been demonstrated to remain constant in spite of the differences between the problems studied. This demonstrates that the coupled criterion can be employed in a structured manner in a wide range of problems. In addition, the physical interpretation of the results is simplified by the common structure. This structure enables even a general computational methodology as is described in Section 8.2.

With regards to showing the suitability of the coupled criterion, the results predicted by the model developed in Chapter 4 have been compared with the predictions obtained by means of a cohesive zone model. Currently, cohesive zone models are considered to be one of the most accepted forms of predicting crack initiation. The comparison shows that the agreement between the cohesive model and the coupled criterion of the finite fracture mechanics is good except at the extreme for small fibers in which the coupled criterion and the cohesive zone model differ in the asymptotic behavior. In addition, a physical explanation of this divergence has been proposed.

The results have also been compared with experiments. The experiments required for a full validation are particularly difficult for the case of the problems studied at the micro scale. At this scale, a first attempt of comparison has been carried out for the problem of crack initiation at the interface between a stiff spherical inclusion and the surrounding matrix. The comparison is based on preliminary results found in the literature. Unfortunately, the results of the comparison are not concluding due to the lack of information provided by the authors about the experimental results and material properties along with the fact that the results are only based on four tests without replication. In spite of this, the theoretical and experimental results seem to agree on the main tendency.

A more satisfying agreement with experiments has been found for the problem of transverse crack initiation in cross-ply laminates. In this case, a first comparison has been carried out with experiments found in the literature. These experiments are well detailed and the elastic, strength and fracture properties required by the coupled-criterion model are reported. The comparison shows a very good agreement in the size effect predicted. However, this comparison is not fully concluding due to two reasons: i) the lack of information about the repetition of the tests carried out for each value of the inner-ply thickness ii) the influence of the residual thermal stresses could not be accounted given the uncertainty about the presence of these stresses due to the particular fabrication method employed by the authors of the experiments. Actually, the influence of the residual thermal stresses has been analyzed and the maximum error introduced has been estimated leading to the conclusion that their influence for the composite employed in these experiments is low in worst-case scenario.

In view of the previous experimental deficiencies, a new set of experiments has been performed in order to confirm the agreement of the predictions given by the coupled criterion with experiments. Several specimens were fabricated for each value of the varying parameter in order to measure the variability of the results. In addition, with the objective to evaluate the size effect, the laminates were designed to be geometrically similar, i.e. with a fixed ratio of the outer- to inner-ply thickness. This avoids to mix the influence of this ratio with the size effect on the experimental results obtained. Moreover, the range of thickness tested is defined to be sufficiently wide to capture the asymptotic behavior. The experimental results show a good agreement with the model within the dispersion observed.

The third objective focuses on the contribution to the knowledge about the initiation of some failure mechanisms in composites. This has been specifically treated in Chapters 4 to 7, where the crack initiation at particular problems, lying in the origin of the different failure mechanisms, has been analyzed.

The initiation of the failure mechanism associated to a transverse loading in long fiber reinforced laminates is analyzed in Chapter 4. It is well known that this failure is initiated at the fiber-matrix interface or very near it. A theoretical model is developed to study the influence of the problem parameters on the critical load leading to the crack onset. This work is a generalization of the model proposed by Mantič (2009). In the context of this thesis, the preference for the asymmetry in the crack initiation has been shown, the asymmetric crack onset requiring a slightly lower critical load. The difference between the critical load given by the symmetric and asymmetric crack initiation depends strongly on the fiber radius. Moreover, the influence of a secondary transverse load has been evaluated. For usual composites as carbon/epoxy and glass/epoxy, a secondary compression decreases the value of the main critical load required for the crack onset. The variation of this influence with the contrast between the elastic properties of fiber and matrix are obtained. In addition, this effect is shown to be stronger for larger fibers. Finally, an experimental procedure, based on the model developed, is proposed to obtain the fracture and strength properties in an indirect form.

An analogous failure mechanism can be found in composites reinforced with spherical particles. The first step of this failure mechanism is studied in Chapter 5. A theoretical model is developed to predict the crack initiation at the interface between a spherical inclusion and the surrounding matrix. The model predicts a strong size effect, particularly for small particles below a certain size.

At meso scale, a theoretical model is developed in order to predict crack initiation in the transverse ply of cross-ply laminates. Initially, this model is based on a 2D approximation, which enables obtaining a very simple model for the prediction of this phenomenon. This model predicts a strong size effect, which can be interpreted as an additional strength for thin laminates not taken into account in the classical failure criteria for composites. In order to evaluate the influence of the 3D effects, a new model has been proposed employing powerful computational tools. The comparison of the results predicted by this new 3D model and the 2D model, whose application is much simpler, shows that the latter is a very good approximation of the general 3D situation. An additional modification of the 2D model is introduced in order to account for the effect of the residual thermal stresses. It is shown that they can be simply introduced by defining a fictitious equivalent strain to be added to the mechanical strain originated by the external loading. The effect of the residual thermal stresses is shown to be very relevant for certain common composites such as carbon/epoxy.

The crack initiation at a reentrant corner in the presence of an interface has been modeled in Chapter 7, obtaining a set of expressions for the critical load. This work enables the generalization of the expressions obtained previously for corners in homogeneous materials in the cases containing interfaces. An equivalent generalized stress intensity factor is defined. The model also gives the conditions under which the crack is initiated either at the interface or in the bulk material as a function of the properties.

8.2 Future developments

Previous conclusions show that the strategy followed in this thesis of combining in parallel the development of the coupled criterion and its application to characterize the failure initiation in composites has led to satisfactory results. The need of combining both processes can also be extended to the future developments. Thus, in view of the results obtained, it is necessary to progress simultaneously in the three general objectives detailed in Section 1.3, i.e. advancing in the techniques associated to the coupled criterion, improving the validation of the coupled criterion as an adequate tool for the prediction of failure initiation in composites, and improving the models generated. In what follows, the future developments proposed are sorted according to the general objectives.

The coupled criterion of the finite fracture mechanics has been shown to be a promising technique to generate very simple models for the prediction of crack initiation. In particular, this is appropiate for modeling and understanding the problems of crack initiation which are at the origin of the first steps of the failure mechanisms in composites. However, it is necessary to progress in certain aspects of the coupled criterion. The main ones are outlined in the following.

The coupled criterion is based mainly on two hypotheses: the assumption of finite length for the crack after the onset and the so-called Leguillon's hypothesis: the condition of fulfilling simultaneously both the stress and energy criteria. Several examples have been presented in the literature showing that the coupled criterion gives good approximations of the results obtained in experiments. However, currently the physical basis and justification of both hypotheses are still not fully clear. Some partial evidences of the requirement postulated in Leguillon's hypothesis have been presented, see a discussion in Section 3.2, but they do not represent a full justification of why this criterion works even in very different material systems. Thus, it is necessary to deepen the understanding of the physical basis of the coupled criterion. A sound physical justification of the coupled criterion would be very useful from a practical point of view as well. For example, it would enable defining range of materials, scales and loads for which the coupled criterion can be applied.

In this thesis, the influence of the fracture-mode mixity in the formulation of the coupled criterion has been widely discussed because the presence of interfaces in composites makes the mixity to be, *a priori*, very relevant in the characterization of the crack initiation in some typical failure mechanisms. In spite of the work carried out, the influence of this mode mixity is still controversial. Thus, it is necessary to generate more knowledge about the influence of the mixity in the onset of a crack of a finite length assumed in the context of the finite fracture mechanics. Given that several forms of generalizing the original coupled criterion to take into account the mixity have been proposed, it would be interesting to design an experiment for measuring this influence in order to evaluate the adequacy of the different methods proposed.

Fracture mechanics and fatigue are traditionally two closely related disciplines. Thus, some of the methods within the context of the finite fracture mechanics described in Section 2.3.4 have been (even previously) applied by other authors for the study of fatigue initiation of cracks in the absence of an initial crack. Some first attempts to generalize the coupled criterion were proposed by Pugno et al. (2007); Leguillon and Murer (2012). Given the relevance of fatigue in the failure initiation process, it would be interesting to work on this generalization, particularly in the experimental comparison of the results predicted by these approaches. In addition, some of the concepts which are employed in the coupled criterion in its static formulation should be revisited for fatigue.

The coupled criterion requires its application to obtain an analytical or approximate solution for a set of elastic problems, as described in Chapter 3. If neither analytical nor approximate solutions in closed form are available, the elastic solutions required are approximated by computational procedures. The numerical solution is extracted from a suitable numerical model and is used in the application of the coupled criterion. However, there does not exist an interaction between the application of the coupled criterion and the numerical model. This strategy limits the complexity of the problems which can be studied, given that for a sufficiently complex problem the number of elastic solutions to be approximated strongly increases, particularly due to the uncertainty about the crack geometry and position. Thus, it is necessary to generate a numerical method able to integrate the numerical solution along with the formulation of the coupled criterion. It would enable the extension of the coupled criterion to more complex problems, which cannot be studied by the coupled criterion currently. It could be argued that the need of a numerical method for the coupled criterion is in opposition to one of the main advantages of the coupled criterion: its ability to generate simple expressions for predicting crack initiation. However, the hypothetical numerical method proposed here could be employed as a first step in the analysis of a global problem providing the preferential solutions. Once the number of solutions to be studied is reduced, the problem can be studied by the usual strategy leading to semianalyticial expressions for the prediction of crack initiation.

With regards to the second objective of verifying that the coupled criterion is a suitable tool for the study of the initiation of failure mechanisms in composites, a set of additional developments are proposed in the following.

The experimental validation of the results predicted by the coupled criterion can be found in the literature for a wide range of problems. However, it is necessary to validate some models for certain problems with no experimental data to compare. This lack of experimental validation is found particularly for problems which are studied at the smallest scales, e.g., the problem studied in Chapter 4. The experimental validation of these models is considered very necessary in order to show the applicability of the coupled criterion beyond certain small scales and certain materials. This is a consequence of the lack of a full physical justification of the coupled criterion discussed previously. Even in those models which have been compared with experiments, the comparison is normally carried out with the variation of one parameter. In these cases, additional experiments are necessary in order to validate the variation of the results predicted when other parameters are varied in order to verify that the coupled criterion account for them adequately.

A particular experiment is motivated by the divergence found between the predictions of the coupled criterion and a cohesive zone model described in Section 4.2 for certain values of the governing parameters. For the problem of crack initiation at the fiber-matrix interface when the composite is subjected to transverse loading, the two criteria diverges in the asymptotic limit of the critical load for small fibers. The coupled criterion predicts that the critical load is proportional to the square root of the inverse of the fiber radius, whereas the cohesive zone model gives a variation which is linear with the inverse of the fiber radius. This divergence is evidence of a deep difference in the physical basis which governs the asymptotic behavior in this extreme, as was discussed in Section 4.2. Thus, new experiments are required to verify which is the actual asymptotic behavior. Given that the divergence between the two models is caused by a divergence in the physical basis, as has been demonstrated, this experiment would also give information about the basis behind the failure initiation in these situations.

Regarding the third objective, which focuses on the characterization of the first steps of some failure mechanisms in composites, a set of particularized future developments are proposed in what follows for each one of the particular problems of crack initiation studied in this thesis.

In the case of the problem of crack initiation at the fiber-matrix interface when the composite is subjected to transverse loading, several future developments are proposed in regards to the conclusions described in Section 4.4. First, the study of interface crack under biaxial loads has to be completed for the case of dominant transverse compression. Then, a full critical curve for biaxial transverse load could be generated. In addition, this curve could be combined with an approach proposed by Carraro and Quaresimin (2014) to generate a model able to evaluate the influence of three of the six independent components of the stress tensor on the initiation of this failure mechanism. The effect of the residual thermal stresses originated by the fabrication process could be introduced in this model using the methodology employed in the problem of transverse cracking in cross-ply laminates. Once the initiation of the crack is characterized, it is necessary to study the subsequent steps of the failure mechanism, in particular the crack growing along the interface and the kinking out towards the matrix. The previous developments are adequate under the hypothesis of low volumetric fiber-content because this enables neglecting the influence of the neighboring fibers. This hypothesis is not accurate for most usual composites, which does not necessarily imply that the results obtained cannot be a good approximation for real composites. To evaluate how the neighboring fibers affect the results obtained by these simplified models, it is necessary to develop multi-fiber models. Given the complexity of the geometry of the crack initiation in these models, this problem must be solved by a numerical method which has to be developed as discussed previously in this section. Besides the new models, it is necessary to verify experimentally the main results obtained by the models already developed, in particular those with a certain practical relevance, e.g. the size effect of the fiber radius or the influence of a secondary transverse load.

With regards to the problem of crack initiation at the interface between a spherical inclusion and the surrounding matrix, some analogous future developments can be proposed. First, the subsequent steps after the crack initiation have to be modeled as an interface crack growth and kink out towards the matrix. The problem of neighboring inclusions is not as relevant as in the case of fibrous composite because the volumetric content is usually lower. Similar to the future development proposed for the previous problem, the influence of the reasidual thermal stresses can be introduced here. In addition, new experiments are required for this problem since the available experiments were not detailed enough to obtain a satisfactory conclusion.

In the problem of transverse crack initiation in cross-ply laminates, the future developments should mainly focus on the extension of the model to take into account other failure mechanisms along with the next steps of the type of failure whose initiation is modeled. A new model can be developed by combining the model developed by Martin et al. (2010) for the delamination due to the singularity at the free edge and the model developed here. These would enable predicting under which conditions each failure mechanism is preferential. Going beyond the initiation of the first transverse crack, the evolution of the crack density with the external strain is also a very relevant problem due to its influence on the permeability of the laminate and even on the global stiffness in some composites. Thus, it would be interesting to extend this model to predict the evolution of the crack density and compare the results with the models generated by employing other criteria. After the transverse cracking, the subsequent step of the failure mechanism is well known: the delamination at the interface between the transverse and the longitudinal ply due to the stress concentration near the crack tip. This step can also be studied by combining the coupled criterion and matched asymptotics. Finally, the combination of the models of transverse cracking, free-edge delamination and interface delamination could provide of a full characterization of this failure mechanism.

For the model developed to predict crack initiation at reentrant corners in the presence of a weak interface between two subdomains, the future developments are mainly based on its experimental validation and its application to problems of practical interest. The experimental validation can be carried out by an Arcan test as that employed to show the application of the model. In addition, it would be necessary to design a set of experiments to estimate the material and interface properties required by the model, as fracture toughness and strength in pure opening and shearing modes. In addition, the model can be extended to account for problems with multi-material corners with any number of materials and interfaces. This extension is almost automatic and enables to apply this model to the failure initiation in adhesive joints.

Bibliography

Alfano, G. (2006). On the influence of the shape of the interface law on the application of cohesive-zone models. *Composites Science and Technology 66*(6), 723–730.

Altenbach, H., S. Smirnov, and V. Kulyk (1995). Analysis of a spherical crack on the interface of a two-phase composite. *Mechanics of Composite Materials 31*(1), 11–19.

Andersons, J., S. Tarasovs, and E. Sparniņš (2010). Finite fracture mechanics analysis of crack onset at a stress concentration in a UD glass/epoxy composite in off-axis tension. *Composites Science and Technology 70*(9), 1380–1385.

ASTM Standard D3039 (2006). Standard test method for tensile properties of polymer matrix composite materials. West Conshohocken, PA: ASTM International.

ASTM Standard D3518 (2007). Standard test method for in-plane shear response of polymer matrix composite materials by tensile test of a ±45° laminate. West Conshohocken, PA: ASTM International.

Aveston, J., G. A. Cooper, and A. Kelly (1971). The properties of fiber composites. In *Conference Proceeding. National Physical Laboratory, Guildford*, Teddington, pp. 15–26. IPC Science and Technology Press.

Bader, M. G., A. Parvizi, and J. E. Bailey (1980). The mechanisms of initiation and development of damage in multi-axial fibre-reinforced plastic laminates. *Mechanical Behavior of Materials, Proceedings International Conference, ICM3 3*, 227–239.

Bailey, J. and A. Parvizi (1981). On fibre debonding effects and the mechanism of transverse-ply failure in cross-ply laminates of glass fibre/thermoset composites. *Journal of Materials Science 16*(3), 649–659.

Banks-Sills, L., N. Travitzky, and D. Ashkenazi (2000). Interface fracture properties of a bimaterial ceramic composite. *Mechanics of Materials 32*(12), 711–722.

Barenblatt, G. and G. Cherepanov (1961). On brittle cracks of longitudinal shear. *PMM* *25*(6). (In Russian).

Barenblatt, G. I. (1962). The mathematical theory of equilibrium cracks in brittle fracture. *Advances in Applied Mechanics* *7*, 55–129.

Barenblatt, G. I. (1996). *Scaling, Self-similarity, and Intermediate Asymptotics.* Cambridge, United Kingdom: Cambridge University Press.

Barroso, A. (2007). *Caracterización de estados singulares de tensión en esquinas multimateriales. Aplicación a uniones adhesivas con materiales compuestos.* Ph. D. thesis, School of Engineering, University of Seville.

Barroso, A., V. Mantič, and F. París (2003). Singularity analysis of anisotropic multimaterial corners. *International Journal of Fracture* *119*(1), 1–23.

Bažant, Z. P. and L. Cedolin (1991). *Stability of Structures: Elastic, Inelastic, Fracture and Damage Theories.* New York: Oxford University Press.

Benallal, A., C. Comi, and J. Lemaitre (1992). Critical damage states at crack initiation. Volume 142, pp. 13–23.

Berthelot, J. (2003). Transverse cracking and delamination in cross-ply glass-fiber and carbon-fiber reinforced plastic laminates: Static and fatigue loading. *Applied Mechanics Reviews* *56*(1), 111–147.

Blázquez, A., V. Mantič, and F. París (2006). Application of BEM to generalized plane problems for anisotropic elastic materials in presence of contact. *Engineering Analysis with Boundary Elements* *30*(6), 489–502.

Blázquez, A., V. Mantič, F. París, and L. McCartney (2008). Stress state characterization of delamination cracks in [0/90] symmetric laminates by BEM. *International Journal of Solids and Structures* *45*(6), 1632–1662.

Blázquez, A., V. Mantič, F. París, and N. McCartney (2009). BEM analysis of damage progress in 0/90 laminates. *Engineering Analysis with Boundary Elements* *33*(6), 762–769.

Blázquez, A., F. París, and V. Mantič (1998). BEM solution of two-dimensional contact problems by weak application of contact conditions with non-conforming discretizations. *International Journal of Solids and Structures* *35*(24), 3259–3278.

Boso, D. P., G. Zavarise, and B. A. Schrefler (2005). A formulation for electrostatic-mechanical contact and its numerical solution. *International Journal for Numerical Methods in Engineering* *64*, 382–400.

Buckingham, E. (1914). On physically similar systems; illustrations of the use of dimensional equations. *Physical Review* *4*, 345–376.

Buehler, M. J. (2008). *Atomistic Modeling of Materials Failure.* Springer.

Buehler, M. J., A. Hartmaier, H. Gao, M. Duchaineau, and F. F. Abraham (2004). Atomic plasticity: description and analysis of a one-billion atom simulation of ductile materials failure. *Computer Methods in Applied Mechanics and Engineering 193*(48-51), 5257–5282.

Camanho, P. P., C. G. Dávila, S. T. Pinho, L. Iannucci, and P. Robinson (2006). Prediction of in situ strengths and matrix cracking in composites under transverse tension and in-plane shear. *Composites Part A: Applied Science and Manufacturing 37*(2), 165–176.

Camanho, P. P., G. H. Erçin, G. Catalanotti, S. Mahdi, and P. Linde (2012). A finite fracture mechanics model for the prediction of the open-hole strength of composite laminates. *Composites Part A: Applied Science and Manufacturing 43*(8), 1219–1225.

Cao, A. and J. Qu (2013). Atomistic simulation study of brittle failure in nanocrystalline graphene under uniaxial tension. *Applied Physics Letters 102*(7).

Cardu, M., L. Chiaravalloti, B. Chiaia, P. Cornetti, and B. Frigo (2008). A coupled stress and energy criterion for natural and artificial triggering of dry snow slab avalanches. In *42nd U.S. Rock Mechanics - 2nd U.S.-Canada Rock Mechanics Symposium*.

Carpinteri, A. (1981). Size effect in fracture toughness testing: a dimensional analysis approach. In G. Sih and M. Mirabile (Eds.), *Analytical and Experimental Fracture Mechanics*, pp. 785–797. International Conference on Analytical and Experimental Fracture Mechanics, Roma, Italy, 1980.

Carpinteri, A. (1982). Notch sensitivity in fracture testing of aggregative materials. *Engineering Fracture Mechanics 16*(4), 467–481.

Carpinteri, A. (1989a). Cusp catastrophe interpretation of fracture instability. *Journal of the Mechanics and Physics of Solids 37*, 567–582.

Carpinteri, A. (1989b). Softening and snap-back instability in cohesive solids. *International Journal for Numerical Methods in Engineering 28*, 1521–1537.

Carpinteri, A., P. Cornetti, N. Pugno, A. Sapora, and D. Taylor (2008). A finite fracture mechanics approach to structure with sharp V-notches. *Engineering fracture mechanics 75*(7), 1736–1752.

Carpinteri, A., M. Paggi, and G. Zavarise (2005). Snap-back instability in microstructured composites and its connection with superplasticity. *Strength, Fracture and Complexity 3*(2), 61–72.

Carpinteri, A., M. Paggi, and G. Zavarise (2008). The effect of contact on the decohesion of laminated beams with multiple microcracks. *International Journal of Solids and Structures 45*, 129–143.

Carraro, P. A. and M. Quaresimin (2011). Un modello per il debonding tra fibra e matrice in presenza di sollecitazioni multiassiali. *40 Convegno Nazionale, de la Associazione italiana per l'analisi delle sollecitazioni.*

Carraro, P. A. and M. Quaresimin (2014). Modelling fibre-matrix debonding under biaxial loading. *Composites Part A: Applied Science and Manufacturing 61*, 33–42.

Catalanotti, G. and P. P. Camanho (2013). A semi-analytical method to predict net-tension failure of mechanically fastened joints in composite laminates. *Composites Science and Technology 76*, 69–76.

Chawla, K. K. (1998). *Composite Materials: Science and Engineering.* Springer.

Chen, D.-H. and S. Nakamichi (1997). Plane problem of cracks generated from the interface of an elliptic inclusion. *JSME International Journal, Series A 40*(3), 275–282.

Cho, J., M. S. Joshi, and C. T. Sun (2006). Effect of inclusion size on mechanical properties of polymeric composites with micro and nano particles. *Composites Science and Technology 66*(13), 1941–1952.

Comninou, M. (1977). The interface crack. *Journal of Applied Mechanics 44*(4), 631–636.

Contreras, J. (2000). Micromechanical modelling of cracks in composites. Applications. (in Spanish). Master's thesis, School of Enginnering, University of Seville (Spain).

Cornetti, P., V. Mantič, and A. Carpinteri (2012). Finite Fracture Mechanics at elastic interfaces. *International Journal of Solids and Structures 49*, 1022–1032.

Cornetti, P., N. Pugno, A. Carpinteri, and D. Taylor (2006). Finite fracture mechanics: A coupled stress and energy failure criterion. *Engineering Fracture Mechanics 73*(14), 2021–2033.

Correa, E. (2008). *Micromechanical analysis of the matrix failure in fibre reinforced composites (In Spanish).* Ph. D. thesis, School of Engineering, University of Seville.

Correa, E., V. Mantič, and F. París (2008). Numerical characterisation of the fibre-matrix interface crack growth in composites under transverse compression. *Engineering Fracture Mechanics 75*(14), 4085–4103.

Cotterell, B. (2010). *Fracture and life.* Imperial College Press, London.

Cottrell, A. H. (1961). Theoretical aspects of radiation damage and brittle fracture in steel pressure vessels. *Iron Steel Institute Special Report 69*, 281–296.

de Borst, R. (2004). Damage, material instabilities, and failure. In E. Stein, E. de Borst, and T. J. R. Hughes (Eds.), *Encyclopedia of Computational Mechanics*, Volume 2, Chapter 10, pp. 274–297. The Atrium: Wiley.

Dlouhy, I. and A. R. Boccaccini (1996). Preparation, microstructure and mechanical properties of metal-particulate/glass-matrix composites. *Composites Science and Technology 56*(12), 1415–1424.

Dugdale, D. S. (1960). Yielding of steel sheets containing slits. *Journal of the Mechanics and Physics of Solids 8*(2), 100–104.

Dundurs, J. (1967). Effect of elastic constants on stress in a composite under plane deformations. *Journal of Composite Materials 1*, 310–322.

Dundurs, J. (1969). Discussion of a paper by D.B. Bogy. *Journal of Applied Mechanics 36*, 650–652.

Dundurs, J. (1975). Properties of elastic bodies in contact. In A. D. de Pater and J. J. Kalker (Eds.), *The mechanics of the contact between deformable bodies*, pp. 54–66. Delft: Delft University Press.

Dunn, M., W. Suwito, and S. Cunningham (1997). Fracture initiation at sharp notches: Correlation using critical stress intensities. *International Journal of Solids and Structures 34*(29), 3873–3883.

Dvorak, G. J. (2013). *Micromechanics of Composite Materials*. Springer.

Dvorak, G. J. and N. Laws (1986). Analysis of first ply failure in composite laminates. *Engineering Fracture Mechanics 25*(5-6), 763–770.

Dvorak, G. J. and N. Laws (1987). Analysis of progressive matrix cracking in composite laminates - II. First ply failure. *Journal of Composite Materials 21*(4), 309–329.

El Haddad, M., K. Smith, and T. Topper (1979). Fatigue crack propagation of short cracks. *J Eng Mater Technol Trans ASME 101*(1), 42–46.

Elices, M., G. Guinea, J. Gómez, and J. Planas (2002). The cohesive zone model: advantages, limitations and challenges. *Engineering Fracture Mechanics 69*, 137–163.

Emburi, J. D. (1986). Plastic flow in dispersion hardened materials. *Metallurgical transactions A 16*, 2191–2200.

England, A. (1966). An arc crack around a circular elastic inclusion. *Journal of Applied Mechanics 33*, 637–640.

Erdogan, F. (2000). Fracture mechanics. *International Journal of Solids and Structures 37*(1-2), 171–183.

Erdogan, F. and G. Sih (1963). On the crack extension in plates under plane loading and transverse shear. *Trans. ASME ser D 85*(4).

Eshelby, J. D. (1957). The determination of the elastic field of an ellipsoidal inclusion, and related problems. *Proceedings of the Royal Society of London. Series A. Mathematical and Physical Sciences 241*(1226), 376–396.

Evans, A. G., M. Rühle, B. J. Dalgleish, and P. G. Charalambides (1990). The fracture energy of bimaterial interfaces. *Materials Science and Engineering: A 126*(1-2), 53–64.

Fisher, J. R. and J. Gurland (1981). Void nucleation in spheroidized carbon steels Part 2: Model. *Metal Science 15*(5), 193–202.

Fracture Analysis Consultants, Inc (2011). *FRANC3D Reference Manual* (6 ed.). Fracture Analysis Consultants, Inc.

Freed, Y. and L. Banks-Sills (2008). A new cohesive zone model for mixed mode interface fracture in bimaterials. *Engineering Fracture Mechanics 75*(15), 4583–4593.

Fu, S.-F., X.-Q. Feng, B. Lauke, and Y.-W. Mai (2008). Effects of particle size, particle/matrix interface adhesion and particle loading on mechanical properties of particulate–polymer composites. *Composites Part B: Engineering 39*(6), 933–961.

Gamstedt, E. K. and B. A. Sjögren (1999). Micromechanisms in tension-compression fatigue of composite laminates containing transverse plies. *Composites Science and Technology 59*(2), 167–178.

Gao, Z. (1995). A circular inclusion with imperfect interface: Eshelby's tensor and related problems. *Journal of Applied Mechanics 62*, 860–866.

García, I. G. and D. Leguillon (2012). Mixed-mode crack initiation at a V-notch in presence of an adhesive joint. *International Journal of Solids and Structures 49*(15–16), 2138 – 2149.

García, I. G., M. Paggi, and V. Mantič (2014). Fiber-size effects on the onset of fiber-matrix debonding under transverse tension: A comparison between cohesive zone and finite fracture mechanics models. *Engineering Fracture Mechanics 115*, 96–110.

Garrett, K. W. and J. E. Bailey (1977). Multiple transverse fracture in 90° cross-ply laminates of a glass fibre-reinforced polyester. *Journal of Materials Science 12*(1), 157–168.

Gent, A. N. (1980). Detachment of an elastic matrix from a rigid spherical inclusion. *Journal of Materials Science 15*(11), 2884–2888.

Gent, A. N. and B. Park (1984). Failure processes in elastomers at or near a rigid spherical inclusion. *Journal of Materials Science 19*(6), 1947–1956.

Gersappe, D. and M. Robbins (1999). Where do polymer adhesives fail? *Europhysics Letters 48*(2), 150–155.

Goodier, J. (1933). Concentration of stress around spherical and cylindrical inclusions and flaws. *Journal of Applied Mechanics 55*, 39–44.

Goods, S. H. and L. M. Brown (1979). The nucleation of cavities by plastic deformation. *Acta metallurgica 27*, 1–27.

Graciani, E., V. Mantič, and F. París (2007). On the estimation of the first interpenetration point in the open model of interface cracks. *International Journal of Fracture 143*(3), 287–290.

Graciani, E., V. Mantič, and F. París (2010). A BEM analysis of a penny-shaped interface crack using the open and the frictionless contact models: Range of validity of various asymptotic solutions. *Engineering Analysis with Boundary Elements 34*(1), 66–78.

Graciani, E., V. Mantič, F. París, and A. Blázquez (2005). Weak formulation of axi-symmetric frictionless contact problems with boundary elements: Application to interface cracks. *Computers and Structures 83*, 836–855.

Griffith, A. A. (1921). The phenomena of rupture and flow in solids. *Philosophical transaction of the royal society of London. Sieres A. Containing papers of a Mathematical or physical character 221*, 163–198.

Griffith, A. A. (1924). The theory of rupture. In *First International Congress of Applied Mechanics*, pp. 55–63.

G'Sell, C., J. M. Hiver, and A. Dahoun (2002). Experimental characterization of deformation damage in solid polymers under tension, and its interrelation with necking. *International Journal of Solids and Structures 39*(13-14), 3857–3872.

Hallbäck, N. and F. Nilsson (1994). Mixed-mode I/II fracture behaviour of an aluminium alloy. *Journal of the Mechanics and Physics of Solids 42*(9), 1345–1374.

Han, R., M. Ingber, and H. Schreyer (2006). Progression of failure in fiber-reinforced materials. *Computers, Materials and Continua 4*(3), 163–176.

Hardiman, N. (1954). Elliptic elastic inclusion in an infinite elastic plate. *The Quarterly Journal of Mechanics and Applied Mechanics 7*, 226–230.

Hashin, Z. (1996). Finite thermoelastic fracture criterion with application to laminate cracking analysis. *Journal of the Mechanics and Physics of Solids 44*(7), 1129–1145.

Hebel, J., R. Dieringer, and W. Becker (2010). Modelling brittle crack formation at geometrical and material discontinuities using a finite fracture mechanics approach. *Engineering Fracture Mechanics 77*(18), 3558–3572.

Henninger, C., D. Leguillon, and E. Martin (2007). Crack initiation at a V-notch—comparison between a brittle fracture criterion and the Dugdale cohesive model. *Comptes Rendus Mécanique 335*(7), 388–393.

Hexply 8852 Product Data (2013). Hexply 8852 epoxy matrix (180°C/365°F curing matrix) UD carbon prepregs product data.

Hillerborg, A., M. Modéer, and P. E. Pertersson (1976). Analysis of crack formation and crack growth in concrete by means of frature mechanics and finite elements. *Cement and concrete research 6*, 773–782.

Hills, D. and J. Barber (1993). Interface cracks. *International Journal of Mechanical Sciencies 35*, 25–37.

Hills, D. A. and D. Dini (2011). Characteristics of the process zone at sharp notch roots. *International Journal of Solids and Structures 48*(14–15), 2177–2183.

Hinton, M. J. and A. S. Kaddour (2013). The background to Part B of the Second World-Wide Failure Exercise: Evaluation of theories for predicting failure in polymer composite laminates under three-dimensional states of stress. *Journal of Composite Materials 47*, 643–652.

Hinton, M. J., A. S. Kaddour, and P. D. Soden (2002). A comparison of the predictive capabilities of current failure theories for composite laminates, judged against experimental evidence. *Composites Science and Technology 62*, 1725–1797.

Hinton, M. J., A. S. Kaddour, and P. D. Soden (2004). *Failure criteria in fibre reinforced polymer composites: the world-wide failure exercise.* Oxford: Elsevier.

Huang, R., K. R. Hebert, and L. S. Chumbley (2004). Microscopic observations of voids in anodic oxide films on aluminium. *Journal of the electromechanical society 151*(7), 379–386.

Hull, D. and T. W. Clyne (1996). *An Introduction to Composite Materials.* Cambridge University Press.

Hutchinson, J. and Z. Suo (1992). Mixed mode cracking in layered materials. *Advances in Applied Mechanics 29*, 63–191.

Irwin, G. R. (1948). Fracture dynamics. *Fracturing of metals. American Society for metals*, 147–166.

Irwin, G. R. (1957). Analysis of stresses and strains near the end of a crack transversing a plate. *Journal of applied mechanics 24*, 361–364.

ISO Standard 15024 (2001). Fibre-reinforced plastic composites - Determination of mode I interlaminar fracure toughness, G1C, for unidirectionally reinforced materials. Geneva, Switzerland: ISO.

Jenq, Y. and S. P. Shah (1985). Two parameters fracture model for concrete. *Journal of engineering mechanics 111*(10), 1227–1241.

Jones, R. (1999). *Mechanics of Composite Materials*. London: Taylor & Francis.

Kaddour, A. S. and M. J. Hinton (2013). Maturity of 3D failure criteria for fibre-reinforced composites: Comparison between theories and experiments: Part B of WWFE-II. *Journal of Composite Materials 47*(6-7), 925–966.

Kamp, N., I. Sinclair, and M. J. Starink (2002). Toughness-strength relations in the overaged 7449 Al-based alloy. *Metallurgical and Materials Transactions A: Physical Metallurgy and Materials Science 33*(4), 1125–1136.

Kobayashi, S. and N. Takeda (2002). Experimental and analytical characterization of transverse cracking behavior in carbon/bismaleimide cross-ply laminates under mechanical fatigue loading. *Composites: Part B 33*, 471–478.

Koyanagi, J., H. Nakatani, and S. Ogihara (2012). Comparison of glass-epoxy interface strengths examined by cruciform specimen and single-fiber pull-out tests under combined stress state. *Composites Part A: Applied Science and Manufacturing 43*(11), 1819–1827.

Krueger, R. (2004). Virtual crack closure technique: History, approach, and applications. *Applied mechanics review 57*(2), 109–143.

Krull, H. and H. Yuan (2011). Suggestions to the cohesive traction-separation law from atomistic simulations. *Engineering Fracture Mechanics 78*(3), 525–533.

Kushch, V.I., S. S. and L. Mishnaevsky Jr. (2010). Elastic interaction of partially debonded circular inclusions. I. Theoretical solution. *International Journal of Solids and Structures 47*(14-15), 1961–1971.

Kushch, V., S. Shmegera, P. Brøndsted, and L. Mishnaevsky (2011). Numerical simulation of progressive debonding in fiber reinforced composite under transverse loading. *International Journal of Engineering Science 49*(1), 17–29.

Labossiere, P. and M. Dunn (1999). Stress intensities at interface corners in anisotropic bimaterials. *Engineering Fracture Mechanics 62*(6), 555–575.

Ladevèze, P., G. Lubineau, and D. Violeau (2006). A computational damage micromodel of laminated composites. *International Journal of Fracture 137*(1-4), 139–150.

Lane, M. (2003). Interface fracture. *Annual Review of Materials Research 33*(1), 29–54.

Lawn, B. (1993). *Fracture of Brittle Solids*. Cambridge University Press, Cambridge.

Lee, H. K. and S. H. Pyo (2007). Micromechanics-based elastic damage modeling of particulate composites with weakened interfaces. *International Journal of Solids and Structures 44*(25-26), 8390–8406.

Lee, H. K. and S. H. Pyo (2008). Multi-level modeling of effective elastic behavior and progressive weakened interface in particulate composites. *Composites Science and Technology 68*(2), 387–397.

Leguillon, D. (1989). Calcul du taux de restitution de l'energie au voisinage d'une singularité. *C. R. Acad. Sci. Paris 309*, 945–950.

Leguillon, D. (2001). A criterion for crack nucleation at a notch in homogeneous materials. *Comptes Rendus de l'Académie des Sciences - Series {IIB} - Mechanics 329*(2), 97–102.

Leguillon, D. (2002). Strength or toughness? A criterion for crack onset at a notch. *European Journal of Mechanics and Solids 21*(1), 61–72.

Leguillon, D. (2013). A simple model of thermal crack pattern formation using the coupled criterion. *Comptes Rendus Mécanique 341*(6), 538–546.

Leguillon, D. and E. Martin (2013a). Crack nucleation at stress concentration points in composite materials - application to the crack deflection by an interface. In V. Mantič (Ed.), *Mathematical Methods and Models in Composites*, Chapter 10, pp. 401–424. London: Imperial College Press.

Leguillon, D. and E. Martin (2013b). The strengthening effect caused by an elastic contrast. Part I: the bimaterial case. *International Journal of Fracture 179*(1-2), 157–167.

Leguillon, D. and E. Martin (2013c). The strengthening effect caused by an elastic contrast. Part II: stratification by a thin stiff layer. *International Journal of Fracture 179*(1-2), 169–178.

Leguillon, D. and S. Murer (2008a). Crack deflection in a biaxial stress state. *International Journal of Fracture 150*(1-2), 75–90.

Leguillon, D. and S. Murer (2008b). Crack kinking out of an interface, influence of the T-stress. In *17th European Conference on Fracture (ECF17), Brno, Czech Republic, September 2-5 2008*.

Leguillon, D. and S. Murer (2012). Fatigue crack nucleation at a stress concentration point. *4th International Conference on crack paths (CP 2012)*.

Leguillon, D., S. Murer, N. Recho, and J. Li (2009). Crack initiation at a V-notch under complex loadings - statistical scattering. In *Proceedings of the International Conference on Fracture, ICF12, Ottawa, Canada, 12-17 July 2009*.

Leguillon, D., D. Quesada, C. Putot, and E. Martin (2007). Prediction of crack initiation at blunt notches and cavities - size effects. *Engineering Fracture Mechanics 74*(15), 2420–2436.

Leguillon, D. and E. Sanchez-Palencia (1987). *Computation of singular solutions in elliptic problems and elasticity.* Paris: Masson.

Leguillon, D. and Z. Yosibash (2003). Crack onset at a V-notch. Influence of the notch tip radius. *International Journal of Fracture 122*(1-2), 1–21.

Leidner, J. and R. T. Woodhams (1974). The strength of polymeric composites containing spherical fillers. *Journal of Applied Polymer Science 18*(6), 1639–1654.

Leite, A., V. Mantič, and F. París (2010). Study of crack onset at holes in PMMA. Difficulties in characterizing the material. In *Anales de Mecánica de la Fractura*, Volume 27. Iberian Conference on Fracture and Structural Integrity 2010, Porto (Portugal).

Lemaitre, J. (1985). A continuous damage mechanics model for ductile fracture. *Journal of Engineering Materials and Technology 107*(1), 83–89.

Lemaitre, J. (1990). Micro-mechanics of crack initiation. *International Journal of Fracture 42*(1), 87–99.

Lemaitre, J. and R. Desmorat (2005). *Engineering Damage Mechanics, Ductile, Creep, Fatigue and Brittle Failures.* Berlin: Springer.

Levy, A. J. (1991). The debonding of elastic inclusions and inhomogeneities. *Journal of the Mechanics and Physics of Solids 39*(4), 477–505.

Levy, A. J. and K. Hardikar (1999). The inclusion pair interaction problem with non-linear interface. *Journal of the Mechanics and Physics of Solids 47*(7), 1477–1508.

Li, D. S. and M. R. Wisnom (1997). Evaluating Weibull parameters for transverse cracking in cross-ply laminates. *Journal of Composite Materials 31*(9), 935–951.

Li, J. and X. B. Zhang (2006). A criterion study for non-singular stress concentrations in brittle or quasi-brittle materials. *Engineering Fracture Mechanics 73*(4), 505–523.

Maccagno, T. M. and J. F. Knott (1989). The fracture behaviour of PMMA in mixed modes I and II. *Engineering Fracture Mechanics 34*(1), 65–86.

Maimí, P., P. P. Camanho, J. A. Mayugo, and C. G. Dávila (2007). A continuum damage model for composite laminates: Part I - constitutive model. *Mechanics of materials 39*, 897–908.

Maimí, P., P. P. Camanho, J. A. Mayugo, and A. Turon (2011). Matrix cracking and delamination in laminated composites. Part I: Ply constitutive law, first ply failure and onset of delamination. *Mechanics of Materials 43*(4), 169–185.

Maimí, P., E. V. González, N. Gascons, and L. Ripoll (2013). Size effect law and critical distance theories to predict the nominal strength of quasibrittle structures. *Applied mechanics reviews 65*(2).

Malyshev, B. and R. Salganik (1965). The strength of adhesive joints using the theory of cracks. *International Journal of Fracture and Mechanics* (1), 114–128.

Mantič, V. (2009). Interface crack onset at a circular cylindrical inclusion under a remote transverse tension. Application of a coupled stress and energy criterion. *International Journal of Solids and Structures 46*(6), 1287–1304.

Mantič, V., A. Blázquez, E. Correa, and F. París (2006). Analysis of interface cracks with contact in composites by 2D BEM. In M. Guagliano and M. H. Aliabadi (Eds.), *Fracture and Damage of Composites*, Chapter 8, pp. 189–248. WIT Press, Southampton.

Mantič, V. and I. G. García (2012). Crack onset and growth at the fibre-matrix interface under a remote biaxial transverse load. Application of a coupled stress and energy criterion. *International Journal of Solids and Structures 49*(17), 2273–2290.

Mantič, V. and F. París (2004). Relation between SIF and ERR based measures of fracture mode mixity in interface cracks. *International Journal of Fracture 130*(2), 557–569.

Marlett, K. (2010). Hexcel 8852 AS4 unidirectional prepreg at 190 gsm & 35% RC qualification material property data report. Technical Report CAM-RP-2010-002 Rev A, National Institute for Aviation Research. Wichita State University, Wichita, Kansas.

Martin, E., D. Leguillon, and N. Carrère (2010). A twofold strength and toughness criterion for the onset of free-edge shear delamination in angle-ply laminates. *International Journal of Solids and Structures 47*(9), 1297–1305.

Martin, E., D. Leguillon, and N. Carrère (2012). A coupled strength and toughness criterion for the prediction of the open hole tensile strength of a composite plate. *International Journal of Solids and Structures 49*(26), 3915–3922.

Martin, E., B. Poitou, D. Leguillon, and J. M. Gatt (2008). Competition between deflection and penetration at an interface in the vicinity of a main crack. *International Journal of Fracture 151*(2), 247–268.

Martin, P. A. (2001). The spherical-cap crack revisited. *International Journal of Solids and Structures 38*(26-27), 4759–4776.

Martynenko, M. A. and I. V. Lebedyeva (2006). Axisymmetric problem for a spherical crack on the interface of elastic media. *Journal of Engineering Mathematics 56*(4), 371–384.

Martyniuk, K., B. F. Sørensen, P. Modregger, and E. M. Lauridsen (2013). 3D in situ observations of glass fibre/matrix interfacial debonding. *Composites: Part A 55*(2), 63–73.

Matzenmiller, A., J. Lubliner, and R. Taylor (1995). A constitutive model for anisotropic damage in fiber-composites. *Mechanics of Materials 20*(2), 125–152.

McCartney, L. N. (1998). Predicting transverse crack formation in cross-ply laminates. *Composites Science and Technology 58*(7), 1069–1081.

McClintock, F. A. (1958). Ductile fracture instability in shear. *Journal of applied mechanics 25*, 582–588.

Michel, J.-B., Y. Kui Shen, A. Presser Aiden, A. Veres, M. K. Gray, J. P. Pickett, D. Hoiberg, D. Clancy, P. Norvig, J. Orwant, S. Pinker, M. A. Nowak, and E. L. Aiden (2011). Quantitative analysis of culture using millions of digitized books. *Science 331*(6014), 176–182.

Miller, A. G. and A. L. Wingert (1979). Fracture surface characterization of commercial graphite/epoxy systems. In R. B. Pipes (Ed.), *Nondestructive evaluation and flaw criticality for composite materials*, pp. 274–297. Philadelphia: American society for testing and materials.

Mower, T. M. and A. S. Argon (1996). An experimental technique to measure the adhesive strength between inclusions and transparent matrices. *Journal of Materials Science 31*(6), 1585–1594.

Murakami, Y. (2001). *Stress Intensity Factors Handbook, Vol 5*. The Society of the Materials Science, Japan, and Elsevier Science Ltd.

Nairn, J. A. (1997). Fracture mechanics of composites with residual thermal stresses. *Journal of Applied Mechanics 64*, 804–815.

Nairn, J. A. (2000). Matrix microcracking in composites. In A. Kelly and C. Zweben (Eds.), *Comprehensive Composite Materials*, Volume 2, pp. 403–432. Oxford: Pergamon.

National Science Board (2014). Science and engineering indicators. Technical report, National Science Fondation, Arlighton, VA.

Nazarov, S. A. (2004). Three-Dimensional formulation of the Novozhilov criterion for Mode I cracks. *Doklady Physics 50*(6), 328–332.

Needleman, A. (1987). A continuum model for void nucleation by inclusion debonding. *Journal of Applied Mechanics 54*(3), 525–531.

Needleman, A. (1990). An analysis of tensile decohesion along an interface. *Journal of the Mechanics and Physics of Solids 38*(3), 289–324.

Neuber, H. (1958). *Theory of notch stresses: Principles for exact calculation of strength with reference to structural form and material.* Springer.

Ngo, D., K. Park, G. H. Paulino, and Y. Huang (2010). On the constitutive relation of materials with microstructure using a potential-based cohesive model for interface interaction. *Engineering Fracture Mechanics 77*(7), 1153–1174.

Novozhilov, V. V. (1969). On a necessary and sufficient criterion for brittle strength (in Russian). *P.M.M. 33*, 201–210.

Nurhaniza, M., M. Ariffin, A. Ali, F. Mustapha, and A. Noraini (2010). Finite element analysis of composites materials for aerospace applications. In *IOP Conf. Series: Materials Science and Engineering*, Volume 11. 9th National Symposium on Polimeric Materials.

Ogihara, S. and J. Koyanagi (2010). Investigation of combined stress state failure criterion for glass fiber/epoxy interface by the cruciform specimen method. *Composites Science and Technology 70*, 143–150.

Orowan, E. (1949). Fracture and strength of solids. *Reports on Progress in Physics 12*(1), 185–232.

Othmani, Y., L. Delannay, and I. Doghri (2011). Equivalent inclusion solution adapted to particle debonding with a non-linear cohesive law. *International Journal of Solids and Structures 48*(24), 3326–3335.

Otsuka, K. and H. Date (2000). Frature process zone in concrete tension specimen. *Engineering fracture mechanics 65*, 111–131.

Paggi, M., A. Carpinteri, and G. Zavarise (2006). A unified interface constitutive law for the study of fracture and contact problems in heterogeneous materials. In *Analysis and Simulation of Contact Problems, Lecture Notes in Applied and Computational Mechanics*, Volume 27, pp. 297–304. Springer-Verlag, Berlin.

Paggi, M. and P. Wriggers (2011). A nonlocal cohesive zone model for finite thickness interfaces – Part I: mathematical formulation and validation with molecular dynamics. *Computational Materials Science 50*(5), 1625–1633.

Paggi, M. and P. Wriggers (2012). Stiffness and strength of hierarchical polycrystalline materials with imperfect interfaces. *Journal of the Mechanics and Physics of Solids 60*(4), 557–572.

París, F. (2001). A study of failure criteria of fibrous composite materials. Technical report, NASA CR-2001-210661.

París, F., A. Blázquez, L. McCartney, and A. Barroso (2010b). Characterization and evolution of matrix and interface related damage in [0/90]s laminates under tension. Part II: Experimental evidence. *Composites Science and Technology 70*(7), 1176–1183.

París, F., A. Blázquez, L. McCartney, and V. Mantič (2010a). Characterization and evolution of matrix and interface related damage in [0/90]s laminates under tension. Part I: Numerical predictions. *Composites Science and Technology 70*(7), 1168–1175.

París, F., J. C. Caño, and J. Varna (1996). The fibre-matrix interface crack - a numerical analysis using boundary elements. *International Journal of Fracture 82*, 11–29.

París, F., E. Correa, and J. Cañas (2003). Micromechanical view of failure of the matrix in fibrous composite materials. *Composites Science and Technology 63*(7), 1041–1052.

París, F., E. Correa, and V. Mantič (2007). Kinking of transversal interface cracks between fiber and matrix. *Journal of Applied Mechanics, Transactions ASME 74*(4), 703–716.

Park, K. and G. H. Paulino (2011). Cohesive zone models: A critical review of traction-separation relationships across fracture surfaces. *Applied Mechanics Reviews 64*(6).

Parvizi, A., K. Garrett, and J. Bailey (1978). Constrained cracking in glass fibre-reinforced epoxy cross-ply laminates. *Journal of Materials Science 13*(1), 195–201.

Peterson, R. (1938). Methods of correlating data from fatigue tests of stress concentration specimens. In *Stephen Timoshenko Anniversary Volume*, pp. 179.

Peterson, R. E. (1959). Notch-sensitivity. In G. Sines and J. L. Waisman (Eds.), *Metal fatigue*, pp. 293–306. McGraw Hill, New York.

Petersson, P. E. (1981). Crack growth and development of fracture zones in plain concrete and similar materials. Technical Report TVBM-1006, Division of Building Materials. Lund Institute of Technology, Lund, Sweden.

Picard, D., D. Leguillon, and C. Putot (2006). A method to estimate the influence of the notch-root radius on the fracture toughness measurement of ceramics. *Journal of the European Ceramic Society 26*(8), 1421–1427.

Priel, E., Z. Yosibash, and D. Leguillon (2008). Failure initiation at a blunt V-notch tip under mixed mode loading. *International Journal of Fracture 149*(2), 143–173.

Pugno, N. M. (2004). Quantized fracture mechanics. *Philosophical magazine 84*(27), 2829–2845.

Pugno, N. M., P. Cornetti, and A. Carpinteri (2007). New unified laws in fatigue: From the Wöhler's to the Paris' regime. *Engineering Fracture Mechanics 74*(4), 595–601.

Quesada, D., D. Leguillon, and C. Putot (2009). Multiple failures in or around a stiff inclusion embedded in a soft matrix under a compressive loading. *European Journal of Mechanics and Solids 28*(4), 668–679.

Reedy Jr, E. and T. Guess (1993). Comparison of butt tensile strength data with interface corner stress intensity factor prediction. *International Journal of Solids and Structures 30*(21), 2929–2936.

Renart, J., N. Blanco, E. Pajares, C. J., L. S., and G. Santacruz (2011). Side clamped beam (SCB) hinge system for delamination tests in beam-type composite specimens. *Composites Science and Technology 71*, 1023–1029.

Rice, J. R. (1988). Elastic fracture mechanics concepts for interfacial cracks. *Journal of Applied Mechanics, Transactions ASME 55*(1), 98–103.

Richard, H. and K. Benitz (1983). A loading device for the creation of mixed mode in fracture mechanics. *International Journal of Fracture 22*(3), 3591–3618.

Rose, J. H., J. Ferrante, and J. R. Smith (1981). Universal binding energy curves for metals and bimetallic interfaces. *Physical Review Letters 47*, 675–678.

Rossmanith, H. P. (1995). An introduction to K. Wieghardt's historical paper "On splitting and cracking of elastic bodies". *Fatigue and Fracture of Engineering Materials and Structures 18*(12), 1367–1369.

Salviato, M., M. Zappalorto, and M. Quaresimin (2013). Nanoparticle debonding strength: A comprehensive study on interfacial effects. *International Journal of Solids and Structures 50*(20–21), 3225–3232.

Sapora, A., P. Cornetti, and A. Carpinteri (2014). V-notched elements under mode II loading conditions. *Structural Engineering and Mechanics 49*(4), 499–508.

Schellekens, J. and R. de Borst (1993). On the numerical integration of interface elements. *International Journal for Numerical Methods in Engineering 36*, 44–66.

Schmauder, S. and M. Meyer (1992). Correlation between Dundurs' parameters and elastic constants. *Z. Metallkd 83*, 524–527.

Seweryn, A. (1994). Brittle fracture criterion for structures with sharp notches. *Engineering Fracture Mechanics 47*(5), 673–681.

Seweryn, A., S. Poskrobko, and Z. Mróz (1997). Brittle fracture in plane elements with sharp notches under mixed-mode loading. *Journal of Engineering Mechanics 123*(6), 535–543.

Shen, W., C. Y. Tang, C. P. Tsui, and L. H. Peng (2002). Effects of two damage mechanisms on effective elastic properties of particulate composites. *Composites Science and Technology 62*(10-11), 1397–1406.

Shi, Y. W., N. N. Zhou, and J. X. Zhang (1994). Comparison of mode I and mode II elastic-plastic fracture toughness for two low alloyed high strength steels. *International Journal of Fracture 68*(1), 89–97.

Soden, P. D., M. J. Hinton, and A. S. Kaddour (1998). A comparison of the predictive capabilities of current failure theories for composite laminates. *Composites Science and Technology 58*(7), 1225–1254.

Spanoudakis, J. and R. J. Young (1984). Crack propagation in a glass particle-filled epoxy resin - Part 1: effect of particle volume fraction and size. *Journal of Materials Science 19*(2), 473–486.

Suga, T., G. Elssner, and S. Schmauder (1988). Composite parameters and mechanical compatibility of material joints. *Journal of Composite Materials 22*, 917–934.

Sun, C. T., B. J. Quinn, J. Tao, D. W. Oplinger, and W. J. Hughes (1996). Comparative evaluation of failure analysis methods for composite laminates. Technical report, DOT/FAA/AR-95/109.

Sun, L., R. F. Gibson, F. Gordaninejad, and J. Suhr (2009). Energy absorption capability of nanocomposites: A review. *Composites Science and Technology 69*(14), 2392–2409.

Tan, H., Y. Huang, C. Liu, G. Ravichandran, and G. H. Paulino (2007). Constitutive behaviors of composites with interface debonding: The extended Mori-Tanaka method for uniaxial tension. *International Journal of Fracture 146*(3), 139–148.

Távara, L., V. Mantič, J. Cañas, E. Graciani, and F. París (2010). BEM analysis of crack onset and growth in composites using the Linear Elastic-Brittle Interface Model. In E. Sapountzakis (Ed.), *Recent Developments in Boundary Elements Methods*, pp. 281–294. WIT Press.

Távara, L., V. Mantič, E. Graciani, and F. París (2011). BEM analysis of crack onset and propagation along fiber-matrix interface under transverse tension using a linear elastic-brittle interface model. *Engineering Analysis with Boundary Elements 35*(2), 207–222.

Taylor, D. (2007). *The theory of critical distances. A new perspective in fracture mechanics*. Oxford: Elsevier B.V.

Taylor, D., P. Cornetti, and N. Pugno (2005). The fracture mechanics of finite crack extension. *Engineering fracture mechanics 72*, 1021–1038.

Toda, H., T. Ueda, K. Toshiro, and A. Takashi (2001). Interfacial debonding criteria under combined stress. *Journal of the Society of Materials Science, Japan 50*(12), 1375–1381.

Toya, M. (1974). A crack along the interface of a rigid circular inclusion embedded in an elastic solid. *International Journal of Fracture 9*(4), 463–470.

Tran, V., D. Leguillon, A. Krishnan, and L. Xu (2010). Strength and fracture energy of adhesive bonding between similar and dissimilar sheet materials. In *Proceedings of the 10th International Fatigue Congress, Prague, Czech Republic, June 6-11 2010*.

Tran, V.-X., D. Leguillon, A. Krishnan, and L. R. Xu (2012). Interface crack initiation at V-notches along adhesive bonding in weakly bonded polymers subjected to mixed-mode loading. *International Journal of Fracture 176*(1), 65–79.

Tszeng, T. (1993). A model of void nucleation from ellipsoidal inclusions in ductile fracture. *Scripta Metallurgica et Materialia 28*(9), 1065–1070.

Tszeng, T. C. (2000). Interfacial stresses and void nucleation in discontinuously reinforced composites. *Journal of Engineering Materials and Technology 122*(1), 86–92.

Tvergaard, V. (1990). Effect of fiber debonding in a whisker-reinforced metal. *Materials Science and Engineering A 107*, 23–40.

Tvergaard, V. and J. Hutchinson (1993). The influence of plasticity on mixed mode interface toughness. *Journal of the Mechanics and Physics of Solids 41*(6), 1119–1135.

van der Meer, F. P. and C. G. Dávila (2013). Cohesive modeling of transverse cracking in laminates under in-plane loading with a single layer of elements per ply. *International Journal of Solids and Structures 50*(20), 3308–3318.

Varna, J., R. Joffe, and R. Talreja (2001). Mixed micromechanics and continuum damage mechanics approach to transverse cracking in [s, 90n]s laminates. *Mechanics of Composite Materials 37*(2), 115–126.

Vaschy, A. (1892). Sur les lois de similitude en physique. *Annales Télégraphiques 19*, 25–28.

Vicentini, D. (2012). *Study of the stress state and failure in adhesive joints with composite materials*. Ph. D. thesis, School of Engineering, University of Seville.

Wagner, H. D. (1986). Some electron microscope observations of longitudinal splitting from surface flaws in Kevlar 29. *Journal of materials scence letters 5*, 439–440.

Wang, A. S. D. (1984). Fracture mechanics of sublaminate cracks in composite materials. *Composites Technology Review 6*(2), 45–62.

Wang, A. S. D., N. N. Kishore, and C. A. Li (1985). Crack development in graphite-epoxy cross-ply laminates under uniaxial tension. *Composites Science and Technology 24*, 1–31.

Wawrzynek, P. A., B. J. Carter, and A. R. Ingraffea (2009). Advances in Simulation of Arbitrary 3D Crack Growth using FRANC3D NG. In *International Congress on Fracture 12, Ottawa*.

Weibull, W. (1939). The phenomenon of rupture in solids. *Ingeniors Vetankaps Akademien-Handlinger* (153).

Weissgraeber, P. and W. Becker (2011). A new finite fracture mechanics approach for assessing the strength of bonded lap joints. *Key Engineering Materials 471-472*, 1075–1080.

Weissgraeber, P. and W. Becker (2013). Finite fracture mechanics model for mixed mode fracture in adhesive joints. *International Journal of Solids and Structures 50*(14-15), 2383–2394.

Wells, A. A. (1961). Unstable crack propagation in metals: cleavage and fracture. In *Proceedings of the Crack Propagation Symposium. College of Aeronautics, Cranfield*, pp. 120–230.

Westergaard, H. M. (1939). Bearing pressures and cracks. *Journal of Applied Mechanics 49*, 49–53.

Wevers, M. and M. Surgeon (2000). Acoustic emission and composites. In A. Kelly and C. Zweben (Eds.), *Comprehensive Composite Materials*, pp. 345–357. Oxford: Pergamon.

Wieghardt, K. (1907). Über das Spalten und Zerreissen elastischer Körper (On splitting and cracking of elastic bodies). *Zeitschrift für Mathematik und Physik 55*(1-2), 60–103. Translation: Rossmanith H.P. Fatigue Fract Engng Mater Struct 1995, 12: 1371-1405.

Williams, M. (1952). Stress singularities resulting from various boundary conditions in angular corners of plates in extension. *Journal of Applied Mechanics 19*, 526–528.

Wisnom, M. R. (2000). Size effects in composites. In A. Kelly and C. Zweben (Eds.), *Comprehensive Composite Materials*, Volume 2, pp. 23–47. Oxford: Pergamon.

Wriggers, P. (2002). *Computational Contact Mechanics*. Wiley & Sons Ltd.

Xie, M. and A. Levy (2007). Defect propagation at a circular interface. *International Journal of Fracture 144*(1), 1–20.

Yang, J. M., S. M. Jeng, and C. J. Yang (1991). Fracture mechanisms of fiber-reinforced titanium alloy matrix composites part I: Interfacial behavior. *Materials Science and Engineering A 138*(2), 155–167.

Yoshinobu, N., M. Yamaguchi, M. Okubo, and T. Matsumoto (1992). Effects of particle size on mechanical and impact properties of epoxy resin filled with spherical silica. *Journal of Applied Polymer Science 45*(7), 1281–1289.

Yosibash, Z., E. Priel, and D. Leguillon (2006). A failure criterion for brittle elastic materials under mixed mode loading. *International Journal of Fracture 141*(1), 289–310.

Zahid, G. H., R. I. Todd, and P. B. Prangnell (1998). Cavitation behaviour of a 2124Al/SiC composite during high strain rate superplasticity. *Superplasticity and Superplastic Forming*, 227–235.

Zappalorto, M., M. Salviato, and M. Quaresimin (2011). Influence of the interphase zone on the nanoparticle debonding stress. *Composites Science and Technology 72*(1), 49–55.

Zeng, X. and S. Li (2010). A multiscale cohesive zone model and simulations of fractures. *Computer Methods in Applied Mechanics and Engineering 199*(9-12), 547–556.

Zhang, H., M. Ericson, J. Varna, and L. Berglund (1997). Transverse single-fibre test for interfacial debonding in composites: 1. Experimental observations. *Composites Part A: Applied Science and Manufacturing 28*(4), 309–315.

Zhang, X. B. and J. Li (2008). A failure criterion for brittle and quasi-brittle materials under any level of stress concentration. *Engineering Fracture Mechanics 75*(17), 4925–4932.

Zhang, Z. X. (2002). An empirical relation between mode I fracture toughness and the tensile strength of rock. *International Journal of Rock Mechanics and Mining Sciences 39*(3), 401–406.

Zwiers, R. I., T. C. T. Ting, and R. L. Spilker (1982). On the logarithmic singularity of free-edge stress in laminated composites under uniform extension. *Journal of Applied Mechanics 49*(3), 561–568.

www.ingramcontent.com/pod-product-compliance
Lightning Source LLC
Chambersburg PA
CBHW080759180526
45168CB00006B/2261